69th Annual Safety in Ammonia Plants and Related Facilities Symposium 2025

AIChE Technical Manual 2025 Vol. 65

Atlanta, Georgia, USA
7-11 September 2025

ISBN: 979-8-3313-2886-3

Printed from e-media with permission by:

Curran Associates, Inc.
57 Morehouse Lane
Red Hook, NY 12571

CURRAN ASSOCIATES INC. proceedings .com

Some format issues inherent in the e-media version may also appear in this print version.

AIChE
120 Wall Street, FL 23
New York, NY 10005-4020

Phone: (800) 242-4363
Fax: (203) 775-5177

www.aiche.org

Additional copies of this publication are available from:

Curran Associates, Inc.
57 Morehouse Lane
Red Hook, NY 12571 USA
Phone: 845-758-0400
Fax: 845-758-2633
Email: curran@proceedings.com
Web: www.proceedings.com

TABLE OF CONTENTS

Seventy Years of AIChE Ammonia Safety Symposium – Lessons Learned 1
Venkat Pattabathula, Taylor Archer, Seshu Dharmavaram

Transfer of Ammonia: Understanding and Reviewing Safety of Transfer Operations 25
Stuart Powell, Dorothy M. Shaffer, Anthony Sarrack

Repair and Investigation into an Ammonia Leak on a Marine Loading Arm 34
Kiran Rampat, Mikyle Moolchan

Lesson Learned Preventing Early Re-Lief in Ammonia Storage 44
Davies Michael, Davide Moncalvo, Brad Luckman, Brian Anderson

Operational Challenges in Ammonia Production: Addressing Water Contamination and Its Impact on Plant Performance 49
Khalid Alyami, Taiebi Mohammed, Abdulrahman Ezzat, Manikandan

High Temperature Exposure of HTS Catalyst at SOR 56
Faisal Jamal, Kamran A. Chaudhry, Maham Bashir, Sheikh A. Ahmad

Problems Encountered in the HTS Reactor When Revamping to Higher Rates 69
Jackson Lee, Karen Braden, Dean Potter

Safe, Reliable and Industrial Operation of SynCOR™ Enable Large-Scale Low Carbon Ammonia Production 76
Thomas S. Christensen, Per J. Dahl, Lari B. Knudsen

Sichuan Lutianhua: A Case History of Operational Excellence and Safety - 20 Years of Proven Reliability with Autothermal Reforming Technology 88
Marco M. Carlucci, Francesco Baratto, Sun Guang

Process Gas Backflow Causes LTS Runaway Reaction 95
Jackson Lee

Prevention of Process Safety Incidents by Establishing Strong Process Safety Leadership and Culture 105
Kamleshkumar Modi

Major Accident Hazard (MAH)! Are You Prepared to Avoid and Mitigate the Consequences of an Unlikely Event? 110
Cristiano Azevedo, Faritha Banu

Analyzing Ammonia Incidents Through the Lens of Process Safety 121
Bruce K. Vaughen, Venkat Pattabathula, Jim Richardson

Nitridation Resistance of High-Temperature Resistant Cast Alloys – The Effect of Alumina Scales on High-Temperature Ammonia Nitridation 131
Benjamin Aymans, Pedro Imizcoz, Pablo Cardin, Dietlinde Jakobi, Matteo Villa, Marcel A. J. Somers

Life Extension of SCC Affected Urea Reactor Through Inspection, Repair & Replacement 144
Syed H. Raza, Hafiz A. R. Awan

Repair Strategies for Tube-Sheets in HP Urea Heat Exchangers 154
Alex Scheerder

Failure Analysis and New Design of Secondary Waste Heat Boiler (102-C) of Ammonia Plant at IFFCO Kalol 165
Bhupen Mehta

Enhanced Safety and Reliability in Fertiberia Palos Ammonia Plant with the Replacement of Obsolete Boilers with Casale-Arvos Double-Tube Design 171
Felipe R. F. Latorre, Heinrich Ormann, Francesco Baratto, Marco M. Carlucci

Nitriding and Crack Propagation in Waste Heat Boiler Lip Seals 180
Abdul A. Al-Shehhi, Saeed Al Mehairbi, Ganesan Balasubramanian

Failure Analysis of Syngas Piping Welds- A Case Study from an Ammonia Plant 189
Tayyab Ejaz

Meta-Analysis of 105-D Ammonia Converter Safety 198
Daragh Blennerhassett, Volodymyr Oliynyk, Sachin Kalatrakkal

Lessons Learned and Mitigation of Risks Associated with Changeout of Internals and Catalyst in North America's Oldest Ammonia Converter 208
Alexis Kimpel, Gabrielle Grills, Taylor Archer, Enrico Masciarelli, Luca Redaelli

Hidden Threat - The Weld Failure Behind a Reverse Flow and Major Unplanned Shutdown 218
Rajindra Benny, Dane Hosein

Use of Risk Based Inspection (RBI) to Demonstrate the Industrial Robustness of the Long Neck Single Nozzle ATR Design, Providing Extended Operating Cycles 230
Matt J. Cousins, Abdulaziz N. Al-Sakran

Catastrophic Tube Failures While Starting Up a Reforming Furnace 240
Ken Wohlgeschaffen

Addressing Reverse Flow Hazards in High Pressure Boiler Feed Water Pumps 246
Mark Schroeder, James McMurphy

Thrust Collar Failure on Integrally Geared Process Air Compressor in Ammonia Plant 256
Mukesh Agrawal, Novi Ariharnomo

Avoiding Fatigue Load of Equipment in Green Ammonia Plants 265
Bernd Mielke, Klaus Nölker, Felix F. Bensing

Ammonia Converter Basket and Its Catalyst Re-Placement Job at CFCL, Gadepan (Kota) 275
Amit Goyal, Akhil Nahar, Vishvajeet Joshi

Split Flow Reactor - Optimized Catalyst Converters for High-Capacity Ammonia Plants 284
Umesh Kudav, Partha P. Chowdhury, Sudesh Kumar, Shailesh Mudaliar

Safety Analysis of PT Green Ammonia Indonesia: Determination of Threat Zone using ALOHA 292
Baskara A. Nugraha, Wiratni, R. B. Cahyono

Lifecycle Integrity Management of Outlet Pigtails 301
Olivia Chung, Nicholas Goodman, Charles Thomas

Author Index

Seventy Years of AIChE Ammonia Safety Symposium – Lessons Learned

This paper discusses the history of the symposium's beginning, the historical perspective of the Ammonia Safety Committee, symposium papers, technical manuals, roundtable sessions, and special awards. A table lists all the safety incidents that have occurred during the last ten years since the symposium's 60th anniversary in Boston, MA, in 2015, and significant incidents reported in the previous decade are highlighted.

Venkat Pattabathula
SVP Chemical Plant Services

Taylor Archer
Clariant

Seshu Dharmavaram
Air Products

Introduction

It has been 112 years since the first ammonia (NH_3) plant based on the Haber-Bosch Process started production at BASF's site in Oppau, Germany, and 70 years since the first ammonia plant safety symposium was held in 1956 as part of the American Institute of Chemical Engineers (AIChE) Annual Meeting in Boston, MA. During this time, tremendous advances have been made in process technology, safety, and reliability of ammonia-related facilities. Unlike some other chemical and refining industries, which are reluctant to share information, the sharing of safety incidents and lessons learned by ammonia producers and technology suppliers at Ammonia Safety Symposia has made the production of ammonia, urea and related products one of the safest chemical processing industries.

Papers on the safety of air separation plants were presented at an "Air and Ammonia Plants Safety" session. Since then, the AIChE "Ammonia Safety Symposium" has been held annually with little change in its format. That it still functions today is a tribute to its effectiveness in reporting accidents, safety improvements, and technology developments related to the international community of fertilizer producers, suppliers, and technology providers.

Seventy years ago, a small group of engineers, chemists, technicians and plant operators came together at the annual Boston, Massachusetts, AIChE Meeting to discuss safety in ammonia and air separation units. In a paper presented at the 2005 Ammonia Safety Symposium celebrating 50 years in Toronto, Canada (**Ref 1**), Gerald Williams from Plant Surveys International, Inc., summed up the first three years of the meetings quite nicely.

Meeting Proceedings and AIChE Ammonia Safety Committee Structure

After the third Symposium (1958), the AIChE felt it should continue these symposia annually. The AIChE also thought that a dedicated subcommittee should be established to provide leadership for the meeting. They asked Norton Walton from ARCO to set up the subcommittee. He subsequently chose 14 other individuals to serve on the subcommittee. William A. Mason from Dow became the first chairperson of the Safety in Air and Ammonia Plants committee.

In late 1959, the first Technical Manual of the Air and Ammonia Plant Safety Symposia was published. The first manual, Volume 1, contained all the papers and roundtable sessions from the first three symposiums. Since then, papers and Q&A sessions have been published in different volumes (while Q&A ceased publication in 2002). Instead, the program chair posted Q&As on the AIChE website at the end of the symposium for a few years, but they were also stopped later. Table 1 summarizes dates, sites, program chairs and the publication volume for each symposium. As can be seen, numbering was not always sequential. For example, proceedings from the first three years are contained in Volume 1. Papers from the fourth symposium are included in Volume 2. After that, volumes were numbered sequentially through the 1980 symposium in Portland, Oregon. Manuals were not published for the 1981 and 1982 symposiums in Montreal, Canada and Los Angeles, California. Publication resumed in 1983 with Volume 24, with each succeeding year numbered sequentially. Electronic copies produced through the coordinated efforts of the NH_3 safety committee and AIChE of papers presented during the first 50 years were available for purchase at the 50th-anniversary meeting in Toronto, Canada. Electronic copies for the first 60 years were available at the Boston meeting in August/September 2025.

During the first few years of the symposium, meetings were chaired by industry volunteers. By 1967, the ammonia committee comprised people worldwide who organized and coordinated yearly meetings at one of the AIChE national meetings. Due to the success of the meetings, the ammonia safety symposium separated from the AIChE annual meetings and has been held as a separate stand-alone meeting since 1992, still with the help of the AIChE.

The Ammonia Safety Committee comprises 16 members representing operating plants, engineering companies, catalyst vendors, and technology suppliers worldwide. Officers include the Committee Chairperson, Secretary, Program Chair and Program Vice-Chair.

Attendance at the first few symposiums consisted of personnel from operating plants and technology suppliers, mainly from North America. To illustrate how international the meeting has become, roughly 50% of the more than 300 attendees at the 2024 San Diego symposium were outside North America. Interest in the symposium topics and information obtained from networking with other participants, among the 50% from operating companies, has been the main driver of the symposium.

The symposiums have almost always been held in North America, with two notable exceptions. In 2013, the symposium was held in Frankfurt, Germany, with a special tour of the BASF plant in nearby Ludwigshafen to see the first commercial reactor to produce ammonia from 1913. This special 100th anniversary of the commercial production of ammonia was recognized during the symposium, and the attendees enjoyed the following tour. In 2023, the symposium returned to Germany, this time in Munich. This was based on feedback from the attendees to try a location in Europe that would enable a higher attendance of local producers and suppliers to attend. It was a successful symposium both times, with high attendance levels and active participation.

Special Awards

The Safety and Health Division and the Ammonia Safety Symposium Committee have occasionally awarded special awards to individuals for their contributions to the ammonia industry.

The AIChE Safety and Health Division's (now known as Process Safety Division) Norton H. Walton/Russell L. Miller Award was presented to Guy S. LeGendre at the Tucson, Arizona, meeting in 1995. This award is for "outstanding chemical engineering contribution and achievements in the Loss Prevention, Safety and Health fields". Mr. Legendre was a former Ammonia Safety Symposium program chairperson and spent most of his career with Monsanto. This award bears the name of the Ammonia Safety Symposium's first Program Chair, Norton H Walton.

The Ammonia Safety Committee Distinguished Service Award was presented to Hays C. Mayo at the Seattle, Washington, meeting in 1999. This award was "in recognition of his contribution to the ammonia industry by way of his leadership in the evolution of the centrifugal compressor concept that has become today's standard in ammonia plant technology used worldwide. Consequently, it became possible to build larger and more efficient plants that have reduced the cost of fertilizer production to benefit the world's food production and supply." Mr. Mayo, a former Ammonia Safety Symposium program chairperson, was employed by M.W. Kellogg Co. (now KBR) while developing their single-train centrifugal ammonia plant design in the 1960s.

At the Tucson, Arizona, meeting in 2000, the Ammonia Safety Committee Distinguished Service Award was presented to Anders Nielsen. The award was "in recognition of his contribution to the ammonia industry through lifelong dedication to fundamental research in catalysts, to the development of improved catalysts for ammonia manufacture, and by his commitment to impart this knowledge through teachings and technical papers, which have contributed significantly to the progress and improvements realized in the ammonia industry". Dr Nielsen, a former Ammonia Safety Symposium program chairperson, was employed at Haldor Topsoe for about 55 years, as Director of R&D. He wrote many technical articles, including one of the standard reference books in the ammonia industry, "An Investigation on Promoted Iron Catalysts for the Synthesis of Ammonia".

Max Appl received the Ammonia Safety Committee Distinguished Service Award at the January 2002 Ammonia Safety Symposium in Montreal, Canada. The award was "in recognition of his contribution to the petrochemical manufacturing industries, particularly to the ammonia industry through his lifelong commitment to enhancing safety in these plants". He continuously exemplified BASF's leadership in the fertilizer and related industries through his extensive contributions to major innovative projects. He was a member of the AIChE Ammonia Safety Committee from 1981-1993 and program chair of the 1988 Symposium in Denver. He was primarily known for making exceptional safety and technology contributions to our industry through his numerous publications. Dr Appl spent most of his career working for BASF at the Ludwigshafen site in Germany. Some of his more prominent publications are the "Modern Ammonia Technology" series of articles in Nitrogen magazine and his book "Ammonia, Principles and Industrial Practice" published by Wiley-VCH.

All the past members of the AIChE Ammonia Safety Committee were presented with plaques recognizing their service to the global ammonia community.

Ammonia Plant Incidents

All the incidents reported from 1956 until 2004 were listed in a table in the paper presented at the 50th anniversary of the AIChE Ammonia Safety Symposium, held in Toronto in 2005 **(Ref 2).** Incidents reported from 2005 to 2014 were presented in the 60th annual symposium held in 2015 in Boston, MA **(Ref 3)**.

Incidents reported during the last 10 years, since 2015, are included in Table 2. Some of the significant incidents from which the industry could learn lessons were as follows:

1. Compressor House Fire

On 6 May 2014, the disconnection of a 12.7 mm (1/2-inch) instrumental tube caused a fire in the compressor room of Pardi's ammonia plant No.2. The failing connection, which was connected to the differential pressure differential transmitter installed between the inlet and outlet of the fourth stage in the synthesis compressor, caused leakage of high-pressure gas at the disconnection point **(Ref 4), s**ee Figures 1 and 2.

This incident created a jet fire with a flame length of approximately 10 meters (32 ft). The plant fire station, with the cooperation of the fire departments of other neighboring petrochemical companies in the Pars Special Economic Energy Zone, controlled the fire in less than an hour. The reconstruction cost was over two million USD, and the shutdown lasted 60 days.

Ammonia synthesis and refrigeration compressors, piping insulations, instrument tools, electrical cables, and part of the compressor room's structure incurred losses.

Because of the damage to the lubricating oil piping flanges and the spreading of oil, both ammonia synthesis and refrigeration compressors were engulfed in the fire and severely damaged.

The incident was caused by a collection of factors such as the improper fitting of the 12.7 mm (½-inch) tube threads, poor arrangement and securing of the tube according to the standard guidelines, possible human error, and physical damage to the instrument tube during unfastening of the scaffolds near the tubing.

The incident caused severe damage to both the synthesis and refrigeration compressors. The ammonia plant was out of service for 60 days to repair and overhaul the compressors.

Figure 1. The view of the fire rising from the roof of the compressor room

Figure 2. Ammonia Synthesis and Refrigeration Compressors were both on fire

Lessons Learned:

- A mist spray system is needed inside the compressor room.
- Increase the foam fire extinguishers and place several high-capacity foam capsules inside the compressor room.
- Low-pressure steam spray nozzles are installed inside the compressor room to extinguish small fires with steam before they spread.
- Having an action plan for fire incidents inside the ammonia plant compressor rooms.

2. Overfiring of Primary Reformer

In early 2015, when a South American ammonia plant was performing a hot restart of its ammonia plant, an uncontrolled and rapid increase in flue gas temperature was noticed in the primary reformer. After introducing natural gas feed to the primary reformer, several catalyst tubes ruptured, and the plant was tripped immediately **(Ref 5),** see Figures 3 and 4.

A thorough Root Cause Analysis (RCA) followed the tube rupture incident. The RCA consisted of a metallurgical examination of both ruptured and remaining catalyst tubes in the furnaces and analyses of the operating data collected from the Distributed Control System (DCS).

The operating data from DCS were analyzed to understand better what went wrong during the hot restart in February 2015. The actual operating data was compared with the maximum allowable parameters derived from a computer simulation of the reformer operation. The firing intensity during the start-up in February 2015 exceeded the recommended practice by the technology licensor.

Following this incident, an automated over-firing protection (OFP) management system was developed that provides four protection elements against the over-firing of primary reformer tubes.

Figure 3. Tube rupture in March 2025

Figure 4. Tubes with longitudinal cracks

Learned:

- Have an inherently safer reformer furnace management system.
- Need for well-trained operators on the potential for significant fire and explosion incidents.

3. Fire in the Synthesis Loop of the Ammonia Plant

A fire incident occurred due to localized cracking/rupture at the weld joint of an inserted spool piece, which carried synthesis gas at 285°C (545°F) and a pressure of 179.5 bar (2603.5 psig). The dislodged spool piece location is marked in Figure 5. The incident occurred on 5 November 2017 at 10:08 AM, and the fire was controlled within 15 minutes. The plant was operating normally before this incident **(Ref 6)**.

The fire damaged the equipment, pipelines, instruments, electrical cables, insulation, etc., in the vicinity, which were assessed and restored. The plant operations were fully normalized by 30 November 2017.

The dissimilar weld joint, along with stressed sites of weld imperfections, gave preferential sites in the Heat Affected Zone (HAZ) and weld for high-temperature hydrogen attack, which eventually led to brittle weld-to-weld failure and, in turn, caused the spool pieces to be torn longitudinally.

Inadequate work controls led to a lack of detection that the wrong material had been poorly welded into a critical service line, which was prone to high-temperature hydrogen attack (HTHA) in this service.

Figure 5. Location of failed spool piece of vertical piece of leg of syngas piping from 123CA exit to 123C inlet.

Root causes

a) Incorrect Pipe Material Used: Positive Material Identification (PMI) of the inserted 1.3-meter (4.25 ft) spool piece revealed that Low-temperature carbon steel A333 Grade 6 (the incorrect construction material for this service) was inserted. Since the existing material was ASTM A335 Grade P22, the 1.3-meter spool piece added should also have been of the same material. LTCS (A333 Grade 6) is not recommended per API RP 941 and has been discontinued since 1991 for such service as synthesis gas at the pipe's operating temperature.
b) Incorrect filler weld metal: The PMI of the failed weld at the welding area was compared with the PMI of other sound welds in the piping system. The analysis of the failed weld showed that welding was done with the 1¼Cr-½Mo electrode (composition like ASTM A335 Grade P11), while healthy welds in the same pipeline showed a 2¼Cr-1Mo electrode (composition like ASTM A335 Grade P22).
c) Defective welds: A visual observation of both welds on the A335 Grade P22 sides indicated a partial or complete lack of fusion at the root pass.

Lessons Learned

- Select the proper material for the operating conditions. For increased temperatures, consider using alloys with higher weight percentages of chromium and molybdenum.
- Use actual operating temperatures to assess HTHA susceptibility and validate that the actual operating temperatures and pressures are below the API RP 941 curve by a defined amount.
- Select inspection methods and frequencies to detect the initial stages of HTHA.
- For any repair/maintenance activities, a PMI test request should be initiated for material identification in case the material is sourced from the site and not withdrawn from centralized stores.
- During any equipment replacement requiring outage/modification, if the new design ratification demands any change in orientation/elevation of any nozzle, the equipment engineer should adequately flag that design as a stop-job hold point. Information should be passed on to relevant areas, such as the Piping Engineer, for proper job design.

4. Mixed Feed Crossover Piping Rupture

A longitudinal weld seam failure of mixed feed crossover piping caused a catastrophic failure of an ammonia plant primary reformer. This weld seam failure was the direct cause of an initial loss of primary containment that induced a reverse flow condition, resulting in the complete failure of other major components throughout the reformer **(Ref 7),** see figures 6 to 9.

On 10 November 2017, the primary reformer mixed feed crossover and other reformer components ruptured in an ammonia plant in Trinidad and Tobago, resulting in multiple fires. The failure occurred in a 1850 MTPD (2,040 STPD) plant. This incident was classified as a Tier 1 process safety incident. No injuries resulted from this event.

Figure 6. Ruptured mixed feed piping

Figure 7. Outlet header tee showing burn-through due to reverse flow

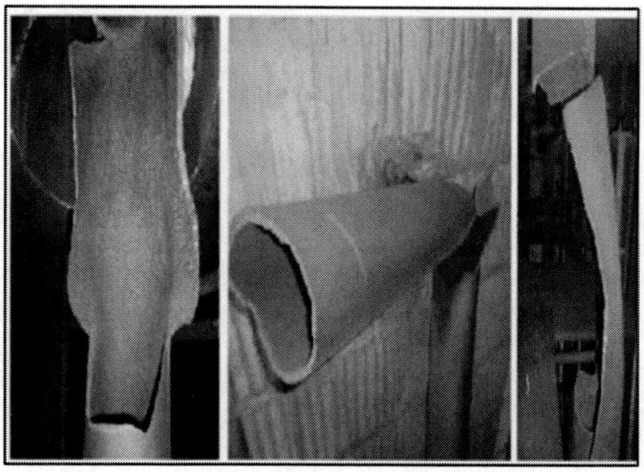

Figure 8. Risers that were ruptured and dislocated

Figure 9. Significant damage to the tunnel walls and covers

The initiating event of this Tier 1 process safety incident was the longitudinal rupture of the mixed feed crossover piping, which resulted in a reverse flow condition. The reverse flow condition caused significant transfer line damage within the radiant section, riser tube and outlet header failures, and accompanying fires and explosions within and under the primary reformer radiant box.

The causal factor was identified as hot/solidification cracking of the longitudinal weld seam during manufacture. The root cause was improper weld geometry, resulting in uneven thermal gradients during welding/cooling, which caused the creation of hot/solidification cracks.

Lessons Learned:

- Implement additional NDE requirements for new pipe procurement to improve inspection confidence
- To ensure serviceability, perform NDE on existing plant piping operating within the creep region.
- Conduct an engineering review of the reforming section to determine the need to enhance the safeguards in this plant section, including mitigation against the reverse flow.
- Conduct a review of all piping in elevated temperature and pressure services to identify the use of similar material in similar services and evaluate each case individually.

5. Dakar Incident

The Dakar ammonia accident in Senegal on March 24, 1992, is the worst ammonia industrial accident ever. This anhydrous ammonia industrial catastrophe claimed 129 lives and injured another 1,150 workers and citizens **(Ref 8), see figures 10 to 12.**

The accident happened at a peanut oil processing facility where ammonia was used to detoxify the product. Anhydrous ammonia was stored in a portable tank commissioned in 1983 and repaired in 1991 before the incident. The weld repairs made were on cracks detected on the tank's surface. Frequent overfilling of the tank ("authorized" to hold 17.7 tons) was one of the primary causes noted in the reports. An overpressure inside the tank led to its catastrophic failure, releasing 22 tons of pressurized ammonia. A heavy white cloud of ammonia aerosol plus vapor spread a significant distance, causing fatalities and injuries.

Around 33 years ago, on March 24, 1992, the accident happened at a peanut oil mill (operated by Sonacos SA) near the Dakar port in Senegal. The debris from the tank truck explosion also pierced process equipment (e.g. hose) containing liquid ammonia under pressure. The release of 22 tons of liquid ammonia was reported. A two-phase flow of ammonia fluid (vapor plus liquid as fine aerosol) formed a dense vapor cloud and spread over a significant distance, resulting in injuries and fatalities. The thick plume settled over the oil mill, nearby offices, and adjacent restaurants where people were present at lunchtime. Forty-one (41) people died immediately, and many others were transported to the nearest trauma center.

The total numbers were determined to be one hundred twenty-nine (129) fatalities and one thousand one hundred and fifty (1150) injuries.

Figure 10. Front of the tank view 1.

Figure 11. Front of tank view 2

Figure 12. The rear of the tank.

Primary Cause: Overfilling
A systematic root cause analysis of the Dakar accident can yield multiple causal factors (related to design, operation, hazard management, etc.) that resulted in the incident. However, one primary cause (overfilling) is apparent and has resulted in and continues to cause numerous incidents worldwide.

Understanding the hazards of overfilling and determining the "Filling Ratio "for various containers (cylinders, tanks, etc.) to avoid incidents like the Dakar accident has been widely recognized. Overfilling high-pressure compressed gases can result in overpressure and loss of containment.

Lessons Learned:

- A detailed analysis of the causal factors can only be done using evidence (preserved/protected) and related data from the day of the accident. After 33 years, it is almost impossible to reconstruct all the details based on the limited data currently available in public literature. However, some general lesson categories (related to technology, operations, management, etc.) and generic causes can still be extracted.
- In addition to the primary cause (i.e., overfilling) noted above, there were many failures in the following categories: technical, operations, facility/corporate leadership, government oversight, and industrial standards/governance. These are all important for the safe operation of ammonia facilities globally.

6. Secondary Reformer Outlet Header Line Rupture

On 17 February 2022, the Dyno Nobel Plant in Waggaman, LA, experienced a rupture of the Secondary Reformer outlet transfer line of its 2300 MTPD Ammonia Plant. Fortunately, no injuries were reported **(Ref 9), p**lease see Figure 13.

The shown loss of containment resulted in a significant fire, damaging adjacent equipment and instrumentation.

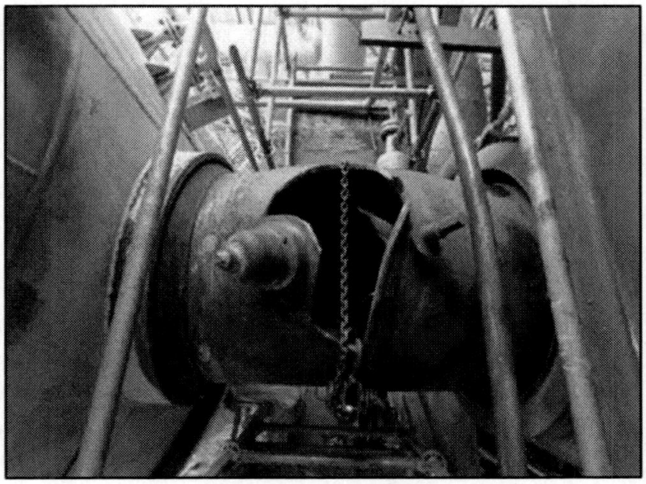

Figure 13. Transfer line rupture

At approximately 4:25 AM, the transfer line between the Secondary Reformer and the Waste Heat Boiler ruptured, resulting in the loss of containment of reformed gas and a subsequent flash fire, which damaged nearby instrumentation and structure. Figure 13 shows this rupture location. This event occurred during the late stages of a plant start-up, as ammonia conversion was observed shortly before the event. Approximately thirty seconds after the rupture, the plant was manually shut down and secured.

An unusual site event was declared, and the site's Emergency Response Team responded to the scene. The necessary notifications were made to outside regulatory agencies and the local community.

Overheating the shell was the most significant factor in the transfer line's failure. The hotspot resulted from refractory degradation and the development of a substantial process gas flow behind the degraded liner.

Lessons Learned:

- Due to the criticality of the transfer line, a robust mechanical integrity program for pressure equipment is vital for the effectiveness of key equipment integrity controls.
- Design and construction controls are considered the primary controls, and process or inspection controls contribute minimal extra integrity. If the design and construction controls are not robust, then the probability of failure of the secondary reformer to waste heat boiler transfer section will become unacceptably high.

7. Fire Beneath a Reformer Furnace

An unexpected fire occurred two weeks after a hydrogen plant's startup following a major plant turnaround. The fire happened beneath the furnace box due to the loss of containment from a radiant section process gas tube **(Ref 10).**

Damage was limited to instrumentation at the bottom of the furnace, including the actuator on the waste heat boiler temperature control (internal bypass) valve, refractory, and the furnace structure (Figures 14 and 15). No one was injured as no one was in the area when the fire occurred, and actions were immediately taken to shut down the plant and extinguish the fire safely.

Root causes:

i. The bottom flanges were not tightened adequately to ensure a tight closure while in operation.

ii. The quality control processes did not identify the compromised enclosure tightening.

iii. This gasket design experiences higher (20%) relaxation than a spiral wound gasket when process temperatures exceed 200ºF (93ºC) due to the softening of the graphite covering layers. This gasket style can mask a potential leak from an untightened flange, using the standard low-pressure nitrogen check for leaks by operators before startup.

Figure 14. Damage to the furnace structure – buckling of I-beams.

Figure 15. Damage to the furnace structure, including bulging of the furnace floor

Lessons Learned:

- Proper training in critical maintenance procedures, such as bolting radiant tube flanges, and specific work permits for this work are essential if company standards or policies do not cover them.
- Company representatives must independently audit contractors' Critical maintenance work to ensure the quality of their work meets company standards.
- Written guidelines should be followed to ensure clear expectations regarding engineering reviews of management of change (MOCs).
- Written standards for gaskets on hydrogen plant radiant tubes should include the need for self-centering gaskets and specify that corrugated metal core gaskets shall not be used in raised face flanges.

8. Syngas Compressor Oil Tank Explosion

On 27 April 2017, at 02:43 a.m., a power outage tripped an ammonia plant. Thirty minutes later, the Syngas compressor oil tank exploded. A thorough investigation identified two independent root causes that led to an unexpected and dangerous situation **(Ref 11),** see Figures 16 and 17.

The nitrogen compressor did not stop because the pressure switch on the suction side was removed. Determining why or even when the pressure switch was removed proved challenging. It probably happened when they upgraded their PLC in 2010. No updated HAZID or PHA was done for the ASU.

The concluding theory was a leak in the 4th-stage compressor end lid. Oil channels in the end lid were routed at angles to get the desired oil flow. The excess holes were plugged after drilling. The management team suspected one of these plugs was corroded and had leaked. This led to a connection between lube oil and process gas. The lube oil return was routed straight back to the tank.

Lessons Learned:

- Conduct HAZID and HAZOP revalidation sessions on all older plants.
- Focus on inherently safer design with a high-integrity safety instrumented system (SIS).

Figure 16. Syngas compressor (3rd and 4th stages) turbine after the explosion and fire

Figure 17. Top view of the Ruptured tank

Summary

The past 70 years have shown us that symposiums, with the presentation of high-quality technical papers addressing the safety concerns of the ammonia industry, question-and-answer sessions after the presentations, and frequent roundtable sessions, have been a winning factor.

The publication of all these papers in the series of Technical Manuals is also a part of this heritage. Its success is due mainly to the authors of more than 1700 technical papers. This would not have been possible without the support of the companies in this industry. Their investment in human and financial resources is a testament to their commitment to promoting safety in this industry.

Will the Symposium continue for another 70 years? If we continue to make ammonia and related products, we must sustain and improve our safety performance. This requires constant vigilance by all industry producers and suppliers.

References

(1) Gerald P. Williams, Plant Surveys International, Inc., *50-Year History of the AIChE Ammonia Safety Symposium,* Ammonia Technical Manual, 2005.

(2) V Pattabathula, B Rani and D Timbres, *The AIChE Ammonia Safety Symposium – 50 Years of Shared Experiences*, AIChE Ammonia Technical Manual, 2005.

(3) V Pattabathula, J Richardson, *Sixty Years of the AIChE Ammonia Safety Symposium*, AIChE Ammonia Technical Manual, 2015.

(4) A Orooji, S Hosseininia, EA Mavlate, *Ammonia plant's compressor room fire – Prevention and Remedies*, AIChE Ammonia Technical Manual, 2016.

(5) PB Jensen, SD Sarkar, G Schulz; *Increase safety by using an automated protection system to avoid overfiring of primary reformer*, AIChE Ammonia Technical Manual, 2018.

(6) MR Sharma, J Sondhi; *Rupture of piping and fire in an ammonia plant synthesis loop*, AIChE Ammonia Technical Manual, 2019.

(7) Z Ramdath, G Davis, *Catastrophic failure of primary reformer due to mixed feed cross-over piping rupture*, AIChE Ammonia Technical Manual, 2022.

(8) S Dharmavaram, V Pattabathula; *The Dakar ammonia accident: Analysis of the worst incident at an anhydrous ammonia user;* AIChE Ammonia Technical Manual, 2022. https://www.aiche.org/sites/default/files/cep/20230747.pdf

(9) B Woods, T Archer: *Secondary reformer outlet transfer line rupture and following effects*; AIChE Ammonia Technical Manual, 2023.

(10) B Ross, K Wohlgeschaffen, *Unusual fire beneath a steam methane reformer furnace,* AIChE Ammonia Technical Manual, 2024.

(11) S Hansen; *Syngas compressor oil tank explosion in an ammonia plant*, Ammonia Technical Manual, 2019

Table 1: 70-Year History of the AIChE Ammonia Safety Symposium Meetings

Meeting No.	Year	Month	Place	Program Chair	Chairperson's Company	Vol. No.
1	1956	Dec.	Boston, MA	Norton Walton	Atlantic Ref (ARCO)	1
2	1957	Sept.	Baltimore, MD	Norton Walton	Atlantic Ref (ARCO)	1
3	1958	Sept.	Salt Lake City, UT	Norton Walton	Atlantic Ref (ARCO)	1
4	1959	Sept.	St. Paul, MN	Bill Mason	Dow	2
5	1960	Sept.	Tulsa, OK	Gordon Weigers	American Cyanamid	3
6	1961	Sept.	Lake Placid, NY	Ray Walton	Rohm & Haas	4
7	1962	Aug.	Denver, CO	Ron Bollen	Dow – Canada	5
8	1963	Sept.	San Juan, PR	Arnold Glass	Monsanto	6
9	1964	Sept.	Las Vegas, NV	Russell Simms	Phillips	7
10	1965	Sept.	Minneapolis, MN	John Warren	DuPont	8
11	1966	Sept.	Atlantic City, NJ	Gary Stockbridge	Kaiser	9
12	1967	Sept.	Mexico City, Mex.	Fred Jones	CIL	10
13	1968	Sept.	Montreal, Can.	John A. Lawrence	CF Industries	11
14	1969	Aug.	Portland, WA	John F. Anderson	Farmland Industries	12
15	1970	Sept.	Denver, CO	Leonard Ball	Air Products	13
16	1971	Sept.	Atlantic City, NJ	Hays C. Mayo	Farmland Industries	14
17	1972	Aug.	Minneapolis, MN	John Cromeans	Consultant	15
18	1973	Sept.	Vancouver, Can.	Joe Stafford	Occidental Petroleum	16
19	1974	Aug.	Salt Lake City, UT	Gene Comeau	Farmland Industries	17
20	1975	Sept.	Boston, MA	James Finneran	Kellogg	18
21	1976	Sept.	Atlantic City, NJ	Philip Ruziska	Esso	19
22	1977	Sept.	Denver, CO	Bill Hoehing	Allied Chemical	20
23	1978	Nov.	Miami, FL	Kenneth Wright	Cominco	21
24	1979	Nov.	San Francisco, CA	Guy Legendre	Monsanto	22
25	1980	Aug.	Portland, OR	Tracy Sandow	IMC	23
26	1981	Oct.	Montreal, Can.	Jack Gearinger	DuPont	None
27	1982	Nov.	Los Angeles, CA	Jan Blanken	UKF	None
28	1983	Aug.	Denver, CO	Harvey Spangler	Farmland Industries	24
29	1984	Nov.	San Francisco, CA	John Livingstone	ICI	25
30	1985	Aug.	Seattle, WA	Gerald P. Williams	Allied Chemical	26
31	1986	Aug.	Boston, MA	Ray LeBlanc	Kellogg	27
32	1987	Aug.	Minneapolis, MN	Jim Chandler	Agrico Chemical	28
33	1988	Aug.	Denver, CO	Max Appl	BASF	29
34	1989	Nov.	San Francisco, CA	William R. Hookey	Chevron	30
35	1990	Aug.	San Diego, CA	Jerry Davis	Triad	31
36	1991	Nov.	Los Angeles, CA	Anders Nielsen	Haldor Topsoe A/S	32
37	1992	Sept.	San Antonio, TX	Phillip Nightingale	ICI	33
38	1993	Sept.	Orlando, FL	James Richardson	UCI	34
39	1994	Oct.	Vancouver, Can.	Theo Huurdeman	DSM	35
40	1995	Sept.	Tucson, AZ	W. J. Delboy	DuPont	36
41	1996	Sept.	Boston, MA	Gunnar Schulstok	Norsk Hydro	37
42	1997	Sept.	San Francisco, CA	Robert W. Clark	ICI Katalco	38
43	1998	Sept.	Charleston, SC	Ib Dybkjaer	Haldor Topsoe A/S	39
44	1999	Sept.	Seattle, WA	Richard L. Johnson	Mississippi Chemical	40
45	2000	Sept.	Tucson, AZ	Ian Welch	PCS Nitrogen	41
46	2002	Jan.	Montreal, Can.	Richard Strait	Kellogg Brown & Root	42
47	2002	Sept.	San Diego, CA	Louis Frey	CF Industries	43
48	2003	Sept.	Orlando, FL	Kevin Vick	Farmland Industries	44
49	2004	Sept.	Denver, CO	Reinhard Michel	Uhde GmbH	45
50	2005	Sept.	Toronto, Can.	Svend Erik Nielsen	Haldor Topsoe A/S	46

51	2006	Sept.	Vancouver, Can	Dorothy Shaffer	Baker Risk	47
52	2007	Sept.	Henderson, NV	Harrie Duisters	DSM	48
53	2008	Sept.	San Antonio, TX	Venkat Pattabathula	Incitec Pivot	49
54	2009	Sept.	Calgary, Can	Danny Franceus	Yara	50
55	2010	Sept.	San Francisco, CA	John Brightling	Johnson Matthey	51
56	2011	Sept.	Montreal, Can	AK Singh	IFFCO	52
57	2012	Sept.	Chicago, IL	David Pierce	Southern Ionics	53
58	2013	Sept.	Frankfurt, Germany	John Mason	Agrium	54
59	2014	Sept.	Vancouver, Can	Bob Collins	KBR	55
60	2015	Aug.	Boston, MA	Ali Jama	QAFCO	56
61	2016	Sept.	Denver, CO	Neal Barkley	CVR	57
62	2017	Sept.	Brooklyn, NY	Michael Warzee	Yara	58
63	2018	Sept.	Toronto, Can	Scott Rodrigue	CFI	59
64	2019	Sept.	San Francisco	Taylor Archer	Clariant	60
65	2021	Sept.	Virtual	Eugene Britton	CFI	61
66	2022	Sept.	Chicago, IL	Klaus Noelker	Thyssenkrupp Uhde	62
67	2023	Aug.	Munich, Germany	Ahmed E Rahmi	QAFCO	63
68	2024	Sept.	San Diego, CA	Mohamad Noueiri	Yara	64
69	2025	Sept.	Atlanta, GA	Marc Gilbertson	EDNF	65

Table -2: AIChE Ammonia Safety Symposium: Incidents/ Fires/ Explosions/ Failures 2015 to 2024

Item	Year Presented	Incident	Root Cause	Mitigation
1	2015	Damages to the horizontal converter basket	Reverse flow	Repaired the basket internals and loaded with new catalyst
2	2015	Synloop WHB failure	Chloride SCC	Replaced with improved materials
3	2015	Explosion in the CO2 removal unit	Wet CO2 corrosion	Installed check valves and replaced elbow materials with SS
4	2015	NH3 converter shell cracks	Hydrogen-induced cracking	Welded nozzle with a blind flange
5	2015	Primary reformer coking	Passing NG feed gas valves	Replaced the catalyst
6	2015	Fouling in process gas cooler	Catalyst and refractory dust	Replaced alumina lumps with corundum balls
7	2016	Failure of the secondary reformer dome	High thermal stresses, caused by the different thermal expansions of the refractory brick dome and the water-cooled shell.	Replaced the dome with a novel design using readily available alumina blocks.
8	2016	Failure of Level Bridle in Potassium Carbonate Service	Localized internal corrosion, as per the materials analysis, had caused a mechanical overload.	Restorative work, including demolition, material procurement, fabrication, NDE, and insulation took seven days
9	2016	Ammonia Synthesis Converter Reaction Loss Followed by Fire	A rapid pressure increase and temperature decrease caused synthesis gas leakage at the ammonia converter outlet flange, resulting in a fire.	Fatima has upgraded the liner material from SS304 to SS321 to avoid brittle damage.
10	2016	Severe Cracks on a New Duplex Syngas Loop Cooler	A combination of wrong welding parameters and insufficient hydraulic expansion of the tubes into the tube sheet caused the welds to fracture badly.	Yara selected mechanical rather than hydraulic expansion based on experiments and calculations.
11	2016	Ammonia Plants Compressor Room Fire: Prevention and Remedies	A half-inch instrument tube became disconnected and caused a fire in the compressor room. The failure of this tube, which was installed between the inlet and outlet of the fourth stage of the synthesis compressor, caused high-pressure gas to leak, creating a jet fire with a flame length of approximately 10 meters.	i. A closed-circuit camera inside the compressor house was installed for online monitoring and recording. ii. A water mist spray system was installed inside the compressor house. iii. High-pressure instrument tubes were replaced with piping. iv. The number of foam fire extinguishers was increased, and several capacity foam capsules were placed inside the compressor room.

				v. Low-pressure steam spray nozzles were created inside the compressor room to extinguish small fires with steam before their spread. vi. A solenoid valve in the steam line was installed in the oil turbine pump to shut it down immediately in case of fire.
12	2016	Failure of reformer inlet pigtails	An interaction of creep, low-cycle fatigue, and high-temperature oxidation after 43 years.	Replacement of pigtails
13	2016	Failure of the riser between the WHB and the steam drum	Sudden overload due to upset process conditions	Improved water treatment and changes in SOPs.
14	2016	Cracks between the reformer/riser tubes and the weldolet of the outlet header	The difference in thermal expansion coefficient between Ni-base weld metal (Alloy 625) and the austenitic Alloy 800H base metal.	Weldolet was repaired.
15	2016	Syngas compressor hydro- dynamic thrust bearing failure	Malfunction of the controlling instrumentation.	A new protection trip on dP between the makeup and recycling stages.
16	2017	Wetting of the Low temp shift (LTS) converter Catalyst.	Leak in BFW Preheater Exchanger	The catalyst was dried slowly with nitrogen, and its activity was fully recovered.
17	2017	Syngas Compressor Failure	Reverse rotation occurred from the fast depressurization of the hot synthesis loop due to the gas passing through synloop shutoff valves.	Compressor labyrinths were replaced. The thrust bearings of the HP case were also replaced.
18	2017	Process Air Compressor Damage	Lack of positive isolation and inadequate LOTO.	The air compressor discharge, XCV702 actuator, was replaced with the proper size one. Positive isolation of the equipment has been incorporated into the work permit. The LOTO (lockout tag out) system became mandatory for any equipment handed over to maintenance.
19	2017	Pool Condenser Tubesheet Damaged	Poor quality of the steam condensate through the U-bundle, resulting in Flow Accelerated Corrosion of the carbon steel tube sheet just underneath the stainless-steel weld overlay	A unique and delicate repair was carried out to restore the integrity of the tubesheet.

20	2017	Failure of Riser Transition Assembly	Insulated riser transition assembly	Heat Treatments to Susceptible Components Non-destructive Evaluation Adherence to MOC Procedures Review of As-Built and Vendor Drawings.
21	2018	Failure of HP Boiler Feed Water Pump	Reverse flow of BFW from the steam drum due to check valve failure.	Modified the ARCNRV and installed XVs on pump discharge lines.
22	2018	Hot Spots in Refractory Lined High-Temperature Equipment	Damage to the refractory over time	External cooling and temperature monitoring until the line was replaced
23	2018	A Rupture in the High-Pressure Drain Line	Stress Corrosion Cracking	a) Use resistant materials for construction. b) When hydro testing, use low chloride content water and dry out thoroughly and quickly. c) Properly applied coatings under insulation. d) Avoid designs that allow stagnant regions where chlorides can concentrate or deposit.
24	2018	Secondary Urea Reactor Top Head and Diaphragm Failure	The diaphragm was constructed with 316 stainless steels rather than the correct 317L.	The MTR (material test requirements) from the OEM shop that repaired the head after the 2010 incident indicated that the diaphragm was made of the correct grade of stainless steel.
25	2018	The Reformer resulted in several tube failures	Overfiring of Primary Reformer	Installed the over-firing protection management system (OFP), which included both 'local' and 'global' to prevent the catalyst tubes from overheating.
26	2018	Reformer Post Combustion Problem	Deviations from the design during the burners' manufacturing and installation resulted in poor combustion in the furnace overall.	Post-combustion issues were resolved after the original revamped design was fabricated and new burners installed, providing significant safe operating margins and improving the reformer's reliability and life.
27	2019	Syngas Compressor Oil Tank Explosion in Ammonia Plant	Oil vapours containing hydrogen were gassed from the oil to a limited extent and ignited as oxygen-enriched ambient air entered the tank.	The low-pressure switch on the suction side of the compressor was reinstalled. Performed a complete HAZOP for the ASU.
28	2019	Process gas cooler tube sheet thermal disintegration	Fibre materials were not resistant to the process gas atmosphere. Bypass online was restricted in axial movement. The gap between the ferrules and the tube sheet was not closed.	Changing the original construction to a welded design eliminates the risk of oversized flanges (chambers to the main PGC body). Improved ferrule and in-liner design, Thermocouples for online tube sheet temperature monitoring.

29	2019	Repair of an HP Scrubber Sphere	Condensation corrosion on liner plates and the leak The leak detection system did not detect a leak in the liner The carbon-steel pressure-retaining part went unnoticed.	Implement a reliable and robust leak detection system. Act immediately upon a leak in the stainless-steel barrier. Apply Safurex® as a protective barrier: Not susceptible to SCC. Low-strength steel is preferred for the pressure shell (mono-wall design). Apply adequate steam tracing to keep the temperature above the dew point.
30	2019	Damage to the Combustion Air Ducting of an Ammonia Primary Reformer	The reverse flow of hot flue gas from the convection section to the radiant section, up through the cold air doors (CAD) to the Hot Air Doors and out towards the atmosphere	Modified Cold and Hot Air Doors Logic Amend the Standard Operating Procedures (SOP's). Install Temperature Indicators (TI's) on all six (6) CADs.
31	2019	Secondary Reformer Loss of Containment Due to Air Pipe Failure	Minimal QA/QC restrained differential thermal expansion leading to thermal fatigue cracks ~ 24 thermal cycles until failure	• Strict QA/QC • reduce thermal cycling if possible • Material considerations • establish a more formal approach to analyzing process abnormalities
32	2019	Weld Damage and Metal Dusting in Ammonia Plant Front End Equipment	A bimetallic weld had suffered from metal dusting. This damage caused a leak in 2006 and a subsequent bundle replacement.	A lot of repair work was carried out on 102c in 2018. A new Inco 800H liner is in the cone, and the Inco 601 is in the riser, connecting to 112D. (100% replacement).
33	2019	Rupture of Piping and Fire in Synthesis Loop of an Ammonia Plant	Incorrect Pipe Spool Material Used: A333 Grade 6 Spool piece used in place of A335 Grade P22 Defective welds - On A335 Grade P22 sides – partial/complete lack of fusion at the root pass Incorrect Filler Weld Metal Visual - Failed spool piece revealed blisters on the surfaces	Select proper material for the operating conditions, validate the actual operating temperatures and pressure limits per the API RP 941 curve, and ensure that the operation is in a safer zone by a sufficient margin. For any repair/maintenance activities, a PMI test request should be initiated for material identification in case the material is sourced from the site and not withdrawn from centralized stores.
34	2019	Poisoning of Synthesis Catalyst	Process gas bypass of the (114C) methanator feed effluent exchanger.	The repair of the 114-C was accomplished safely.

35	2019	Methanator Vessel Overheating Why Still a "Hot Topic"?	The low flow trip for the lean solution bulk stream was not set high enough to prevent the increase in CO2 content.	Ensure that the CO2 removal and methanator trip systems are working properly Trip thermocouples are correctly located, working and verified Maintenance and inspection routines are correctly executed on trip systems, trip valves, solenoids, relays, and thermocouples.
36	2019	Cracked dissimilar welds in outlet headers of a primary reformer after 34 years of operation	The combination of the dissimilar field weld (base metal & filler alloy 625) and the operating conditions in the material's creep range initiated damage from within the header wall. The difference in thermal expansion of both materials results in very high stress levels, which causes damage.	Matching welds for the field welds connecting the radiant harps Elimination of weld-o-let Upgraded metallurgy and sizing to minimize stress because of the difference in elongation from the thermal expansion of the catalyst pipes compared to the risers Eliminate several critical welds that are hard to inspect.
37	2021	Failure of the hot-walled ammonia synthesis converter	Hydrogen-assisted cracking resulted from a combination of nitriding and hydrogen-induced embrittlement	Adjust inspection planning for all converters. Make sure TOFD and PAUT are performed to find cracks in all orientations. Establish MTP curves for all converters and adjust operating procedures if required.
38	2021	A major failure of the ammonia synthesis converter internals	Overpressure of the converter during startup	Minimum stop of 15% on Valve #3. Installation of DCS indications with alarms for the differential pressure between the inlet and outlet of Interchanger #1.
39	2021	Corrosion failure of a new CO2 stripper reboiler	High heat transfer at one end of the exchanger led to excessive boiling and corrosion of the carbon steel shell & bolting.	The shell was repaired, and a butterfly valve was installed to throttle the flow to the cooler side of the reboiler.
40	2021	Failure of the HP steam superheater header	Stress relaxation cracking	The header was replaced.
41	2021	Urea reactor shell rupture	Leakage in the liner and blockages of weep tubes during operation resulted in process media entering the shell in interlayers, exerting pressure on the outer layer. Thickness reduction due to corrosion under insulation and a localized reduction in thickness due to a pre-existing crack made the outer layer unable to withstand the operating pressure. This led to bulging of the outer shell layer and subsequent shell rupture.	Approved contractors should perform liner fabrication and welding. Liner welding should be performed using fixtures to restrict the lifting of the liner during welding and create excessive air gaps. The optimum weld sequence should be developed to reduce the liner shrinkage during installation. The weld process parameters and filler wire selection should be optimized to restrict the heat input during pass partition plate welds. All attachments on liners shall be full penetration welds.

				Vent holes should be inspected and verified as open following the erection of lined vessels and during each turnaround. As the process licensor specifies, periodic quality checks for the weep tube leak determination system are mandatory.
42	2021	An explosion inside the urea reactor	The most likely cause is the formation of a combustible mixture that explodes due to self-ignition or ignition caused by a static discharge. •High hydrogen content in carbon dioxide •Accumulation of inerts	• A hydrogen analyzer has been installed in the carbon dioxide feed to the reactor. •The startup and shutdown procedures have been adapted to prevent accumulation. The vent valve will not be fully closed anymore during these periods to allow the inert to be vented to the downstream sections.
43	2021	Corrosion under insulation	Chloride Stress Corrosion Cracking (Cl-SCC) is caused by water containing chlorides.	The use of thermal insulation for services that required insulation just for personnel protection was stopped. Improved practices around preparing metal surfaces before coating application and selecting a coating for new insulation applications and repairs. The use of piping for rigging and support during maintenance activities was abandoned to prevent damage to insulation. Insulation craftsmanship and caulking methods were reviewed Annual insulation surveys by the operations team to identify deficiencies and avoid water ingress by repairing insulation damage. Repositioning of supports and brackets where insulation is replaced, and avoiding couplings under insulation lagging Using organic coatings on carbon steel equipment, use aluminium foil to protect stainless steel.
44	2021	LTS harsh conditions	Under the risk of condensation (wetting), Sulphur poisoning, and high temperatures (due to poor temperature control).	Damage to wetted LTS catalyst can be minimized by using an appropriate dry-out procedure before restarting production.

45	2021	Repair of the second ammonia converter	Failure was caused by a combination of factors, including operating in hydrogen service, high temperature, and high pressure. The enhanced F-22 material used in the original R-0502 was more brittle and susceptible to cracking than standard F-22 material.	The reactor was replaced with a new vessel.
46	2021	Urea reactor failures	Generalized deterioration of the liner base metal and seam welds could have led to potential catastrophic failure in the worst-case scenario.	The repair process consisted of installing liner bands in 25Cr22Ni2Mo material on top of the existing circumferential liner seam welds, installing new tray support cleats, and refurbishing the remaining corroded liner seam welds by weld build-up. The new liner bands were connected to the existing leak detection system behind the liner to convey unforeseen leaks toward the weep tubes.
47	2022	Catastrophic Failure of Primary Reformer Mixed Feed Crossover Piping Rupture	Due to incorrect weld geometry, the failure mechanism was determined to be hot/solidification cracking during fabrication.	• Implemented additional NDE requirements for new pipe procurement to improve inspection confidence. • Performed NDE on existing plant piping operating within the creep region to ensure serviceability. • Conducted an engineering review of the reforming section to determine the need to enhance the safeguards in this plant section, including mitigation against reverse flow.
48	2022	Primary Reformer Air-Steam Coil Alloy 800HT Tube Failure	Creep and high-temperature degradation on the Alloy 800HT Air-Steam Coil tubes in a Primary Reformer Convection Section	Schedule and conduct inspections of these tubes during turnarounds to ascertain their integrity. This material is susceptible to creep and high-temperature degradation over a prolonged period. Based on inspection and metallographic examination results, aging facilities should be planned for replacements.
49	2022	Ammonia Release During Ammonia Import Activity	The quick connect/disconnect coupler disconnecting from the ship manifold flange	• Procedural checks throughout the job cycle, • Preventive maintenance is designed in conjunction with the OEM and • Project design reviews utilizing human factors analysis.

50	2022	Methanator Temperature Runaway Results in a Fire	The solvent's water content was not checked to verify that it was low. More importantly, the reactor was not bypassed when the outlet temperature rose above 850°F (454°C).	Plants are now equipped with highly instrumented trip systems that, using a voting system, quickly bypass methanation at high bed temperatures.
51	2022	The Dakar Ammonia Accident: Analysis of the Worst Incident at an Ammonia User	The reports noted frequent overfilling of the tank ("authorized" to hold 17.7 tonnes) as one of the primary causes.	Continue developing and implementing standards for safe designs, operations, and governance, which will improve process safety performance at anhydrous ammonia storage and handling facilities.
52	2023	Secondary Reformer Outlet Transfer Line Rupture and Following Effects	The overheating of the shell was the most significant factor in the failure of the transfer line.	The transfer line was rebuilt to the intended design with more rigorous QA/QC controls to ensure a higher liner standard and refractory installation.
53	2023	Failure of Secondary Reformer due to Top Dish Head Rupture, its refurbishment and Root Cause Analysis	The root cause of the failure was weakness in the design, which the signer improved in later years at new plants.	A robust action plan was developed to address vulnerable areas and mitigate risks associated with ageing plants. It includes improved liaison with the process licensor and document sharing, implementation of RBI and PSM systems, and establishment of an Asset Reliability and Integrity department.
54	2023	Mixed Feed Reheater 304H Stainless Steel Outlet Manifold and Piping Failures	304H stainless steel is susceptible to sigma phase embrittlement after prolonged high-temperature operation.	The failed sections were repaired and partially replaced to return the plant to production, with the intent of a complete replacement during the next turnaround (TAR).
55	2023	Damaging Effect of Two-Phase Flow on High-Speed Centrifugal Air Compressor	Due to liquid carryover from the third-stage cooler to the fourth-stage suction, the high-pressure rotor was subjected to heavy erosion on the suction impellers. This caused a deep cut in the inlet blade tips and an imbalance, damaging the rotor.	The mechanical trap was replaced with a larger capacity (1 inch to 2 inches), and a sight glass and electronic sensor connected to the DCS ensured it was always in service.
56	2023	Hydrogen Plant Relief Line Failure Due to Acoustic Vibrations	The investigation determined that the failure mechanism was acoustic-induced vibrations (AIV), a fast-acting damage mechanism that cannot be identified through usual long-term inspection/monitoring.	To raise awareness of AIV, the operator training program was updated to include AIV awareness and the risk associated with venting full process flow through atypical vent locations. Cautions were added to the normal startup and shutdown procedures to warn against routing the complete process flow through these vent lines. These lines were also locked closed (due to the low frequency of these valves being used) and tagged with the same warning.

57	2023	Unprecedented Catalyst Migration Problem of 2000 MTPD Ammonia Plant	The ammonia converter basket breakdown was causing unprecedented constant migration of catalyst to some synthesis-loop equipment and increasing pressure drop, most notably on the Unitized Chiller tube annulus. The plant rate was reduced to 75%, and was at risk of getting lower than the turn-down rate.	The Ammonia Converter basket was repaired, and many methods and efforts were made to remove the migrated catalyst from the equipment, especially the unitized chiller tube annulus.
58	2024	Primary reformer riser transitional assembly failure	It was determined that insufficient quality assurance and control (QA/QC) during the installation of the 2009 transition cone were the primary causes of the failure.	Temporary repairs were carried out to return the plant to service during the 2023 outage, and it was planned to replace all riser transitional assemblies (RTAS) during the next scheduled turnaround.
59	2024	Fire beneath a reforming furnace	The bottom flanges were not tightened adequately to ensure a tight closure while in operation. The quality control processes were inadequate in identifying the compromised enclosure tightening.	The contracting company has developed a certified training program to familiarize all levels of workers with contractors' procedures, including a bolting training program and competency training assessment for its workers. The operating company has also. enhanced its in-house Quality Assurance process with defined roles and responsibilities.
60	2024	Hydrogen-assisted cracking (HAC) in an ammonia synloop	Hydrogen-assisted cracking (HAC) may occur in low-alloy steel welded components in hydrogen service. It is typically seen in the heat-affected zones of equipment operating at high hydrogen partial pressure (>7 MPa/1015 psi) and operating wall metal temperatures (>250°C).	A detailed analysis of the operating conditions and the definition of specific minimum pressure station temperature curves, considering the amount of available hydrogen, the geometry of the equipment, and the Operating loads, have been identified as the best approach to mitigating the risk of HAC in combination with a proper insection plan involving advanced NDT techniques.
61	2024	Secondary reformer waste heat boiler tube rupture	Short-term overheating during startup; insufficient water flow in boiler tubes, corrosion, scaling, high process gas temperature, tube side overpressure,	The startup has emphasized the importance of controlling the temperature rate increase at the WHB's inlet, and a new temperature rate indicator has been installed there to prevent similar failures in the future.
62	2024	Catalyst overheating due to feed preheat coil failure.	The failure of the feed preheat coil resulted from a disruption in the vacuum system, subsequently influencing the speed of all compressors.	Several programs have been implemented to prevent similar incidents in the future, including adding a low-flow feed gas sensor downstream of the feed preheat coil and replacing the feed preheater coil, ensuring that all pressure gauge indications in the vacuum system can be read correctly and regularly maintained, routine checks of piping in the vacuum system, and reviewing the existing interlock system for potential unmitigated risk.

63	2024	Failure of primary reformer tubes	Root cause analysis indicated that the rupture of the primary reformer tube was caused by creep initiated by a hot spot in the primary reformer tube. Field measurement results showed a significant increase in tube wall temperature, which reached a peak of more than 922 °C. High temperatures on the tube catalyst induced carbon formation on the catalyst's active sites, causing metal sintering. This led to further catalyst deactivation, which worsened the tube wall temperature.	The most effective method to prevent primary reformer tube failure is to thoroughly examine the composition of natural gas entering the primary reformer and adjust operating parameters based on the data. The tube wall temperature of the primary reformer tube needs to be checked to verify the burner adjustment.
64	2024	Syngas compressor oil-seal failure with fire	A failure analysis study was carried out using the fault tree method. The results indicated no damage to the compressor and seals, and it became clear how the fire had occurred by comparing the operational data under normal conditions versus upset conditions and linking the main events from the blackout until the end of the emergency. With the loss of the barrier fluid and the pressurization of the oil system with synthesis gas coming from the compressor's internal pressure, the gas escaped with oil through the atmospheric vents, creating conditions conducive to the start of a fire.	Short-term remedial actions included rerouting atmospheric vents containing synthesis gas to suitable locations, i.e. from oil reservoirs and coupling covers and degassing the entire oil inventory with thermo-vacuum equipment. Longer-term actions included the installation of a water mist firefighting system connected to a fire detection system, a hydrogen detection sensor and a nitrogen-smothering system in the oil tank, installation of an automatic surge control system, and re-evaluation of steam and power supply during different upset and emergency scenarios.

Transfer of Ammonia: Understanding and Reviewing Safety of Transfer Operations

Transferring ammonia is a concern for our industry both inside and outside the plant fence. Ammonia clouds can be toxic (the Immediately Dangerous to Life or Health, IDLH, for ammonia is 300 ppm), impacting operators for even small releases, but large releases and their easily noticed odor can impact an entire community. Ammonia transfer operations happen regularly—even many times a day for truck loading operations--which increases the risk of loss of containment. The sheer number of transfers can also make the operation seem routine, inviting complacency over time that increases the potential risk of human errors. Robust systems to minimize the risks to personnel and the surrounding community are essential. A recent paper by Yara provided a good review of the root cause of some of those failures and introduced a practical approach to mitigating the associated risks. This paper provides a wraparound review of transfer operations, historical trends, and challenges to mitigation of those risks. Key Safety Practice Safeguards that ammonia and other industries are using to identify and prevent incidents are discussed.

Stuart Powell, Dorothy Maxwell Shaffer, Anthony Sarrack Baker Engineering and Risk Consultants, Inc. (BakerRisk®)

Introduction

This document investigates historical data on the release frequency during transfer operations from ammonia manufacturers, and discusses the value of modeling loss of containment of potential ammonia releases to compare, on a theoretical basis, the extent of ammonia clouds during transfer. Finally, the results of a conversation by several hazardous materials experts, including Exxon—a major player in hazardous chemical operations—explores the factors that can truly make a step change improvement in addressing challenges with transfer operation safety, and presents good practices that could be considered and/or adopted by the ammonia industry to provide an extra measure of safety.

1) Reported Incidents Review

It is well known in the processing industry that loading and unloading operations are some of the most risk-prone operations that are performed at a processing or storage facility. The risk comes from the inherently manual nature of the work and the frequency at which the work is performed. To reduce the risk, loading personnel are extensively trained. Additionally, advances in leak detection technology and improvements in Mechanical Integrity (M.I.) processes make the detection of equipment degradation more accurate, which can help ward off issues before they harm others. Advancements in tools and operating procedures have also made the operation less risky than in the past. Despite all this, there is still a human element involved in the connection and disconnection process and required maintenance of equipment and safety systems that require diligent and continuous attention for a safe and efficient operation.

In the US, environmental regulations require a company to report releases of certain regulated substances including ammonia, when they occur, to the Environmental Protection Agency (EPA)[1].

The release reports are publicly available online through the EPA Risk Management Program (RMP). Data has been collected for loss of containment events in the last 30 years that have involved ammonia production companies (primarily North American Industry Classification System, NAICS, 325311 Classification, to be specific) for transfer related equipment such as hoses, railcar valves, relief valves on trucks/rails, loading arms, etc. Data from a total of 27 transfer related events were identified and compiled after reviewing data from all 50 states, and the number of events reported, regardless of the actual consequence onsite or offsite, was compiled into Figure 1 below.

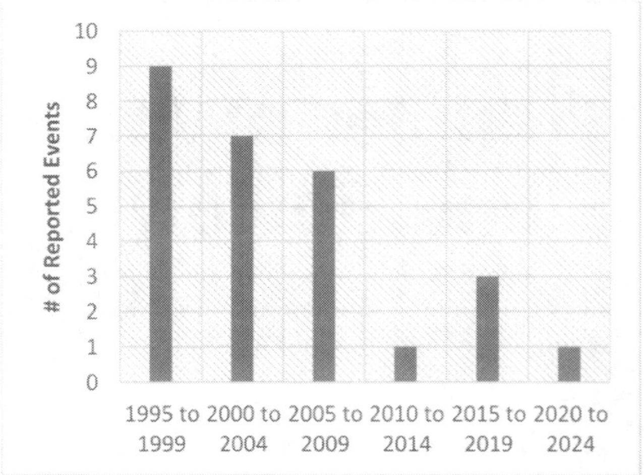

Figure 1. EPA RMP-site Nitrogenous Manufacturing Incident Reported Releases – Transfer Related

The incidents reported to the EPA also require information such as the number of people onsite and offsite were affected, what the initiating event was thought to be, and other contributing factors. Additional research can be done to further dissect the information in EPA RMP to determine the impact that ammonia releases have had on populations, although that was not the intent of this paper.

Looking at the data above, assuming that companies are abiding by the reporting requirements from EPA, the ammonia producers' attention is clearly improving safety during transfers. The entire industry can benefit from producers with deep experience sharing their expertise and learnings to maintain safety along with those new to the ammonia industry with the increased push for cleaner energy. Another key point is that the current ammonia industry can benefit from the expertise of new producers with extensive experience with other chemical hazards, who have extensive experience in developing standards and successful hazard control practices as well.

[1] "Reporting Requirements for Continuous Releases of Hazardous Substances, A Guide for Facilities on Compliance", https://www.epa.gov/epcra/reporting-requirements-continuous-releases-hazardous-substances-guide-facilities-compliance

2) Modeling Loss of Containment

When it comes to the initial design of a chemical plant, locating the area where loading/unloading will occur is of vital importance in regard to where occupied buildings should be located, or even areas outside the fence line where the public could be located. If design engineers know how far a toxic cloud could potentially travel from a loss of containment scenario, then they can be better equipped to seek to locate the loading/unloading operation away from potential occupied locations, both onsite and offsite.

One incident that highlights the need to understand how far toxic clouds can go was the Tanner Industries anhydrous ammonia release back in 2009.[2] During offloading of anhydrous ammonia, operators did not make it a point to check the proper hose to use, leading to a loss of containment that resulted in 1 offsite fatality on a nearby highway.

Studies such as Process Hazard Analysis (PHA) and Facility Siting Studies (FSS) aid in this analysis as they look at different aspects of a process and seek to answer questions such as:

1. Based on the current design, what could go wrong?
2. If something does go wrong, what could the impacts be to personnel nearby or the environment?
3. What safety systems can be added to the design to either prevent the scenario from occurring, or to mitigate the impact if it occurs?
4. What emergency response can be used to mitigate the impact of the release at the site and local area?

When performing an FSS, the recommended practice that is usually utilized is the guidance provided in American Petroleum Institute (API) 752,[3] 753,[4] and 756.[5] These recommended practices address "occupied" structures where personnel are planned to be located, and the hazards that could impact them. With that determined, risks beyond the company's acceptable criteria can be mitigated either on a consequence or risk basis. These measures can relocate the building, relocate personnel, choose a different building construction type, apply mitigations at the building to prevent toxic air ingress, or apply additional mitigations at the release source to prevent, or at least minimize, the probability that the full event occurs.

As part of the guidance in the APIs, the scenario selection is critical to determining the hazards to model. Care must be taken to consider what a 'credible' event is. Historically, a 2-inch diameter hole size has been applied for 'credible events' in consequence-based studies. That said, industry has seen from history[6] that larger than 2-inch releases have occurred, and thus consideration should be given for modeling greater than 2-inch releases when performing an FSS in accordance with API. In addition, with the development of risk-based approaches that consider the frequency and probability of factors such as release size, ignition, wind direction, etc., more and more studies are being done using a larger hole size that

[2] "Cargo Hose Rupture and Release of Anhydrous Ammonia
During Offloading of a Werner Transportation Services Cargo Tank Motor Vehicle at the Tanner Industries Plant Swansea, South Carolina", published by the National Transportation Safety Board, April 2012.

[3] API RP 752: API (American Petroleum Institute) "Management of Hazards Associated with Location of Process Plant Permanent Buildings," API Recommended Practice 752, 4th Ed., Washington D.C, 2024.

[4] API RP 753: API (American Petroleum Institute) "Management of Hazards Associated with Location of Process Plant Portable Buildings," API Recommended Practice 753, 2nd Ed., Washington, D.C., 2024.

[5] API RP 756: API (American Petroleum Institute) "Management of Hazards Associated with Location of Process Plant Tents," API Recommended Practice 756, 2nd Ed., Washington D.C., 2024.

[6] "Facility Siting - Negotiating the Minefield of "Maximum Credible Event", Graham, K., et. al., 17th AIChE Annual Global Congress on Process Safety, April 2021.

accounts for the lower frequency of release relative to smaller release sizes. Regardless, API does let the user dictate the 'credible release' scenarios to model when performing the analysis.

The distance a toxic plume travels is largely dependent on the process conditions at which the transfer operation is occurring, and the weather conditions present at the facility. One of the major inputs to perform this dispersion modeling is the hole size of the hose/loading arm that is used in the operation. Usually, for barge/ship loading, a loading arm is used that might have a larger diameter than a transfer hose for a truck/railcar, in order to achieve higher flow rates. Because of the larger diameter and flow rates, toxic clouds from a loss of containment at a ship/barge could be much larger than what is produced by a truck/rail car as a result. Understanding of this is critical for owners to implement necessary safeguards to mitigate a release, and to work to locate these operations away from potential receptors.

Additional considerations are needed for assessing the air ingress into the potentially occupied building, but knowing how far a nearby toxic release could go can help engineers and designers implement adequate spacing of occupied buildings from potential release sources during the design phase of a potential chemical plant. Doing so helps to minimize impacts to occupants of buildings and is generally a lower capital cost than designing buildings with toxic gas detection and shutdown capabilities.

3) Transfer Operations: Safeguards

Many papers have been published on the factors that lead to the loss of containment of process fluid during transfer operations. One recent paper highlights several safeguards to consider during this operation that has a relatively high likelihood of an accidental release, as compared to fixed equipment.[7] In addition, standards such as the Occupational Safety and Health Administration (OSHA) 1910.111 – "Storage and Handling of anhydrous ammonia" for those outside ammonia manufacturing, US Department of Transportation (DOT) MC331 Regulations on transfer of compressed gas, Fertilizer Europe's Guidance for transporting ammonia in rail, and CGA G-2.1-2021 from the Compressed Gas Association (CGA) provide additional resources for key safeguards for those involved in the operation.

Using failure rate data from known industry sources, it is possible to identify the causes of an accidental release during a transfer and to estimate the likelihood of each of these causes. Because this data is often difficult to attain, BakerRisk is collecting the failure rate data for transfer operations equipment to help ensure that risk mitigation decisions can be effectively evaluated. Doing so can help operating companies identify methods of reducing the likelihood of accidental releases and the risk associated with transfer operations, which in turn, contributes to the potential for making a cost/benefit analysis once the consequences, risk, risk reduction, and cost information have all been obtained.

For instance, in the case of truck loading/ unloading of NH_3, unintentional pull away while the transfer line is pressurized is a concern that gets raised in PHAs. A site or company may review its own historical data on these events and determine that, historically, this has occurred once every 10,000 transfers (1E-4 per transfer). The typical safeguards that can be applied to these transfers to reduce the probability of loss of containment include: (1) driver controls /wheel checks, and (2) Interlocks/hose replacement policies. Assuming these controls are not currently in place, a site may elect to give 1 credit, similar to a Layer of Protection Analysis (LOPA), for each of these items such that if driver controls/chocking is implemented, the probability of occurrence drops to 1E-5 per transfer, and if protective interlocks are added that prevent the

[7] "Experiences from ammonia loading incidents", by Mathieu Tagliaferri (YARA), Vincent Duponchel (YARA), published by Ammonia Technical Manual, 2023.

vehicle from pulling away without disconnecting as well, then the probability drops to 1E-6 per transfer. Risk calculations can be performed with such changes to quantify the safety benefit they would afford. The cost of implementing the changes can be compared to the safety benefit they would afford to determine if the changes are practical. A similar exercise can be utilized for other failure mechanisms, utilizing the safeguards that are in place as the baseline, and considering additions to reduce the likelihood of occurrence of loss of containment.

The causes of accidental releases during ship/barge loading of NH_3 differ from those of loading NH_3 onto railcars/trucks, as do the safeguards that can be implemented to reduce the likelihood of an accidental release. Care should be taken to understand the different causes of accidental releases for the type of loading operation being performed. For instance, for some truck/rail loading operations, hoses could have a higher susceptibility to being dragged and thus hose degradation for truck/rail can be higher than for ship/barge, if hoses are utilized. In addition, transfer to ship/barge has to consider the movement of the vessel from waves that may strain the hose or transfer arm.

Other causes of an accidental release during transfer of NH_3 to a truck, railcar, barge, or ship include:

- Overpressure
  - Is the receiving vessel susceptible to overpressure?
  - Is there a PSV to prevent overpressure and how reliable is it?
  - Does it have a rupture disc or other protective device to prevent vessel failure?
- Corrosion
  - Is the transfer equipment susceptible to corrosion for this operation?
  - What protections are in place?
- Noncompatible material transfer
  - Is the receiving vessel solely used for ammonia?
  - Are incompatible materials unloaded in the same area and could be connected to the wrong line?
- Pull away
  - What type of protections are available for the hose connection?
  - What controls are in place for wheel checks?
  - What are the procedures for drivers regarding preventing pull away?
- Improper connection
  - What type of coupling is used?
  - How well trained are the operators?
  - Is the hose pressure tested prior to transfer?
- Overfill
  - Are safeguards in place to detect high level and shutdown feed?
  - Have overfill lines been considered in the design?
- Line rupture
  - Are excess flow valves present?
  - How often are they functionally tested?

4) Key Conversation Takeaways:

While many of these considerations would be addressed by "normal" mitigations, what does it take to go beyond the normalizations of loading hazards to reduce incidence? A conversation between process safety experts from different backgrounds resulted in these points.

Exxon—Two process safety experts from Exxon contributed , both having deep experience in dealing with a variety of toxic materials and hazardous materials. They emphasized a need for truly "critical safeguards" that call for a higher level of attention and verification as will be discussed in this section.

Dan Walters—Forty-two years of boots on the ground safety consultant/site safety manager with several companies, now an independent consultant.

Dorothy Shaffer—BakerRisk Ammonia SME and former Director of Risk and Reliability for Terra Industries with 40 years of experience.

Stuart Powell— Senior Process Safety Consultant at BakerRisk.

➢ **Third party risks**: Working with third parties can introduce unknown and uncontrolled risks, as were present in the Tanner incident. What is the strategy to manage those risks?

- Exxon ensures that vendor contracts are written with key safety requirements identified. A new vendor will undergo a review. Audits are not often performed but are expected if issues arise.
- Use of standard equipment sets are expected. (See more under "Challenges & Opportunities")
- Site personnel should be able to verify the transport equipment and for import, the product quality. Every year, ammonia site loading personnel check all trailers at their first visit for compliance with the DOT MC331 form[8], verifying PSV checks, data plate, and current hose inspections. Those not complying are turned away. Import ammonia should have a certification of quality that meets the specifications, as well as documentation on where it is from. Water content is critical when ammonia is added into atmospheric tanks.

➢ **Site mitigation:**

- Most ammonia manufacturing sites reduce the risk of releases by using loading arms rather than hoses which—even if tested on the regular basis-- can deteriorate quickly if dragged or damaged.
- Exxon's own equipment is built to consistent specifications. A good practice is to have consistency in ammonia transfer safety expectations that are actively supported at all company levels.
- Understand what equipment conditions and materials are required and what should be rejected and ensure PMI and training is in place.
- Exxon uses an umbilical cord that trips the feed and incoming isolation systems in the case of a decouple.
- Ammonia transfers only take place when two loaders are present, to support and activate emergency operations if the other is incapacitated.

➢ **Operations and Human Factors**: The group discussed the Hazard and Operability Studies (HAZOPs), Job Safety Analyses (JSAs), Operating Procedures and Training, which have become such a norm that it is difficult to emphasize the critical items in operations.

Exxon's strategies include:

- Exxon Critical Safeguards: Truly critical safeguard steps in operating procedures are in bold and must be verified on the form with initials from a second set of eyes. An example may be verification that the coupling between a ship and the loading arm is properly secured with the interlock in place. Only the most important steps are included, as when everything is critical, nothing is critical.
- Move from "fast brain" to "slow brain": Exxon recognizes the experienced hazard of thinking that faster is better ("fast brain"). Through a careful review of the timing to do all steps correctly, a minimum time for loading is determined, and they set a maximum number of loads in the day based on that information. Verification steps for critical safeguards also supports "slow brain" work.
- Taking time to do loading correctly is emphasized by supervisors and at least once a year by a visit from the plant manager.
- Operating procedure development and HAZOP review not only include the more experienced personnel, but a new operator is also included. The experienced personnel

[8] https://www.govinfo.gov/app/details/CFR-2011-title49-vol3/CFR-2011-title49-vol3-sec178-337

will tend to "wash over things" while the real goal is providing procedures so that everyone—new or experienced—performs the tasks and questions the hazards competently.

➢ **Challenges & Opportunities:**

- Standard equipment: Standardizing equipment on site and with outside vendor equipment can reduce operations errors and provide better recognition of equipment that does not meet a standard. Exxon identified that the vendor loading equipment designs are typically limited to a few standard types. However, this is not the same for ammonia truck transports: one ammonia production site audited incoming ammonia trucks from outside companies, finding 3 different setups on the first 5 trucks checked, including different fill designs that affected operations.
- Clear inspection criteria and proper tools: After an ammonia release, a site found that its method of tightening couplings with a hammer caused damage to the coupling. The correct equipment was obtained. However, operators were finding that hammers are being used by drivers coming into the site, resulting in damaged couplings. Education and a Go/NoGo inspection of devices are necessary for both parties.
- Industry standards are in place from several chemical sectors, including chlorine, LPG, etc. As ammonia becomes a clear method for transport fuel, similar standards for loading and transport will help fill the deficiencies that may arise.

5) Transfer Operations: Positive Human Factors, Vendor/Operator Relationship

When 'Human Factors' are referenced in a process safety discussion, it is usually in the context of highlighting the human elements of a process that are susceptible to fail because of the human performing the task and the inherent 'failure' of the person to perform the task right, every time. That said, humans can be the strongest link in a process if they are setup for success. Attention should be given to positive 'human factors' that can be considered in each stage of the transfer operation to reduce the probability that a loss of containment or process upset would occur.

One of the areas for 'positive human factors' is the relationship between the operating company and the transport company as the attention of both will determine success or failure of a transfer. The safety and quality commitment of both groups needs to be established and maintained for the safety of personnel and the community.

The sections below divide the transfer operation into a few different steps, starting with long before the transfer even occurs, and ending with after the transfer has taken place. Some of these 'key practices' are currently in operation in the ammonia producers standard work practice, but implementation industry wide and for new companies that are entering the ammonia space, adoption should be considered to help improve operations.

Before Arrival

- Development of vendor/operator relationship through detailed vendor screening process and working hand-in-hand to take ownership of the transfer operation together. Items to consider include:
    - Development of detailed Standard Operating Procedures (SOPs) with safety critical steps clearly labeled.
    - Including junior operators on both the vendor and operator side to review and provide comments on SOPs
    - Performing a HAZOP on the transient loading operation and SOPs to improve documentation
    - Seeking consistency across all containers/trucks to be used in the transfer operation. One recent find was

different valve types between trucks that would arrive onsite for transfer, leading to issues with connection operations.
  - Development of plans for hose maintenance, and other standard procedures to handle situations that would arise. One example is what to do if one of the transfer tanks is damaged. What is the procedure to return to the tank to service? Who will be notified? Where will this be documented?
- Detailed checklist developed with input from vendor, including:
  - Verification of hose documentation showing hose expiration date
  - Who will perform the transfer, what role will the vendor play, and what role will the operating company play? This can be tricky and difficult depending on how many vendors are employed by the operating company. Diligence and procedures based on industry best practice are critical to keep this organized.
- Detailed documentation on all aspects of the operation and defining what 'success' and 'failure' look like for the various components, i.e. – mechanical integrity program, critical safeguard list.
- Timing the operations team on how fast they can routinely perform the transfer. Using the data to set safe limits for the scheduling of number of transfers for a given day.
- Detailed training for new operators, from both vendor and operating company. Routinely, the new operators at a facility get assigned to the loading/unloading position. Care must be taken to ensure they are prepared to perform the task sufficiently.
- Establishing dedicated loaders from the vendor company for each operating facility, such that the vendor team and the operating team develop a working relationship and gain experience together in handling the transfer operation.

During Transfer

- Independent verification from shift supervisor or operator to confirm line up for transfer ('4 eye review').
- Latest and greatest tools are readily available, and operations/vendors are trained on them for connection/disconnection.
- Site management visits and reviews field loading operation, with transfer vendor company management, on a regular basis to show importance and lead discussions on safe work practices reviews and improvements.
- Development of solid emergency response policies and conducting drills to ensure operators know the correct course of action to take. Some sites are implementing systems similar to 'amber alerts' to quickly inform public of the need to shelter in the event of a toxic release.
- Development of clearly defined exclusion zones and muster areas, should a release occur.
- Ensuring that procedures, training materials, and other critical documents are available in multiple languages, based on the vendor and operating company team.

After Transfer

- Review of near-misses and incidents, with vendor and operating company present
- Periodic reviews of SOPs and critical documentation for safety systems
- Soliciting feedback from new operators and vendors and including feedback into the overall workflow.

6) Future Considerations

While industry has taken large actions to reduce the number of ammonia releases, there is still work to be done towards continued improvement.

It has been noted that some avenues worth pursuing include working with transportation companies to standardize the setups used in truck loading cases, as sometimes different trucks are used with different setups, leading to confusion and possible deviations from approved SOPs.
In addition, working with other industries such as hydrogen fluoride and chlorine, to share best practices and compare standards, to ensure that what is working in one industry is applied to another industry. As part of this, review of European standards on transfer of toxic materials could be undertaken and recommendations applied to US based sites.

References

1. "Reporting Requirements for Continuous Releases of Hazardous Substances, A Guide for Facilities on Compliance", https://www.epa.gov/epcra/reporting-requirements-continuous-releases-hazardous-substances-guide-facilities-compliance
2. Cargo Hose Rupture and Release of Anhydrous Ammonia During Offloading of a Werner Transportation Services Cargo Tank Motor Vehicle at the Tanner Industries Plant Swansea, South Carolina", published by the National Transportation Safety Board, April 2012.
3. API RP 752: API (American Petroleum Institute) "Management of Hazards Associated with Location of Process Plant Permanent Buildings," API Recommended Practice 752, 4th Ed., Washington D.C, 2024.
4. API RP 753: API (American Petroleum Institute) "Management of Hazards Associated with Location of Process Plant Portable Buildings," API Recommended Practice 753, 2nd Ed., Washington, D.C., 2024.
5. API RP 756: API (American Petroleum Institute) "Management of Hazards Associated with Location of Process Plant Tents," API Recommended Practice 756, 2nd Ed., Washington D.C., 2024.
6. "Experiences from ammonia loading incidents", by Mathieu Tagliaferri (YARA), Vincent Duponchel (YARA), published by Ammonia Technical Manual

Repair and Investigation into an Ammonia Leak on a Marine Loading Arm

This paper examines the repair and failure investigation of a double ball valve for a marine loading arm. The failure resulted in a tier 2 process safety event and inspection checks revealed significant damages to the Valve Body. Root Cause incident investigation was used to determine the causal factors and establish reliability improvement recommendations while metallurgical analysis was done to characterize the failure mechanism. A novel repair method vetted by a level 3 fitness for service assessment was applied and the repair process was validated with a complement of non-destructive examination and functional checks.

Kiran Rampat, Mikyle Moolchan
Point Lisas Nitrogen Limited

Introduction

Point Lisas Nitrogen Limited, PLNL, manufactures 1850 MTPD of 99.6wt% anhydrous ammonia with trace amounts of oil and moisture. Additionally, the company manages a 5.8km Ammonia pipeline which runs from the Plant storage area to a shared international port facility, PLIPDECO. The line runs through a nearby fishing community and transfers ammonia product to the loading arm which is sited at the Port.

On February 04th 2024, at approximately 10:50am, during Ammonia product loading of the LPG/C Marcella vessel, a leak developed at the Emergency Release System (ERS) on the Loading Arm. Rates were adjusted from 600MT/hr to 300MT/hr to minimize the ammonia release and loading activities were completed at the reduced rate. The leak was later observed to be due to failure of the Double Ball Valve assembly (DBV).

The failure resulted in Loss of Primary Containment (LOPC) with the release being confined to the immediate area. There were no injuries sustained and the incident was classified as a Tier 2 Process Safety Event (PSE).

Following the event, checks were done on the compromised joint and from this, there were indications of seal failure due to deformation of the mating flanges between the upstream and downstream ball valves. Additionally, a crack on the external surface of the valve body was also identified. These were subsequently repaired using a novel method.

Root Cause Incident Investigation using a Human & Organizational Performance (H&OP) approach was applied to review this failure event and determine areas for improvement. Inspection of the affected components was also done to confirm the failure characteristics of the crack formation. Finite Element Analysis (FEA) modelling and a Level 3 fitness for service assessment was done to validate the novel repair method.

Asset Description

The Marine Loading Arm is a hydraulically operated Rotating Counterweighted Marine Arm (RCMA) with piggyback vapor return line, manufactured by Technip FMC and installed in 2012. The vapor arm and the liquid loading arm form a common assembly that is operated in unison. The Marine Loading Arm is used to transfer liquified ammonia product and recover ammonia vapors generated during ship loading operations. The design pressure, temperature and loading rate of the liquid line is 265psi (18.3Bar), -31°F to 120°F (-34°C to 49°C) and 7,798GPM (1,209MT/hr) respectively. The Loading Arm has freedom of movement to adjust to ship position due to low and high ocean tides, changes in ship weight as cargo is loaded and drift of the ship as per Figure 1. It is also equipped with a two stage over travel alarm system and a two stage Emergency Shutdown System (ESD1 and ESD2) in the event of excessive motion.

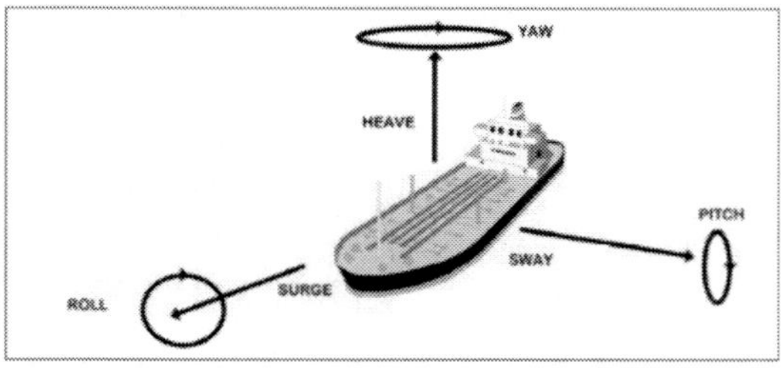

Figure 1. Ship movements

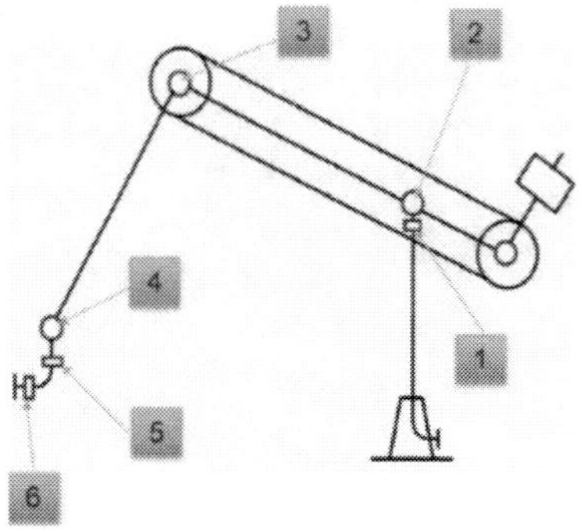

Figure 2. showing swivel joint positions

The material of construction for the Loading Arm assembly and DBV is low temperature carbon steel and ASTM A351 Type CF8M respectively. Dimensionally, the liquid loading line is 12 inches in diameter and 60 feet in length. The arm possesses six degrees of freedom between ship and shore and the movements are governed by the swivel joints listed 1 to 6 in Figure 2. These swivel joints are independently controlled and articulated to allow the arm to connect to the ship.

Near the terminal end of both Loading and Vapor Return Arms, there is an Emergency Release System (ERS) with dry breakaway couplings that is activated during an ESD2 scenario. The ERS is designed to prevent damage to the loading system, ship, bulk transfer arms etc. and to operate with minimum release of Ammonia when activated. Activation of the ERS can be automatic or manual and is achieved by the use of a hydraulically operated Powered Emergency Release Coupling (PERC) and DBV.

Each DBV comprises two mating flanges. However, instead of having them bolted together, they are clamped via a PERC system. When the PERC activates, the two ball valves close and the clamp is released, allowing the valves to separate as per Figure 3. One valve remains connected to the loading arm whilst the other valve remains connected to the ship.

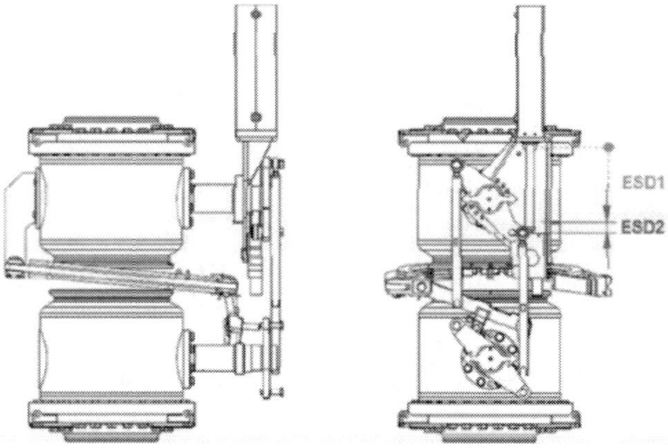

Figure 3. showing ERS during ESD2 activation

Asset Operation

Prior to commencement of the loading process, an initial cool down of the ammonia loading line is carried out using the Ammonia Recirculating Pump, 2106-J as per Figure 7. showing schematic of Loading Process. During this process, a small volume of liquid ammonia from the storage tanks is circulated though the loading line, and back to the tanks via a 4" recirculating line.

A summary of the Standard Operating Procedure (SOP) is provided in the following steps for the loading process:

1. A safety briefing and information exchange are done between the ship crew, PLNL operators and the team of fitters.
2. The Loading Arm is then placed in an appropriate position to perform loading operations. This process requires careful articulation of the arm to ensure alignment with the necessary connection points on the ship to be loaded.
3. The DBV is placed in the closed position and the covers for the liquid loading line and vapor return lines as shown in Figure 4 are removed.
4. Both liquid and vapor return lines are connected to the ship by the team of fitters as per Figure 5. The DBV is placed in the opened position immediately afterwards. N.B. the duration that the DBV is left in the closed position is minimized as much as possible.
5. Various manual ESD1 checks are performed to verify functionality.
6. A pressure test is done to verify that there are no leaks. The pressure test is done by starting one (1) of the Loading Pumps, 2100JA or 2100JB located at the Plant site facility and partially opening the pump discharge valve together with the Loading Arm main isolation valve by approximately 5%. For this test, the liquid loading line valve, XV8016 is left fully open and the ship tank isolation valve is left in the closed position. The ship manifold pressure is checked and verified to hold at 60psi (4bar).

Figure 4. showing the liquid and vapor lines covered

7. On pressure test acceptance, the ship's Chief Mate is informed that the loading process is ready to commence and the ship tank isolation valves are opened.
8. The Loading Arm main isolation valve is opened fully and the minimum flow of 200MT/hr is established by opening the Loading Pump discharge valve as required. N.B. The product at the commencement of loading is relatively hot at approximately 6.8°F (-14°C).

Figure 5 showing Loading Arm connected to ship

The increase in loading rates to the steady state design condition of 1050MT/hr is determined by the ship's tank pressure and temperature. Since one (1) Loading Pump provides approx. 3900GPM (605MT/hr) liquid ammonia of flow, the second pump is placed in service for additional flow requirements.

Under steady state conditions during normal loading operation, with both loading pumps in service, liquid ammonia is transferred to the ship at a temperature of -28°F (-33°C), pressure of 35 psig (2.4Bar) and constant loading rate of 6,773GPM (1050 MT/hr). While the ship is being loaded, ammonia return vapor that cannot be liquified by the ship's refrigeration system, is sent through the vapor return arm and returned to the Ammonia Storage Tanks, 2101-F/2102-F or routed to the flare. At times, the ship may require a reduced loading rate, and this is achieved by having the loading pump's discharge valves throttled. As conditions are stabilized and the ship requests increased loading rates, the valve will be opened further.

The ship chief mate advises on the progress of the loading operations and provides notice when the loading should end. On completion of Loading Operations, the following steps are taken:

1. Loading rates are reduced to 200MT/hr and the Loading Pumps are then taken offline.
2. The main isolation valve for the Loading Arm liquid line is closed.
3. The liquid arm is allowed to drain to the ship tank for approximately 30 minutes resulting in an estimated volume of 0.5MT being transferred. Draining is done in combination with de-icing by means of washing the arm with sea water until no noticeable ice formation is observed and the ship manifold pressure reading is less than 5psi.
4. When ice formation stops and pressure indications are as required, the ship tank isolation valve and liquid line DBV are closed.
5. The enclosed volume of ammonia between the valves are depressurized by means of a 1" flexible rubber hose to a trough with water.
6. Both the liquid line and vapor return lines are disconnected from the ship and the respective covers are reinstalled.
7. The 1" SS flexible hose connecting the liquid loading line to the vapor return line as per Figure 7 is lined up by opening both 1" ball valves
8. The Liquid DBV is opened immediately.
9. On the vapor return line, MOV8011 is opened fully and on the liquid loading line, XV8016 is closed to depressurize the remaining vapor in the loading line.
10. The Loading Arm is placed in the parked position as per Figure 6 and final verification checks are done to ensure that all depressurization valves are in the appropriate position.

Figure 6. showing Loading Arm in the 'parked' position

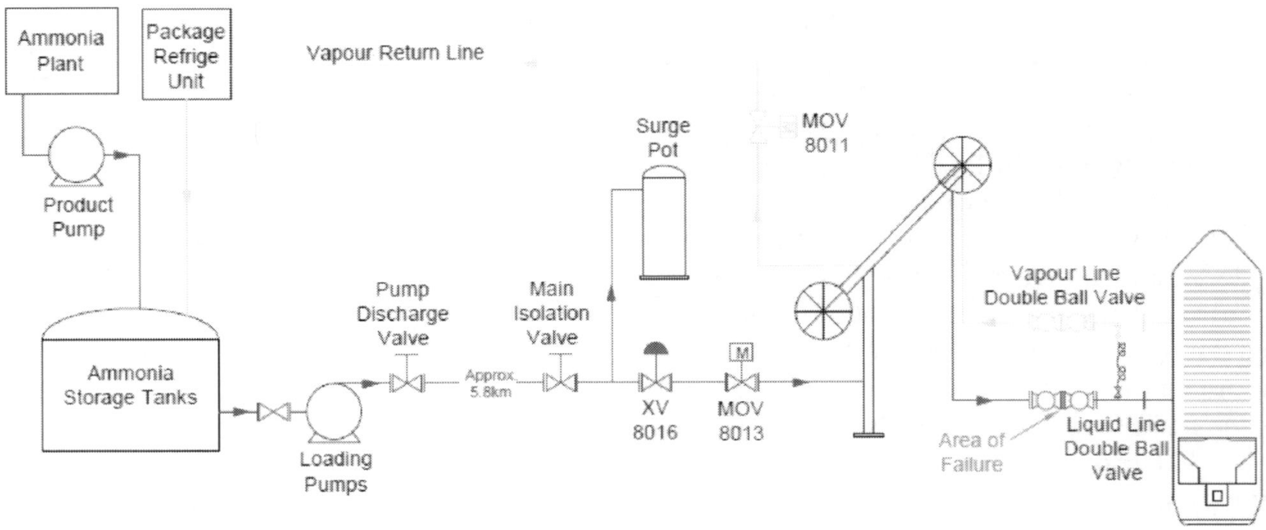

Figure 7. showing schematic of Loading Process

Incident Details

The sequence of events leading up to the failure event are listed below:

Jan 02nd 2024: Loading of Trammo Cornell. No issues were identified, and the volume loaded was 31,650.1MT.

Jan 03rd 2024: Loading of Hellas Eagle. The volume loaded was 25,234.3MT.

Jan 15th 2024: Loading of Libramont. The 1" SS hose used to depressurize the Liquid loading line to the vapor return line began to leak after disconnection of the loading arm. The volume loaded was 15,000.4MT. The DBV was left in a closed position due to the leaking hose.

Feb 03rd 2024: The bleed hose was replaced, and loading commenced at 11:05pm.

Feb 04th 2024: At 10:45am, the vessel indicated 1hr notice for completion of loading. 10:50am the vessel requested a reduction in loading rate to 600MT/hr from 1050MT/hr. At 10:51am the vessel reported a leak on the vapor line but on closer inspection, the leak was confirmed to be coming from the liquid loading line at the ball valve location. The rate was further reduced to 300MT/hr and the leak stopped after 20 minutes. Since the leak had stopped, loading continued at the 300MT/hr rate and was completed at 12:10pm.

Macro inspection checks were carried out on the Loading Arm DBV in situ after the failure event, prior to valve disassembly. From these checks, flange distortion was observed circumferentially around the mating face between the upstream and downstream valves as per Figure 8.

Figure 8. showing flange deformation during initial inspection

Additionally, a crack was observed at the outer surface of the valve body as per Figure 9.

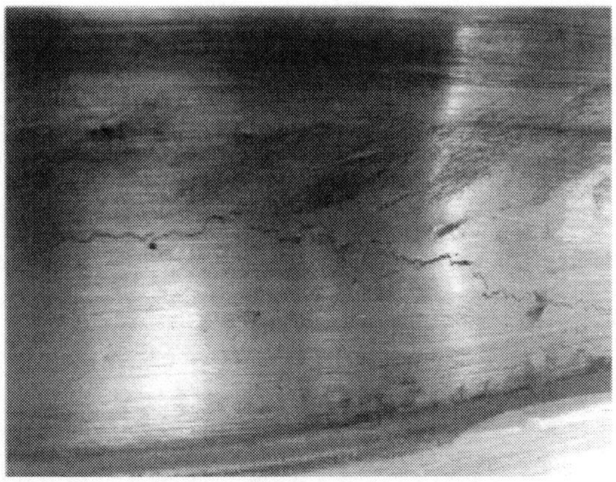

Figure 9. Crack observed during initial inspection

The valve was taken to the repair shop for disassembly and further Non-Destructive Examination (NDE).

Repair Details

Initial repair plans were developed to correct the observed flange distortion and crack indications. These plans included grinding out of the cracks and weld-building all affected areas with subsequent re-machining to original dimensions. Dimensional checks were conducted on the flanges of both upstream valve, 'A' and downstream valve, 'B'. These values were recorded before any machining/welding commenced. From this, flange 'A' measured Ø12.925" maximum and 12.950" minimum while flange 'B' measured Ø12.930" maximum and 12.965" minimum. Radiographic testing was then conducted to determine the extent of defect and 0.020" of material was removed radially from the flanges of both flange 'A' and 'B'. Dye penetrant test (PT) was conducted on the machined areas and indications of bleed out were visible. A further 0.020" was removed from the same areas and PT was conducted once more revealing indications.

Phased Array and Ultrasonic Shearwave testing were performed on flange A and B which revealed linear indications typical of cracking. Field Metallographic Replication Testing was done on the machined areas on flange A and B and from this, several locations were considered unfit for continued use.

The replicas revealed the presence of carbides indicative of sensitization. This was corroborated by higher than expected hardness values as high as 268HB compared to an expected hardness of 200HB. In one area, the morphology of the cracks resembled Stress Corrosion Cracking. It is postulated that due to continual exposure to chlorides from sea water, chloride induced Stress Corrosion Cracking played a role in the embrittlement of the cast CF8M valve body.
The process of machining 0.020" radial off the diameter on flange A and B followed by dye penetrant test was repeated until no indications were visible after conducting the dye penetrant test. A total of 0.200" radial was removed from flange A and 0.240" radial was removed from flange B. After machining flange, A and B, radiographic testing was conducted, and results were found to be acceptable.

Based on the significant volume of material removed and significant warping experienced during welding, several novel repair approaches were undertaken. At first, welding was done in accordance with an approved Weld Procedure Specification for the material using ER316L filer rods and the Gas Tungsten Arc Welding (GTAW) process. However, due to the warping and solidification cracking experienced during welding, several weld parameters were adjusted. This included utilizing smaller filler rods (3/32" instead of 1/8"). A staggered sequence was used in depositing weld material on alternating areas of the valve body. Tie-in weld passes were done using a backstep sequence. Additionally, the heat input was reduced by lowering the current by approximately 30-50amps from the original range of 150-200amps.

Consideration was given to slow cooling to reduce the residual stress and possibility of solidification cracking by using ceramic fiber blankets. However, this was rejected since slow cooling could have led to increased sensitization and increased embrittlement of the microstructure.

This prompted the consideration to use a novel repair method involving a reinforcement 'backing' ring. The effectiveness of this repair approach was validated using FEA stress analysis through a level 3 Fitness for Service evaluation. FEA modelling was done to aid in the design of the reinforcement 'backing' ring. Figure 10 shows the Critical Flaw Size assessment without the backing ring and Figure 11 shows the simulation with backing ring.

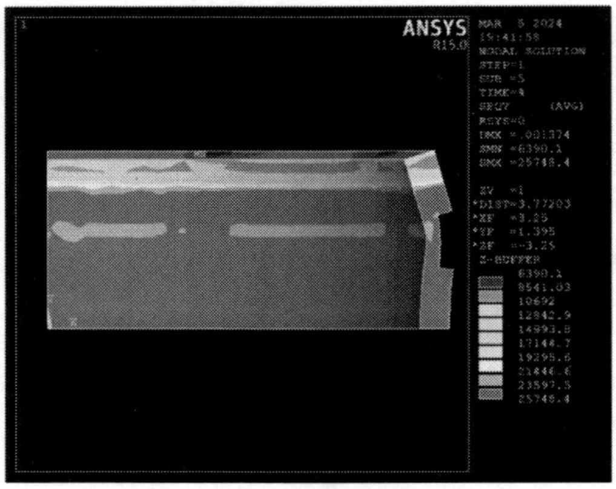

Figure 10. Simulation of stress without backing ring design

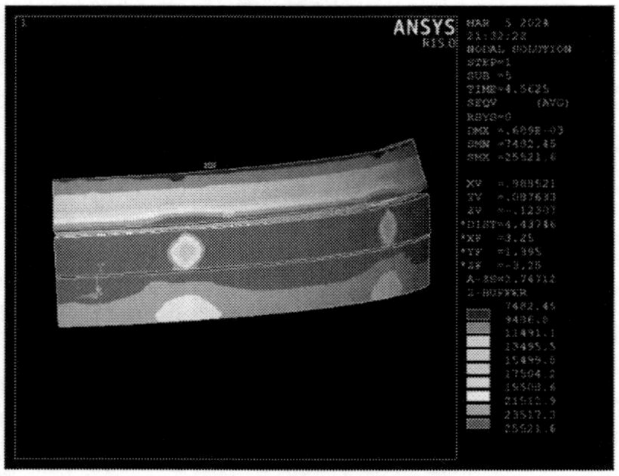

Figure 11.Simulation with backing ring design

Two (2) reinforcement rings were manufactured from 316 stainless steel as per Figure 12 to accommodate the material removed from flange A and B. The ring for flange 'A' had a thickness of 0.200" and thickness of ring for flange 'B' was 0.240". The rings were designed to have a 0.001" interference fit to allow a snug fit with small compressive forces which counteracts the dominant hoop stress due to internal pressure. The installation of the rings consisted of several plug welds in addition to the top and bottom circumferential fillet welds. Notably, the residual stress associated with these welds were also considered in their design. The rings were then split in halves and welded onto flange A and B respectively as per welding procedures, WPS FTT-22 and PQR FTT-11.

Figure 12. The constructed backing ring

Upon completing the welding process, a dye penetrant test as well as a radiographic test were conducted on the welds. The face of flange A and B were also weld build up because of distortion on the faces. Both flanges were final machined as per OEM drawings specification and final dimensional checks were conducted. Final Penetrant test, Radiographic test and Phased Array Ultrasonic Tests were conducted on both flanges 'A' and 'B' and the results were found acceptable.

Figure 13. Pressure and function testing at the repair shop

Further to this, a trial assembly of the valves with the PERC was conducted at the repair shop to verify fit-up. The valves were then sandblasted to SSPC SP-10 specifications and coated with two (2) coats of PPG HI-TEMP 1027 paint to act as a barrier from the constant exposure to chlorides from seawater. Upon completion of the repairs, final checks included pressure testing and function testing of the PERC ERS system as per Figure 13.

Incident Investigation

Root Cause Incident Investigation was done using a Learning Teams approach that was based on the principles of H&OP.

Based on this approach, discussions were carried out with personnel directly involved in the incident and was not restricted solely to the valve failure but generally encompassed causal factors as well as related issues. Problem statements and recommendations were subsequently developed. The boundaries of the incident under investigation spanned the period from the first similar incident which took place in 2019, and then this incident which occurred in February 2024. It considered previous loadings done in 2024 but not all loadings between similar or contributing incidents. And finally, it considered the damage mechanism found during laboratory testing, NDE of valve and the characteristic of the crack indications to help determine the root cause of the failure. Several key factors were identified from the investigation carried out.

The pressure developed within the void space between the ball valves were found to exceed the allowable design limits based on the expansion of trapped liquid ammonia flashing to vapor. With the void completely full of liquid ammonia and there is no empty space for liquid expansion the pressure would increase at the rate of 1.5 x 106 Pa/°C (i.e. 217.56 psig/°C). Assuming a 2°C increase in the void space between valves, the pressure increase would be approximately 435 psig as per Figure 14 showing the ammonia phase diagram.

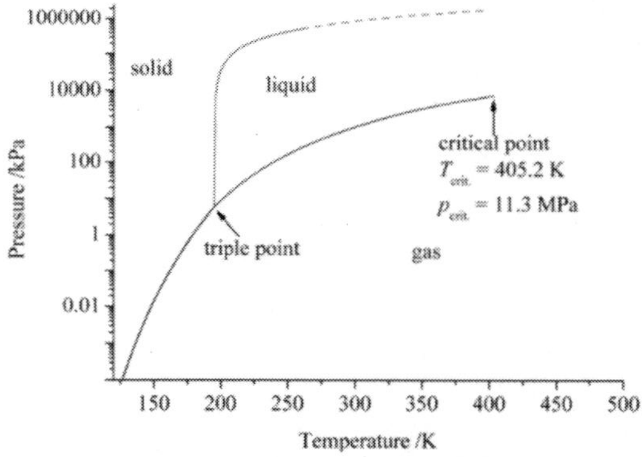

Figure 14. Ammonia phase diagram

Training on current procedures and revisions to cater for situations where the DBV goes close was found to be insufficient. In both incidents, the ball valves were allowed to remain close causing the pressure build up. It was explained that in this particular event there were no major anomalies and previous to this shipping there would have been three shipping cycles for the month of January 2024.

At the preceding ship loading in January 2024, a leak was detected on the 1" SS braided drain hose. As a result, both ball valves were left in the closed position with the arm parked and the drain hose isolated. Normally, the DBV is left open and the drain hose lined up to the vapor line. However, in this instance as the DBV was left closed there would have been trapped ammonia vapors and liquid within the cavity between the two ball valves. As a result there was an increase in pressure with increasing ambient temperature over the time that the Loading Arm was parked. The overpressure scenario was calculated for trapped liquid ammonia and found to be sufficient to result in the deformation of the flange faces.

The investigation identified a lack of awareness on the need to drain the ball valves and methods to rectify a closed DBV in the shortest possible time. In the case of the damaged drain hose, a spare hose could have been installed urgently. There may be other scenarios that could cause the same result and need to be discussed and included in training/procedure where possible.
Environmental factors were considered also. The valve morphology as well as swivel joint pitting pointed to degradation by seawater and therefore remediation/prevention is a possible area for improvement. Contamination from oil was also noted when removing the arm for maintenance. From the in-situ metallography performed, the failure mechanism shown in Figure 15 was consistent with Chloride-induced Stress Corrosion Cracking.

Figure 15. Crack indication using NDE

A total of nine (9) replicas were performed on various locations on the top and bottom sections of the 2102L valve body. Hardness was also completed at each location and PMI testing was completed at 7 locations on the valve.

Replica was performed on the trough of the clamp groove, at a location known to have crack-like indications (via PT), on the bottom half of the 2102L Loading Arm Valve as per Figure 16. This location was selected to characterize the cracking and to ascertain the microstructural integrity of the material at this location. Cracks were observed close to regions of weld repair and the weld repair material revealed a microstructure of fine dendritic weld material as per Figure 17. As microcracks were observed at this location, the material was considered unfit for continued service. These areas were skimmed by machining and reinforced by use of the backing rings and weld build-up.

Figure 16. Location of Replica testing

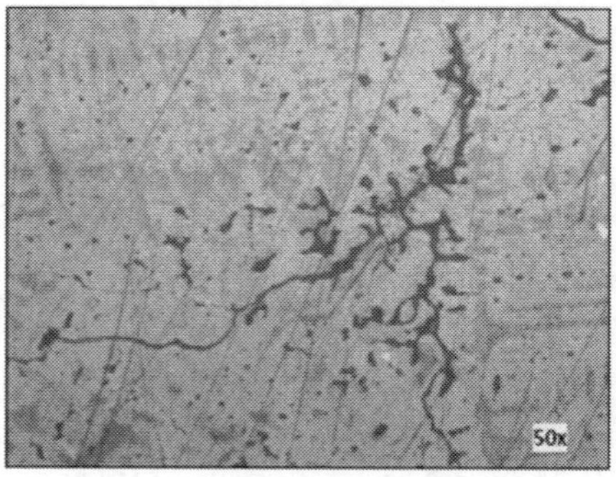

Figure 17. Replica micrograph

Post-Repair Performance

After field installation, there was a total of three (3) vessel loadings amounting to 75,557.88MT. The DBV would have been in service for 10 months and suffered a single over pressure scenario which occurred in Jan 2025. After this overpressure scenario, from subsequent inspection checks, it should be noted that there were no indications of further cracking at the area of repair thereby giving a true indication of the success of the repairs performed. The over pressure resulted in deformation of the seal resulting in ammonia leakage.

Improvement Recommendations

Several recommendations were considered to help improve the reliability of the DBV.

1. There is an action plan for revising the SOP for draining the arm to allow for improved decontamination of the arm and DBV. The drain down time should be increased from 30mins to 1hour and the cover installed on the DBV immediately after disconnection from ship to allow for reopening ball valves.
   Ensure that the 1" SS hose used to depressurize the liquid arm to the vapor return line is connected and lined up as per SOP.
2. There is an action plan for installing a freshwater connection at the loading arm for deicing purposes. This would replace the existing practice of using sea water for de-icing the arm and mitigate chloride-induced stress corrosion cracking.
3. There is an action plan for upgrading the coating system to withstand the low temperature and marine environment. The proposed coating system being investigated is Thurmalox 70.
4. Based on the incident investigation, it was decided that the DBV may be outfitted with a pressure relieving system as an additional safeguard. Engineering is currently in progress in collaboration with the DBV manufacturer.

Conclusion/Summary

The failure event which occurred on 04th Feb 2024 was due to an over pressure scenario due to trapped liquid ammonia in the DBV assembly and a novel repair approach vetted by a level 3 fitness for service assessment was applied. The repair yielded favorable results with over 10 months of continued service without reoccurrence of failure mode. Future initiatives are being considered to prevent an overpressure scenario together with combatting stress corrosion cracking.

References

Bertrand, M. (2024). Replica Metallography, Hardness Testing and PMI on 2102L Ammonia Loading Arm Ball Valve (9-off). Chaguanas, Trinidad, In-Corr-Tech Limited.

Du, W. (2024). Critical Flaw Size Assessment of Double Ball Valve Body. Houston, TX, Vysus Group.
Motilal, R. (2021). Procedure to disconnect ammonia loading arm. Point Lisas, Trinidad, Point Lisas Nitrogen Limited.

Neeranjan, N. (2024). LOPC of the Double Ball Valve (Loading Arm). Point Lisas, Trinidad, Point Lisas Nitrogen Limited.

Technologies, F. (2012). Installation, Operating and Maintenance Manual. Cedex, France, FMC Technologies.

Young, J. (2024). Ammonia Loading Arm Double Ball Valve - Job Completion Report. Chaguanas, Trinidad, Mustapha Engineering Works Limited.

Lesson Learned Preventing Early Relief in Ammonia Storage

This paper describes on-site events when early relief from pilot-operated pressure relief valves protecting cryogenic storage tanks was experienced due to clogging of the sensing line. Some of those events were the consequence of weather extremes, while others resulted from a lack of maintenance.

This paper presents solutions validated from testing pilot-operated pressure relief valves in a climate chamber, addressing various extreme weather conditions, and discusses lessons learned to avoid mistakes during maintenance and startup, as well as due to poor engineering.

Dr. Davies Michael, Dr. Davide Moncalvo,
Braunschweiger Flammenfilter GmbH (PROTEGO)

Brad Luckman, Brian Anderson
PROTEGO (USA), Inc.

Introduction

Near-atmospheric storage tanks for cryogenic applications are very large reservoirs with typically both large diameter and height. Like every other storage tank, the tight requirements on the tank design pressure and vacuum make them extremely susceptible to small changes in the storage pressure. The most common sizing standards, like ISO 28300 [1] and API 2000 [2], provide detailed guidance about the most common causes for overpressure and vacuum for storage tanks in general and for cryogenic applications in particular. Despite its high level of detail, the guidance in the standards does not release operators from performing a complete risk assessment and hazard evaluation. Both standards point out that many causes for overpressure and vacuum are shared by all storage tanks, like changes in the pump flow rates or in the barometric pressure or due to external fires. Others, like weather cooling, may have a different effect in cryogenic tanks due to the low storage temperatures. In other words, in many cryogenic applications – including ammonia – the nearby environment is warmer than the cryogenic liquid, leading to a heat inflow into the stored liquid. Any heat input into the stored liquid, which includes also thermal dissipation from submerged pumps, causes pressurization by evaporation of boil-off gases, which must be recondensed or released from the tank.

The common layers of overpressure protection also apply to cryogenic storage tanks. During operation, a vapor recovery unit (VRU) condenses boil-off gases, thus reducing pressure accumulation in the headspace. The vapor recovery unit can be seen as the first independent protection layer (IPL) and part of good process design. The vapor recovery units can also be connected to a safety instrumented system (SIS) and activated on demand. In both ways, the vapor recovery unit is a preventive solution that permits continuous operation of the cryogenic storage tank without interruption.

When the safety instrumented system or the vapor recovery unit fails, the next protection layers are mitigative solutions like pilot-operated pressure relief valves (POPRV). Pilot-operated pressure relief valves are a common solution for releasing clean gaseous streams with negligible or barely traceable impurities like liquid or solid formations. A detailed description of the working principle of the diaphragm POPRV is provided in Annex C of both API 2000 [1] and ISO 28300 [2]. In a diaphragm POPRV, the dome, the valve body volume above the diaphragm, is connected to the storage tank by remote sensing. The sensing line ensures that the tank pressure is communicated to the dome so that a careful diaphragm design can exert sufficient force to keep the disk on the seat during normal storage operation. The membrane seating force, which increases with the dome pressure and hence with the tank pressure, guarantees premium seat tightness if compared with spring-loaded pressure relief valves (PRV) with the same settings. When the pressure in the storage tank reaches the set pressure, the pilot releases the dome pressure so that the membrane seating force is insufficient to keep the disk on the seat, and the valve opens. When the tank overpressure falls below the blowdown, the pilot closes, and the dome pressure is restored, causing the disk to reseat. According to the standards, the typical blowdown of POPRV is below 7%.

POPRV can operate as a second layer of protection for storage tanks by conveying the boil-off gases to an incinerator or a flare. Unlike the vapor recovery unit, the boil-off gases are burned and lost irreversibly. Despite the economic loss, the environmental impact is modest, consisting of NO_X and steam and a small, unburned fraction (< 2%) of the inlet flow in the case of ammonia. When the POPRV to the flare fails, a third and final layer of protection is direct ambient release of the boil-off vapors, which may lead to a possible local emergency response caused by pollution and toxicity.

An example of a good design to avoid conflicts in operation between the POPRV to flare (2nd IPL) and the POPRV to atmosphere (3rd IPL) is presented in Table 1.

A good design of the protection layers avoids the unexpected opening of the POPRV and boil-off discharge at pressures below the POPRV set pressure. This event is often addressed as early relief in the following chapter.

Pressure	Set	Relieving
	[psig]	[psig]
POPRV to Flare	1.7	1.9
POPRV to Atm.	2.0	2.2

Table 1: Settings of the POPRV discharging to flare (2nd IPL) and to atmosphere (3rd IPL)

Early relief related to extreme ambient temperatures

This chapter addresses an early relief tank from a near-atmospheric storage tank storing normal butane (CAS 106-97-8).

In that occurrence, the ambient temperature dropped to -20°C (-4°F), which is below the boiling point of butane around -0.5°C (31.1°F) [3]. The sensing line was clogged by the condensed butane vapors, which resulted in loss of dome pressure and opening of the main valve. In case of ammonia, condensation in the sensing lines is less probable due to the low boiling point of ammonia at -33.3°C (-28°F) [4], but it is still a credible scenario for a risk analysis for tanks located in the Arctic region, where ambient temperatures can be as low as -52°C (-61.6°F).

One solution to avoid early relief of cryogenic tanks in cold weather is to heat trace the whole assembly of the pilot unit and main valve and envelop it within an insulation. The heating enclosure must not impair device venting and pilot function. An example is given in Fig. 1, where enough room between the pilot unit and the insulation is provided, and the inlet and outlet connections are accessible by the operators, see Fig. 2. This rather voluminous structure has been designed to resist wind loads at installation.

The enclosure has been tested in the climate chamber, an externally insulated metal room where the target ambient temperature can be set or programmed, as shown in Fig. 3, to simulate existing ambient conditions. For this test routine, three sensors have been mounted on the sample: the first one on the inner membrane housing, the second on the main valve body, and the third one on the inner jacket but outside of the POPRV. The test conditions in a climate chamber are independent from the operating conditions of the storage tank and the POPRV settings.

Fig. 1: Example of a heating enclosure to protect a POPRV from cold weather

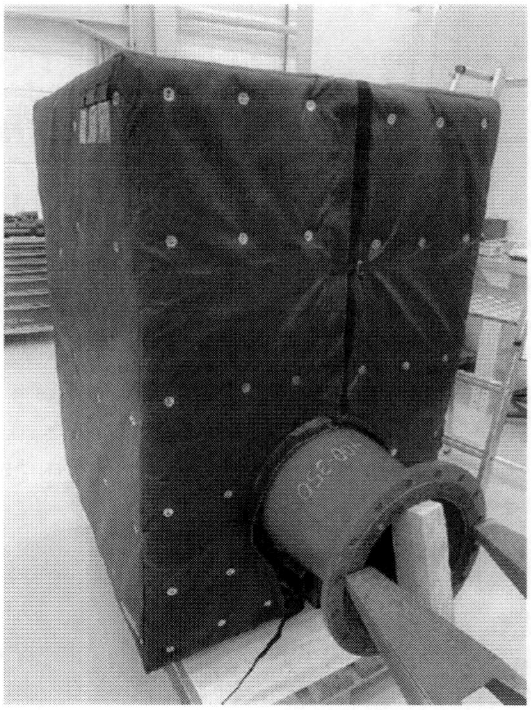

Fig. 2: POPRV inside a heating enclosure with accessible inlet and outlet flanges

Fig. 3: Testing of the heating enclosure within a climate chamber

In this climate chamber, the temperatures of the heating enclosure and the POPRV have been measured when the ambient temperature drops from the initial -20°C (-4°F) to -52°C (-61.6°F). At this lower ambient temperature, early relief in ammonia storage tanks can be a credible scenario. Fig. 4 clearly shows that at an ambient temperature of -20°C, all three sensors measure temperatures above the target temperature of

+10°C (+50°F), but fail when the ambient temperature drops to -52°C. This figure also shows that the temperature at the membrane is most sensitive to changes in the ambient temperature. The lesson learned from this figure is that the heating enclosure may be a good solution to prevent early relief due to clogging of the sensing line when designed and tested to work within the range of ambient temperatures of the installation.

The climate chamber also permits addressing the behavior of the POPRV in case of prolonged exposure to hot temperatures or oscillatory temperature profiles. For this second round of tests, the POPRV has been traced by electric heating and mounted on a vessel inside the chamber, subjected to temperature oscillations between -70°C (-94°F) and 60°C (140°F). The tank temperature and pressure in Fig. 5 show that the pressure profile oscillates between 1.8 kPa (0.26 psi) and 2.45 kPa (0.355 psi) gauge, leading over temperature with a similar interval. The POPRV presented some plastic deformation of the membrane, which unbalanced the force distribution in the dome and led to disk offset, with leakage and early relief as consequences.

Seismic vibrations, as well, can lead to plastic deformation or even rupture of the membrane.

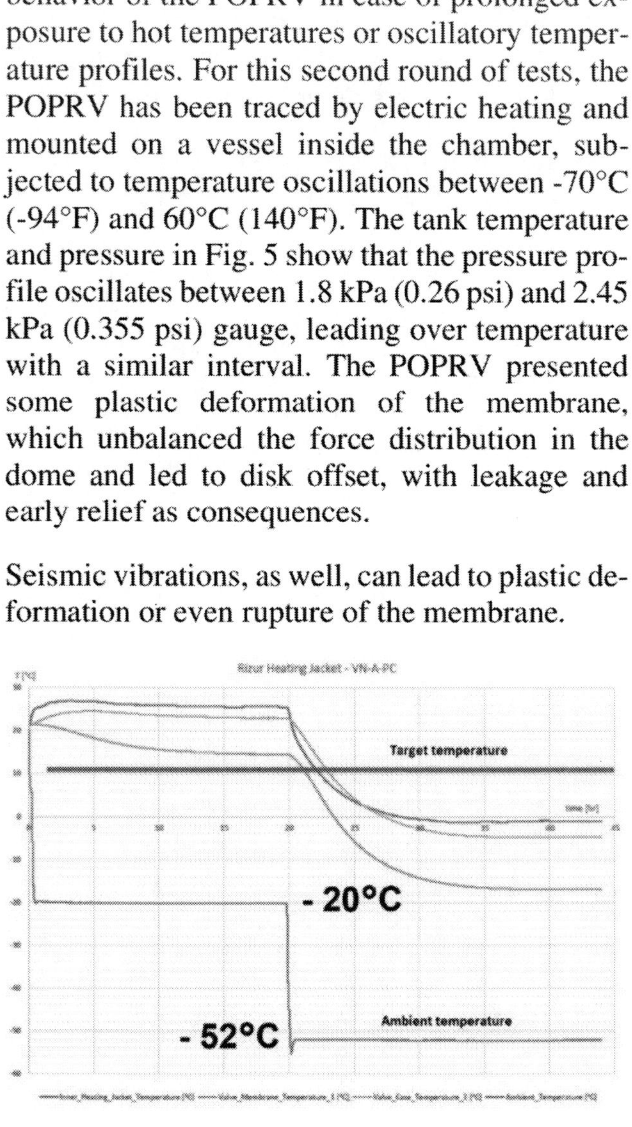

Fig. 4: Temperature profiles in the POPRV and the enclosure due to drop in ambient temperature

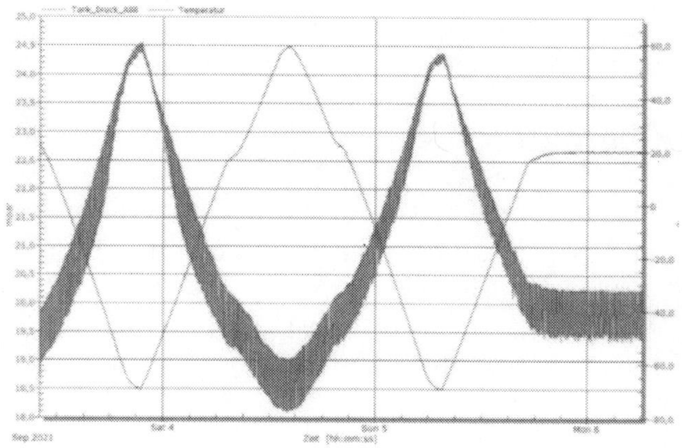

Fig. 5: Oscillatory temperature and pressure profiles in the enclosure in the heating chamber

Early Relief During Startup and Maintenance

Early relief during startup can eventually occur in cryogenic storage tanks at any latitude and climate. During startup, several batches of cold cryogenic liquid are injected into the empty storage tank. Good practice involves a preliminary purge of ambient air from the storage tank using nitrogen to avoid steam condensation and stress corrosion cracking issues on the walls during cooling. The injected liquid batch cools down the storage tank from ambient temperature and pressurizes it. The warmer gas moves to the top of the headspace and is vented out. The injection is done in several rounds to minimize thermal and structural stress in the tank walls.

It is common to protect the POPRV nozzle and the sensing line during startup against clogging. Good operating procedure calls for the removal of the protection before operation. When the protection removal is forgotten, the sensing line is suffocated and may lead to early relief. Similarly, clogging of the sensing line can occur when steam freezes during the startup cooling procedure if nitrogen purge is avoided.

Early relief during maintenance is worth mentioning. Typically, it involves operators who forget to actuate the ball valve on the sensing line after maintenance. Easy accessibility to the sensing line should be guaranteed even when the POPRV is enclosed in the heating jacket.

Conclusions

This paper discusses early relief from pilot-operated pressure relief valves protecting ammonia and other cryogenic storage tanks. Our paper proves that a heating jacket tested within the ambient temperature extremes helps preventing early relief caused by a clogged sensing line. Sensing line clogging is a common cause of early relief, caused by freezing vapors and particulates or packing material. Crucial to this concern are also maintenance shutdowns and startup phases. Excessive heating or seismic vibrations can generate plastic deformation of the membrane and an offset at the disk seat, leading to leakage or early relief.

Literature

[1] API 2000, Venting of atmospheric and low-pressure storage tanks, 7th Ed., 2014.

[2] ISO 28300, Petroleum, petrochemical and natural gas industries — Venting of atmospheric and low-pressure storage tanks, 1st Ed., 2008.

[3] NIST Chemistry Webbook, SRD 69

Operational Challenges in Ammonia Production: Addressing Water Contamination and Its Impact on Plant Performance

This paper investigates an incident at Saudi Arabian Mining (Ma'aden) Ammonia Plant 3, where outsourced water with elevated conductivity led to out-of-specification boiler feed water (BFW), resulting in contamination of the steam system. The high conductivity of water caused chloride contamination, catalyst deactivation, and significant corrosion in key mechanical components. A root cause analysis (RCA) was performed, identifying high chloride levels as the primary issue. Remedial actions included the installation of higher-capacity reverse osmosis units, resin replacement in water treatment systems, turbine repairs, and the addition of a chloride guard along with the catalyst. These measures successfully restored plant operation and efficiency.

Khalid Alyami, Taiebi Mohammed
Saudi Arabian Mining Company (Ma'aden)

Abdulrahman Ezzat, Manikandan
Johnson Matthey

Introduction

Saudi Arabian Mining (Ma'aden) operates three large scale ammonia plants, each with a nameplate capacity of 3300 metric ton per day (MTPD) (3638 STPD). Ammonia Plant 3 was commissioned in 2021, featuring a unique water system due to its location. Water was sourced from a nearby utility provider and processed within the plant to produce boiler feed water (BFW). Several months after commissioning, the outsourced water had a conductivity level higher than the design limit, leading to out-of-specification BFW. This contamination affected the entire steam system. This paper explores the causes of the incident and its impact on mechanical equipment, catalysts, and process safety.

Plant Description

Plant 3 is a standalone facility, located separate from Plants 1 and 2. During the design phase, it was decided to outsource potable water to a third party supplier, with maximum specifications of total dissolved solids (TDS) level of 110 mg/l and a conductivity of 175 µS/cm. The water was then processed through a demineralization unit to reduce conductivity to below 2 µS/cm, followed by a polishing unit to further reduce conductivity to below 0.2 µS/cm and sodium and chlorides to less than 0.02 mg/l. This system was specifically designed to treat water with a conductivity of 175 µS/cm. The processed water was used to produce boiler feed water (BFW), which in turn was used to generate high pressure (HP), medium pressure (MP), and low pressure (LP) steam.

Incident and Timeline

Plant 3 was commissioned at the end of 2021 while the first droplet of ammonia was produced at the beginning of 2022. A few months after start-up, the outsourced water supply conductivity was higher than expected and reached 450 µS/cm. This exceeded the capabilities of the demineralized and polished water unit, which was designed to handle 175 µS/cm. This higher conductivity led to polished water having slightly higher conductivity than design of 0.2 μS/cm. This water was then sent to boilers to produce steam.

In March 2023, and as a first remedial action, two portable reverse osmosis (RO) units were rented to reduce the high conductivity and bring it closer to design specification. The RO units reduced the conductivity slightly, but it was not sufficient to solve the problem. The RO membrane required cleaning on a regular basis. On several occasions like membrane cleaning, chemical dosing, and membrane unit trips, the plant team was forced to bypass the RO system to sustain operation.

In August 2023, the Plant 3 reformer tripped on low steam to carbon ratio (S/C). During the plant restart, a failure occurred on a dry gas seal in the refrigeration gas compressor (RGC). This resulted in tripping of back end of the plant while the front end was kept running (70% of plant rate) with venting downstream of the CO_2 removal unit while the RGC was handed over to maintenance for seal replacement work. During this period with only the front end running, polished water conductivity exceeded the limits due to poor performance of RO units. This resulted in high polished water conductivity above the normal range for a few days. Also during the period the high temperature shift (HTS) reactor CO slip increased from ~3.0 mol% to 7 mol%. The HTS reactor inlet temperature was increased to improve the CO slip and it was observed to be very marginal. Due to the reduced plant rate, the CO slip from the low temperature shift (LTS) reactor was kept within normal levels.

In the meantime, the repair work on the RGC seal was completed and the back end was lined up for operation. Soon after resuming production, the plant faced a HP steam superheater coil failure which forced the whole plant to Trip.

It was decided to inspect the HP steam superheater coil and inspect the HTS reactor due to the higher CO slip.

HP Steam Superheater Coil Inspection

This HP steam superheat coil is located in the flue gas convection duct of the primary reformer. The inspection found the liner sleeve located in the header after the desuperheater was detached (Figure 1) and carried over to the HP steam superheater coils, thus plugging the two coils. This sleeve protects the line from erosion which is likely after the desuperheater spray. These two coils failed due to the choking of the coil with the sleeve material, which resulted in flow starvation and overheating failure. During the inspection it was found that commissioning debris (Figure 2) did block some of the tubes which may have also contributed to the failure.

Figure 1: Header Liner sleeve detached.

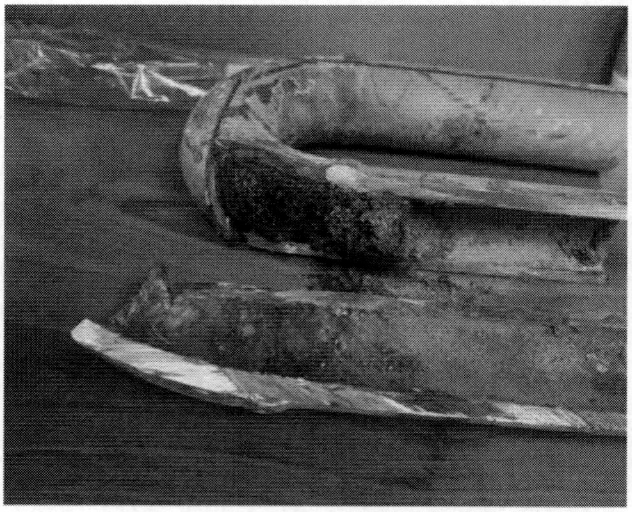

Figure 2: Foreign debris inside the tube.

During inspection of the HP steam coil, traces of white powder were found inside the coil (Figure 3).

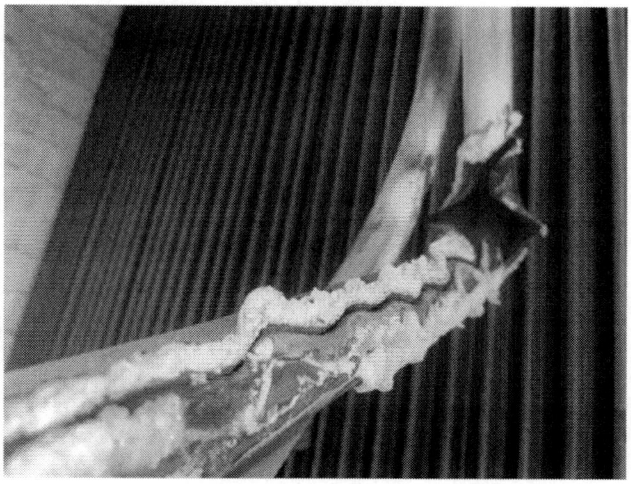

Figure 3: Failed HP steam superheater coils.

It was decided to open the HTS reactor for inspection. Upon opening the HTS reactor, a significant deposit of white material was found on top of the catalyst, which was later analyzed and found to contain chlorides. Due to the unavailability of a full charge of catalyst and uncertainty regarding the source of the chlorides at that time, it was decided to remove approximately 30% of the catalyst. Spent catalyst samples were collected for further analysis.

Due to the location of the superheater coil and challenges to replace damaged tubes, it was decided to cut these tubes from the header and install plugs at the header.

After the plant restarted the CO slip was significantly reduced and performance returned to normal levels.

Investigation and Root Cause Analysis (RCA)

A root cause analysis (RCA) was conducted to identify the cause of the HP steam coil failure and the higher CO slip from the HTS reactor.

Below are the inspection details along with the result of the spent catalyst analysis:

Mechanical Inspections and Findings

The investigation revealed significant mechanical damage caused by chloride contamination, which had a profound impact on key plant equipment. The following mechanical findings were observed:

1. **Corrosion of Steam Turbines**

As the HP steam coil inspection revealed chloride deposits, the impact of HP steam on the machinery and associated equipment was reviewed. Based on this, it was decided to inspect the syngas compressor turbine and RGC turbine as they are driven by HP steam. Significant pitting corrosion was identified on the steam turbines in both the syngas and refrigeration systems. (Figure 4) The corrosion, believed to be caused by chloride- presence in the system, led to reduced efficiency in the turbines, ultimately affecting their performance and reliability. The presence of chlorides in the steam system accelerated the degradation of turbine materials, necessitating repairs and affecting the overall mechanical stability of the plant. Both syngas compressor turbine and RGC turbine rotors were replaced.

Figure 4: SynGas Turbine Rotor

As the chloride contamination was likely due to poor BFW quality, the steam drum internals were inspected and found to be normal with no significant deposit formation. However as a precaution, the steam drum was flushed and rinsed several times to remove any chlorides and effectiveness of washing was measured by chloride analysis during that step. Peak chloride level during flushing was observed to be 65 ppm (max) which progressively dropped to < 0.02 ppm over a period of 24 hours indicating the removal of chloride from the system.

The Reformed Gas Boiler was also inspected and tube conditions were found to be normal.

2. **Failure in Auxiliary Boiler Tubes**:

The auxiliary boiler system, which is crucial for steam generation, exhibited severe corrosion in the tubes(Figure 5). Chloride contamination likely causes the degradation of the metal surfaces, compromising the heat transfer efficiency. This corrosion not only threatened the structural integrity of the boiler but also increased the risk of boiler failure and contributed to system inefficiencies. Further inspection confirmed that the corroded components needed to be replaced to ensure continued safe operation.

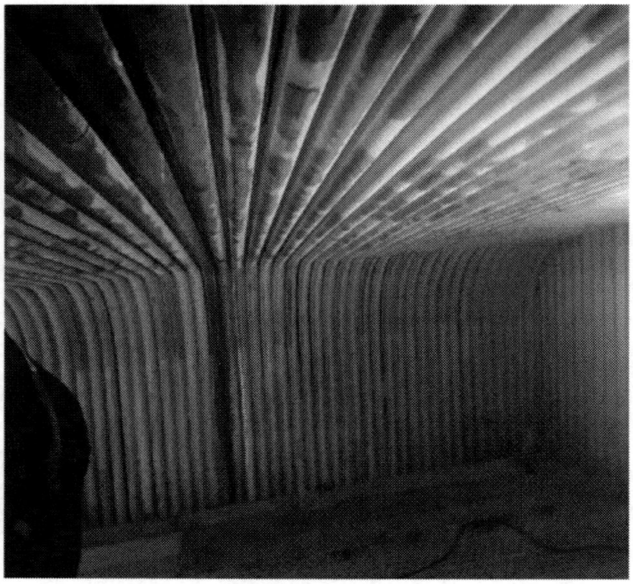

Figure 5: Auxiliary Boiler Tubes.

The mechanical findings highlight the severe impact of chloride contamination on plant components, particularly on turbines, boilers, and critical steam system equipment. The corrosion damage resulted in decreased efficiency, forced shutdowns, and increased repair costs, underlining the importance of controlling water quality to preserve the integrity and functionality of plant equipment.

Skimmed HTS Analysis Findings and Test Methods

Samples of the HTS spent catalyst was sent to Johnson Matthey for further analysis. Below is the list of samples (refer Table 1) with indication of their position inside the reactor:

Sample No.	Sample Name	Location in Bed	Materials
1	L1-C	Top	Alumina Support Balls
2	L1-E	Top	
3	L1-W	Top	
4	L2-C	200 mm depth	HTS Catalyst (KATALCO 71-6)
5	L2-N	200 mm depth	
6	L2-S	200 mm depth	
7	L3-C	700 mm depth	
8	L3-E	700 mm depth	
9	L3-W	700 mm depth	
10	L4-C	1205 mm depth	
11	L5-C	1205 mm depth	
12	L5-N	1405 mm depth	
13	L5-S	1405 mm depth	

Table 1: List of HTS spent samples.

Various analytic methodologies were used to provide physical and chemical analysis of the spent HTS catalyst.

Visual Observation

Pellets from samples taken from the top of the bed (L2) contained a considerable amount of grey dust in comparison to other samples. This suggests poisoning or a considerable amount of carryover from upstream processes. The amount of dust decreased as the bed depth increased. Discharged Katalco 71-6 pellets are typically matte black in appearance, however pellets from the top of the bed were slightly grey in appearance. Further down the bed the pellets were much darker, however none of the samples were matte black as would be expected. The alumina support balls were intact, but some grey dust was visible. Figure 6)

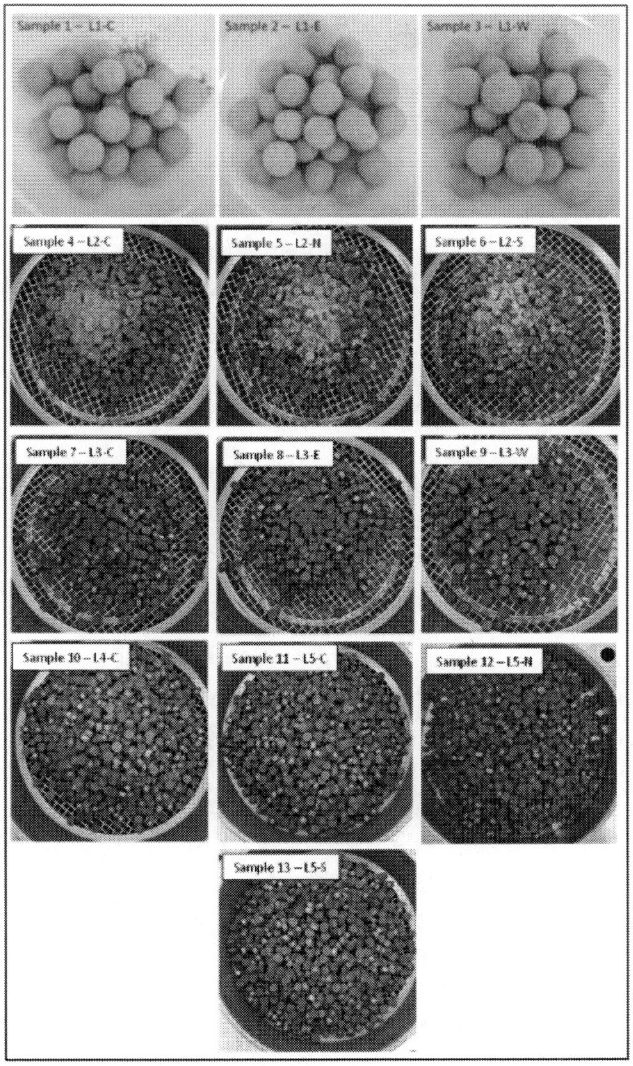

Figure 6: Photographs of all samples, including ceramic support balls (Samples 1-3) and Katalco 71-6 (Samples 4-13).

Brunauer-Emmett-Teller (BET) Surface Area

The BET Surface Area Test measures the surface area of porous materials. The BET surface area for all samples was similar, but slightly lower than expected. The porosimetry graph median pore diameter data revealed the samples had different pore structures with a shift to smaller average pore sizes moving deeper into the bed. Entrapment is typically close to 100 %v/v for discharged Katalco 71-6. The samples towards the top of the bed had lower values than this, indicating blockage of the pores, which could account for the observed lower performance and increased CO slip.

Ion Speciation

An Ion Chromatograph was used to determine elemental composition of non-metal components in samples. Specifically, to determine the presence of Halides, Sulphate and Nitrogen compounds. Several of the findings for halides are shown in Table 2.

Sample No.	F (ppm)	Cl (ppm)	Br (ppm)
Sample 4 – L2-C	4	4468	14
Sample 5 – L2-N	3	4712	2
Sample 7 – L3-C	3	3850	12
Sample 8 – L3-E	3	3222	10
Sample 9 – L3-W	3	3801	18
Sample 10 – L4-C	8	2395	<5
Sample 11 – L5-C	6	1349	<5
Sample 13 – L5-S	6	1114	<5

Table 2: Ion Chromatograph results

A small amount of fluoride was detected throughout the catalyst bed. Bromide was observed at the top of the bed but was undetectable in the samples at the bottom of the skimmed section of the bed. There are extremely high levels of chloride contamination throughout the entire sampled area of the bed. The chloride levels in these samples are concerning as there was almost 0.47 wt% Cl in samples taken from the top of the bed. Higher chloride levels, as found, can lead to bulk formation of iron and copper chlorides, causing a severe loss in activity and pellet integrity. The samples at the top of the bed contain greater than >4500 ppm Cl, which could account for the observed deactivation of the catalyst.

X-Ray Fluorescence (XRF)

XRF analysis is used to determine the bulk elemental composition of samples. Results of this analysis on the spent HTS catalyst is shown in Table 3.

	Na (wt. %)	K (wt. %)	Al (wt. %)	Si (wt. %)
Sample 4 – L2-C	0.32	0.13	0.28	0.35
Sample 5 – L2-N	0.32	0.14	0.32	0.34
Sample 7 – L3-C	0.27	0.10	0.25	0.31
Sample 8 – L3-E	0.24	0.08	0.18	0.28
Sample 9 – L3-W	0.27	0.09	0.17	0.29
Sample 10 – L4-C	0.29	0.04	0.06	0.02
Sample 11 – L5-C	0.21	0.03	0.03	<0.01
Sample 13 – L5-S	0.20	0.02	0.04	<0.01

Table 3: X-Ray Fluorescence (XRF)

The spent HTS catalyst samples were found to contain extremely high levels of Cl, which can significantly affect catalyst performance and lead to loss of pellet integrity. As well as Cl, the catalyst samples contain high amounts of potassium (K), sodium (Na), aluminum (Al), and silicon (Si) which can also affect performance of the catalyst. The level of poisoning was high enough to account for the observed deactivation and CO slip.

Summary of Findings

From the results of the RCA investigation, the main findings were:

1. High Water Conductivity: The outsourced water had a conductivity level of 450 µS/cm, which exceeded the design limit of 175 µS/cm, leading to out-of-specification BFW.
2. Chloride Contamination: The water's high conductivity contributed to significant chloride contamination in the system, resulting in catalyst poisoning, mechanical corrosion, and damage to various plant components.
3. Catalyst Deactivation: High chloride levels in the HTS reactor caused a loss in catalyst activity, which contributed to increased CO slip and operational inefficiencies.
4. Corrosion Damage: Pitting corrosion was observed on steam turbines, especially in the syngas and refrigeration systems, and corrosion was also found in the auxiliary boiler tubes.
5. Effect on Process Safety: The contamination and mechanical issues compromised plant safety, leading to operational disruptions and potential risks to the steam system and reactor performance.
6. Root Cause of Issues: The primary cause of the problems was traced back to the high conductivity outsourced water, which led to chloride contamination and subsequent damage to catalyst, equipment, and processes.

Remedial Actions

The remedy actions taken to address the issues caused by high conductivity outsourced water and its impact on the plant included:

1. Replaced the two-reverse osmosis (RO) units with five higher-capacity RO units (120% of the plant load) to better control the outsourced water's conductivity.
2. Replaced the resin in both the demineralized and polished water treatment units.
3. Installed new rotors for the syngas and refrigeration turbines.
4. Replaced corroded parts in the auxiliary boiler.

These measures aimed to restore the plant's operational stability and prevent further damage from high conductivity water.

Conclusion

The analysis reveals that the high conductivity of outsourced water was the primary cause of operational failures at Plant 3. This issue, stemming from water quality exceeding design limits, led to contamination, catalyst deactivation, and corrosion, which hindered the ammonia production process.

Corrective actions, including installing additional reverse osmosis units, replacing critical components, and upgrading the water treatment system, have been implemented to address these challenges and restore stability. The plant is now better equipped to handle water quality variations, ensuring smoother operations.

Moving forward, continuous monitoring and system refinements are necessary to maintain plant performance. The lessons learned highlight the importance of planning, monitoring outsourced resources, and taking prompt action to prevent infrastructure damage, ultimately improving operational efficiency and safety.

High Temperature Exposure of HTS Catalyst at SOR

A fresh charge of HTS catalyst was exposed to high temperatures of 680°F (360°C) for 18 days. Normal SOR inlet temperature is recommended to be around 620-630°F (326-332°C). This high temperature exposure resulted in shifting of the exotherm from the top to the middle of the catalyst bed and a decline in the performance of the HTS catalyst as indicated by the carbon monoxide slip at the outlet and ATE (approach to equilibrium) of the reactor. This paper describes the root cause and contributory factors resulting from an obstruction of the front-end waste heat boiler bypass valve leading to this incident.

Faisal Jamal, Kamran Arif Chaudhry, Maham Bashir
Technology Division, Fatima Group Head Office, Lahore, Pakistan

Sheikh Adeel Ahmad
FatimaFert Limited, Sheikhupura, Pakistan

Introduction

Fatima Group (FG) is one of the largest and most progressive business conglomerates in Pakistan, operating in diversified sectors of Fertilizers, Textiles, Sugar, Energy, Packaging, Mining and Trading commodities. Over the last 3 decades, FG has progressed to play a significant role in Pakistan's economy.

FG is one of the leading manufacturers of quality fertilizers in Pakistan. FG owns and operates three large scale fertilizer manufacturing complexes located strategically in the fertile belts of Pakistan, with a diverse product range and a combined fertilizer capacity of 2.8 million MTPA. The fertilizer manufacturing facilities include Fatima Fertilizer Company Limited in Sadiqabad, PakArab Fertilizer Company Limited in Multan and FatimaFert Limited in Sheikhupura. Our fertilizer brand, Sarsabz, was launched in 2011 as Pakistan's only specialized nitrate-based premium fertilizer.

Health, Safety and Environment (HSE) is the highest priority of our operations. FG continuously thrives in HSE and Operational Excellence with stringent and voluntary compliance of OSHA PSM standards reaching combined safe man-hours of 140 million.

FatimaFert limited (FFT) was formerly owned by Dawood Hercules Chemical Limited (DHCL). FFT became a subsidiary company of FG in 2015. It operates an Ammonia – Urea complex near Lahore in Pakistan. The Kellogg designed ammonia plant was commissioned in 1971 with a nameplate capacity of 625 MTPD and was upgraded to 815 MTPD in 1991 with a decrease in energy index by 10%. Most of the equipment remained the same, including the bayonet primary waste heat boiler (101-C) and typical shell and tube secondary waste heat boiler (102-C) installed downstream of the Secondary Reformer.

The urea plant was originally a Mitsui Toatsu Japan designed Total Recycle process with 1100 MTPD capacity that has been revamped to 1350 MTPD by UTIC USA in 1991.

Historical operation by DHCL, as in many similar plants around the world, failure of the bayonet Waste Heat Boiler (WHB) has been a major cause of production loss due to frequent shutdowns. The DHCL's ammonia plant experienced a total of 30 failure incidents of the Primary Waste Heat Boiler (101-C) out of which only two were before the revamp in 1991. These failures resulted due to operational upsets, water chemistry, and spacer / nail (design) issues. The primary and secondary waste heat boiler were vertical exchangers that suffered from known corrosion issues resulting from solids deposits at the bottom of the exchangers. After frequent failures, the boilers were replaced with higher capacity single Borsig Fire Tube WHB with a dedicated steam drum (1101-C and 1101-F, respectively).

Alongside this a dedicated steam drum (1101-F) was also installed and commissioned in 2006. Since then, the WHB has operated without any leakage or major problem resulting in a significant improvement in the service factor and cumulative energy consumption numbers [1].

Figure 1: Steam Drum 1101-F and WHB 1101-C with Bypass valve (TRCe-610)

Ammonia production was further increased from 815 to 890 MTPD by capacity utilization and the replacement of the primary reformer tubes with thin-walled tubes along with the Borsig waste heat boiler and dedicated steam drum.

Process Description

FFT Ammonia plant is a conventional KBR technology design equipped with water jacketed secondary reformer and transfer line. Figure 2 explains the block flow diagram from secondary reformer and shift conversion reaction. Partially reformed natural gas (CH4 Slip 10-11%) from the primary reformer is directed towards the secondary reformer (103-D). Compressed process air and reformed gas from the primary reformer are burned together in the secondary reformer to complete the desired reforming reaction using a nickel-based catalyst (CH4 Slip 0.3-0.5%). The stoichiometric amount of process air is controlled to maintain the hydrogen to nitrogen ratio (H/N) required for the ammonia synthesis reaction. Reformed gas leaves the bottom of the Secondary Reformer at 1807°F (986°C) and a significant amount of energy is recovered by steam generation which is later used to drive the ammonia plant's major machines and provide process steam required for the reforming reaction. Reformed gas at exit from 103-D enters a waste heat boiler (1101-C) where its heat is recovered by producing 1500 psig (10.3 MPa) steam. The temperature at the exit of 1101-C is controlled in the range of 620 to 680 °F (326-360°C) by using the WHB bypass temperature control valve (TRCe-610) depending on the health and life of High Temperature Shift Convertor (104-D) catalyst. At FFT, the high temperature shift convertor (HTS) temperature is maintained at around 625°F (329.4 °C) at the start of run (SOR) of the new catalyst charge and later 1101-C exit temperature is gradually increased over the years, by controlling the opening and allowing mixing of the 103-D high temperature exit gas through the 1101-C internal bypass line, while the low temperature gas exits through the tubes on the outer side of the exchanger by exchanging heat with the BFW (boiler feed water) on the shell side to produce steam. The reformed process gas enters the HTS (104-D) where the water gas shift conversion reaction takes place over the iron-based catalyst, converting the inlet CO to CO_2 and giving the desired CO slip. This gas then enters a series of three exchangers 103-C, 1104-C and 1112-C, respectively in order to achieve the desired inlet temperature at the Low Temperature Shift Convertor (LTS) 108-D.

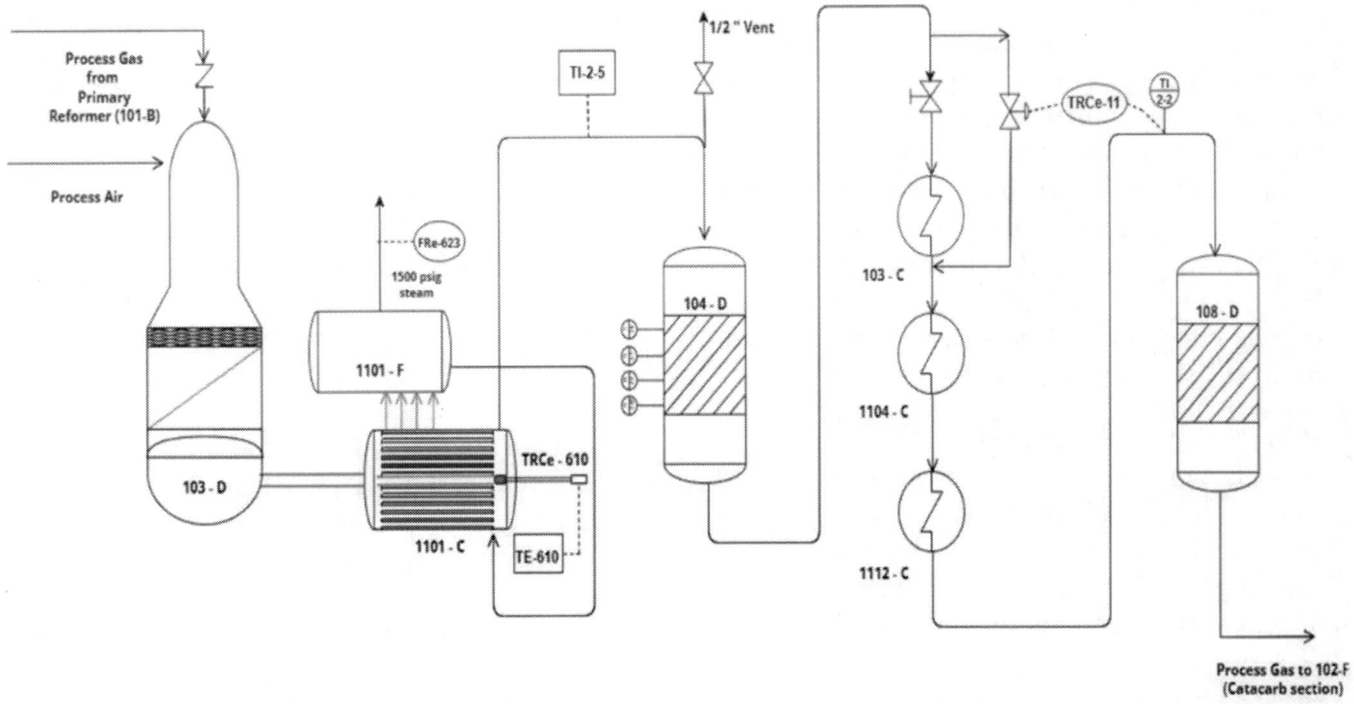

Figure 2: FFT Ammonia plant Block flow diagram – Primary Reformer outlet to Shift conversion Section

Problem Description

The plant was restarted in March 2023 after a major Turnaround during which the HTS Catalyst was replaced after completion of its useful service life of 8 years. Plant start-up was normal, and production was resumed on 29th March 2023. The HTS inlet temperature was maintained at 625 °F (329.4°C) as per OEM recommended range at SOR to gain maximum life with optimized conversion.

On 29th April 2023, 1 month after the plant start-up, an increase in CO slip at the LTS exit was observed (0.18 to 0.21 dry mol%). The WHB By-pass valve (TRCe-610) was opened to maintain the CO slip at LTS exit by increasing HTS inlet from 625 °F to 650 °F (329 to 343 °C). Later after normalization of operating parameters, despite the complete closure of TRCe-610 bypass valve, HTS inlet temperatures remained on the higher side 635 °F (335°C).

On 21st May 2023, plant load was reduced by 14 % to attend the Hydraulic turbine casing drain line leakage. Plant was normalized on the 22nd May 2023 after rectification of leakage. During 22nd May to 12th June 2023, the HTS inlet Temperature at TRCe-610 gradually increased from 635 °F to 680 °F (335-360 °C). Consequently, the steam production from steam drum (1101-F) reduced by 5%.

The operations team tried to maintain HTS inlet temperature by closing the TRCe-610 valve, and the instrument team also checked the control valve which was found fully closed. This was also confirmed by observing the protruding valve stem in the field and the percentage valve opening on the Distributed Control System (DCS). Efforts were also made to tighten the valve through hand jack, but no improvement was observed. Figure 3 depicts orientation and Figure 4 shows TRCe-610 hand jack location. As the temperature at the HTS inlet reached a maximum of 680 °F (360 °C), an emergency shutdown was taken to prevent risk of prolonged high temperature exposure to the fresh charge of the HTS catalyst, to find the root cause of the incident and to rectify the issue.

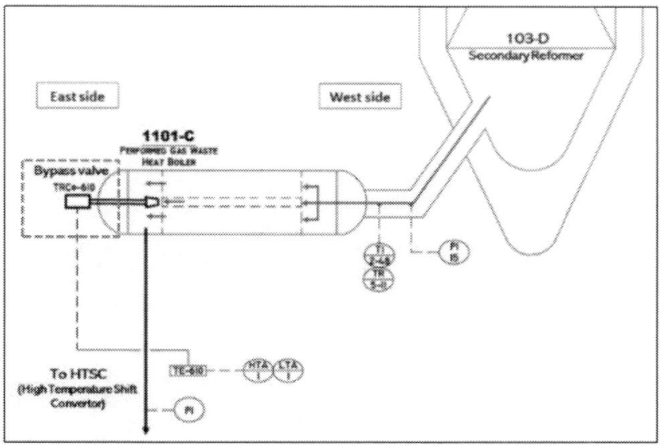

Figure 3: 1101-C downstream of 103-D (East and West sides) along with TRCe-610

Figure 4: TRCe-610 valve installed at the East side of the WHB 1101-C

Sequence of Events

22nd May to 12th June, 2023

HTS inlet temperature gradually increased from 635°F to 657°F (335 - 347 °C) from 22nd May to 5th June and further increased to 680 °F (360 °C) until 12th June. Steam generation reduction by 5% was also observed resulting in a higher Energy Index of the Ammonia plant. Steam drum pressure was reduced to mitigate the high temperature condition by reducing the boiling temperature, but no fruitful results were achieved.

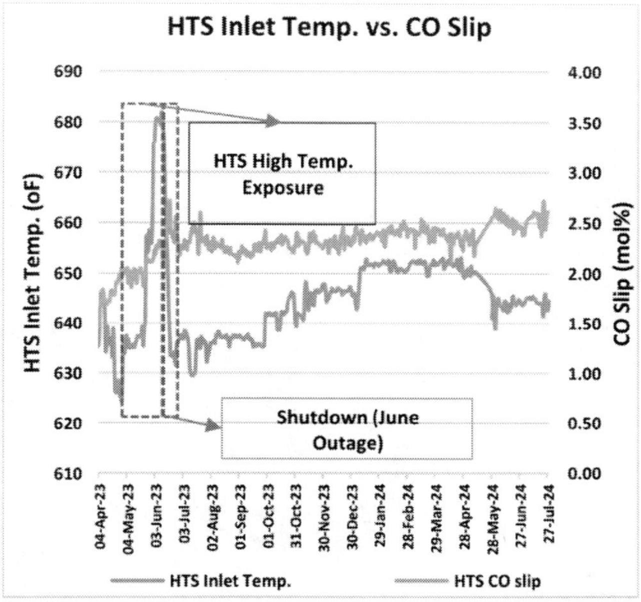

Figure 5: HTS Inlet Temperature and HTS CO slip pre and post temperature exposure

13th June, 2023

After multiple efforts, it was decided to take an emergency plant shutdown to identify and address the root cause of 1101-C higher outlet temperature. Plant shutdown was initiated on 13th June as per SOP (standard operating procedure).

14th and 15th June, 2023

After cooling, West and East side manways were opened along with the removal of the internal bolted head cover of the WHB on the TRCe-610 (East side) for initial inspection.

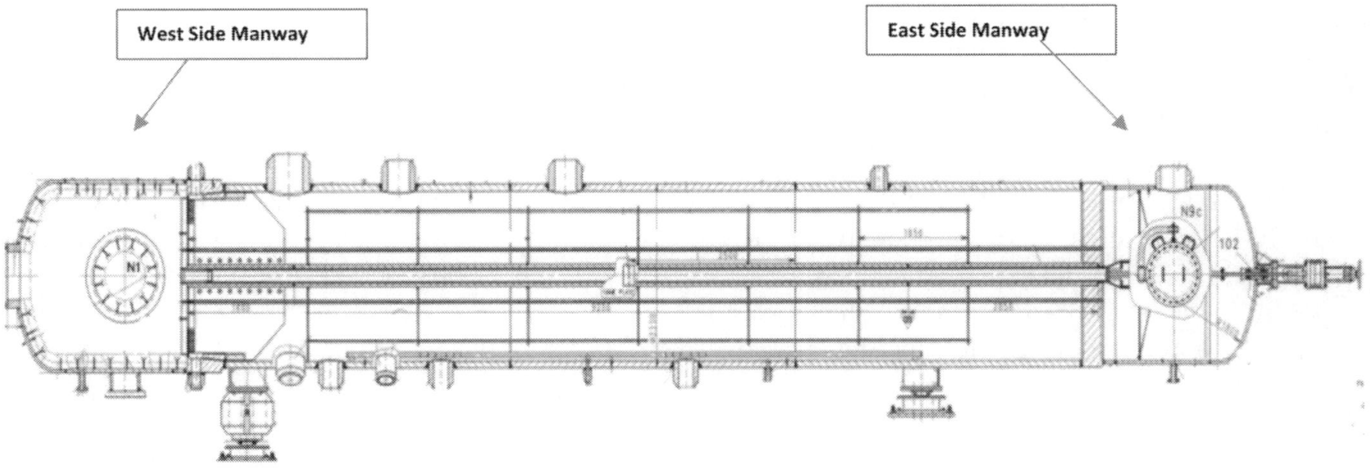

Figure 6: WHB 1101-C drawing depicting east side and west side manways

During initial inspection from the east side manway the control valve assembly and shaft were found intact.

Figure 7: TRC-610 initial inspection from east side man-way of 1101-C

Borescope inspection of the bypass valve disc and seat cone revealed that the conical portion of bypass valve has been ruptured (Figure-8).

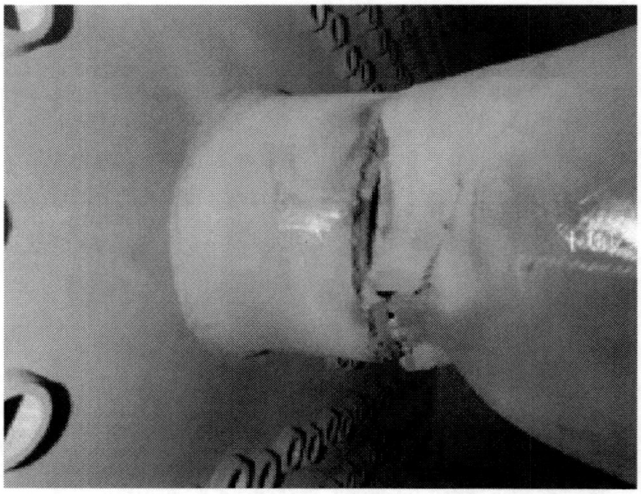

Figure 8: Boroscopy: Ruptured bypass valve seat cone weld joint and base metal

Refractory pieces were found stuck between the valve disc and seat causing the valve to be partially open even at full closure and tightening of valve by hand jack.

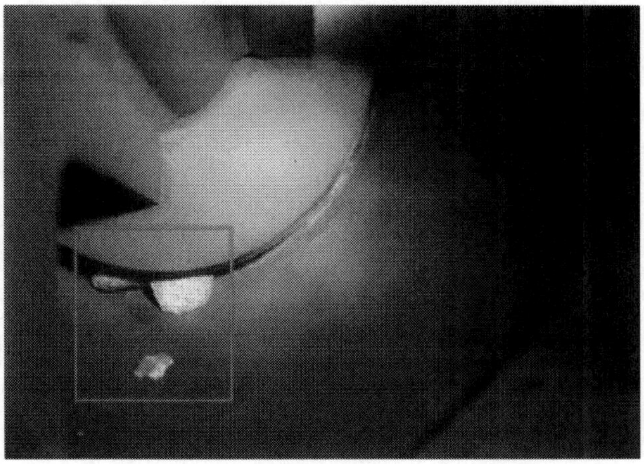

Figure 9: Refractory pieces found between valve disc and seat

Internal bolted head cover was removed for the first time since commissioning for detailed inspection of valve assembly and seat. Earlier valve internals were inspected using Boroscopy technique.

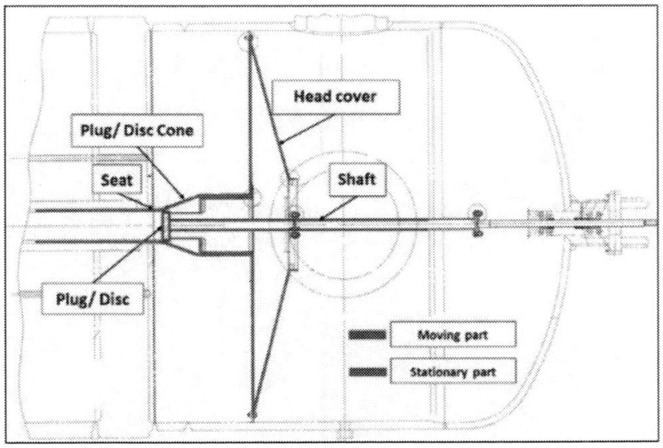

Figure 10: Internals of Temperature Control Valve Assembly

Detailed internal inspection of the Waste Heat Recovery Boiler along with its transfer line from both ends, 103-D (West) and TRCe-610 (East), sides was carried out. The following were the major observations;

East Side:

Bypass valve plug cone was found ruptured from base metal and its weld joint (Figure-8). Bypass valve disc was found slightly worn off and 1" longitudinal crack was also found on the circumference. The valve disc (valve plug) was found cracked from top and side wise as well as shown in Figure 11.

Figure 11: Crack at bypass valve disc

Cracks in structural supports of bypass valve shaft were observed. All structural supports of bypass valve were also found cracked (Figure 12). Flanged connections were observed to be misaligned. The bypass valve shaft was also found slightly misaligned.

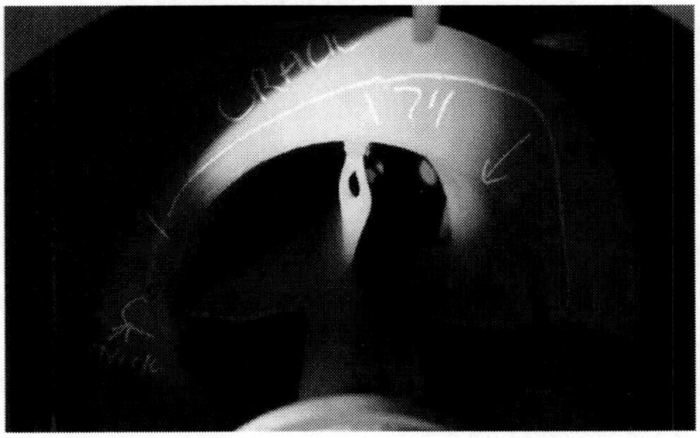

Figure 12: Cracks observed in structural supports of valve assembly

A crack was observed in the bypass valve seat cone base metal.

Figure 13: Crack in by-pass valve seat cone

West End:

Refractory from the bulged portion of the transfer line liner was found missing as per the initial inspection.

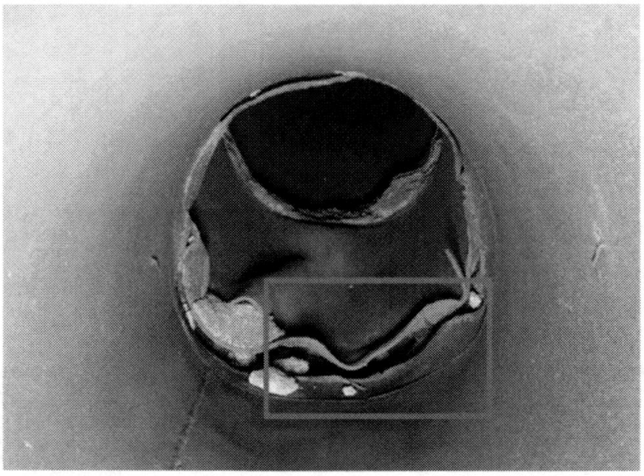

Figure 14: Missing refractory in the bulged section of transfer line as viewed from the East end

Upon inspection of 1101-C inlet side (West) tube sheet, approximately 60-70% of the tubes (ferrules) were observed partially choked with refractory pieces. Refractory dust and pieces were also observed at the bottom of the channel head.

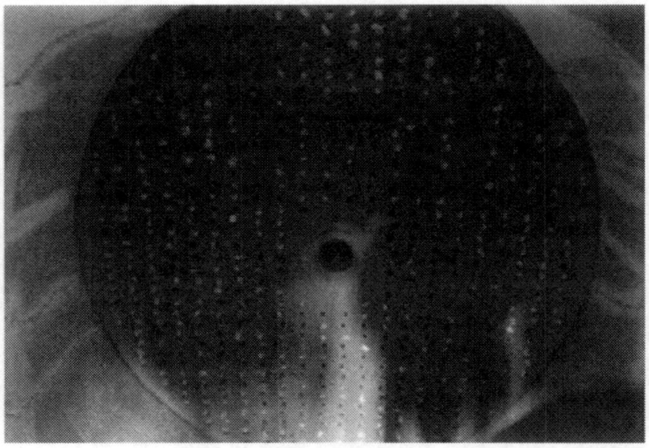

Figure 15: Tube sheet Condition from West Site

16th June, 2023

Corrective actions were taken covering welding repairs as per standard Welding Procedure Specification (WPS). Refractory pieces and debris found in the transfer line and tube ferrules were removed from both channel heads. The valve assembly was installed and its opening/closing was calibrated and confirmed with DCS commands.

A thorough final inspection was conducted and found satisfactory for box up.

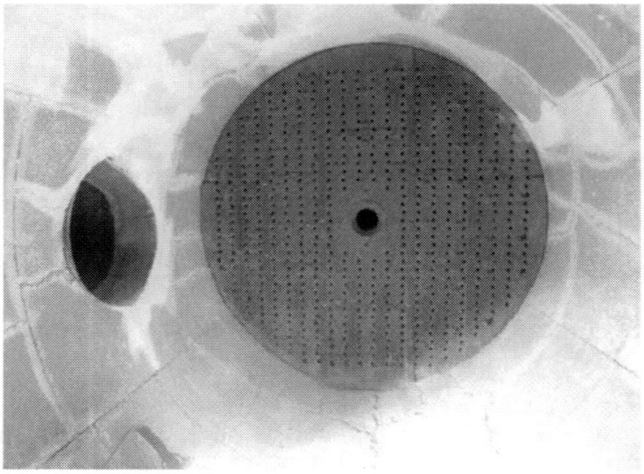

Figure 16: Inlet Tube sheet and channel head after cleaning

17th and 18th June, 2023

WHB and secondary reformer transfer line were boxed up. Startup was initiated after ensuring all pre-startup checks were completed. Production was resumed after an outage duration of 6 days.

18th June, 2023

Normal operating conditions of the plant after start-up were achieved where HTS CO slip and dP were 2.36 mol% and 4.7 psi (32.4 kPa), respectively at an inlet temperature of 632 °F (333.3 °C).

Key findings

To identify the source of refractory, which clogged the tube inlets and obstructed the bypass valve disc, detailed inspection of 103-D Bottom Plug and Transfer line between 103-D and 1101-C was carried out.

During the 2023 turnaround, a 1.5ft portion of the liner was observed bulged severely and a gap was created beneath it which could have led to high temperature process gas penetrating into the refractory and with time further refractory degradation can lead to exposing the pressure wall to extremely high temperatures >1650°F (899 °C). The refractory damage did not result in substantial increase in heat loss from the transfer line and the jacket water consumption remained normal. The complete liner repair is a difficult job due to access limitations, congested area, slope of transfer line and time constraints. It was decided to pour refractory under the gap created as per procedure until next TA before permanent replacement of transfer line liner. During startup, considering it a minor repair, a formal dry out procedure was not carried out resulting in compromised bonding of poured refractory.

During the June 2023 Outage while performing a detailed inspection it was found out that the refractory repairs carried out in Transfer line liner piece from the bulged portion (during Turnaround 2023) did not remain sustainable after operation and refractory beneath this bulged portion was observed removed.

Bypass valve disc was found slightly worn off and a crack was observed, possibly due to over tightening of the valve with hand jack during efforts to close the valve completely. All vertical supporting strips of the bypass valve disc were found cracked, which are observed in every TA, and were repaired each time as per WPS. The valve disc was not repaired and was installed as found.

The bypass valve seat cone weld joint was repaired to ensure effective closure and with respect to plant startup, and cone base metal was not repaired due to multiple ruptures and plastic deformation. Repaired weld joint was checked with Die Penetrant Testing (DPT) and was found satisfactory.

Figure 17: Bypass valve disc deformed from one end

Detailed Evaluation and Analysis

Root Cause:

Detailed investigation revealed that a refractory repair job was carried out in TA-2023 based on observation of excessive refractory loss under deformed / bulged portion of liner (Figure 14.

Whereas, transfer line liner bulging occurred during commissioning and since then it was operating satisfactory with minor loss of refractory over the years. During this outage, increased bulging was observed in 1.5ft portion of liner and a gap was created beneath it; accordingly, refractory repair was deemed mandatory to avoid HTHA (High Temperature Hydrogen Attack) and high temperature in jacketed transfer line. Refractory repair methodology beneath the bulged liner was found to be inadequate with respect to quality control measures and application due to time limitation and area access constraints. Catalyst dust was present beneath the liner which restricted the bonding of old refractory with newly poured refractory.

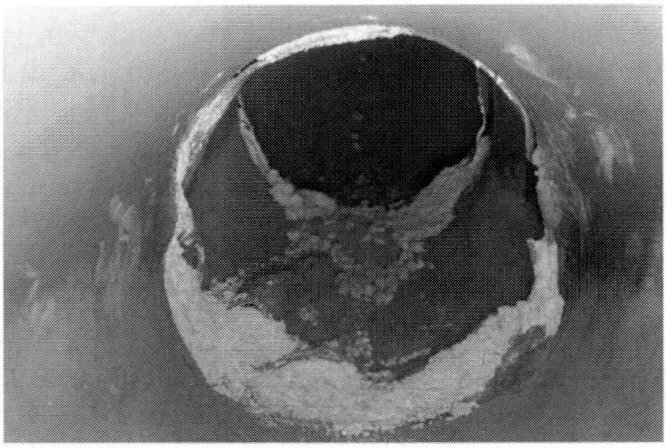

Figure 18: Refractory Repair job during TA-23

Furthermore, dry out of the new refractory was also not carried out in view of it being considered a minor refractory repair job. After start-up in March 2023, castable refractory, which was repaired (poured) in TA 2023 in the transfer line liner bulged section, detached and was carried downstream to the 1101-C tube sheet and to the bypass tube resulting in a partial blockage of the tube inlet ferrules. Refractory pieces travelled through 1101-C bypass tube towards the valve seat, and got stuck during valve partially opening on 29th April 2023 during an HTS inlet temperature adjustment to arrest the CO slip of LTS. Bypass valve (TRCe-610) remained completely closed until 29th April 2023, with no margin for refractory to get stuck between the valve seat and disc. Refractory pieces stuck between the valve disc and seat (Figure 9 hindered its complete closure, which resulted in bypassing of hot gas and the higher-than-normal HTS inlet temperature. Later, as the percentage blockage of inlet tube ferrules increased due to continued carry over of refractory pieces, the flow of hot gas increased through the bypass line resulting in the gradual increase in temperature until the temperature reached its maximum value of 680 °F (360 °C).

Contributory Factors:

Inspections Regime:

As per inspection regime since commissioning of WHB (1101-C) in 2006, inspections were carried out with respect to the health of the transfer line refractory and liner on the west side. Valve shaft and bolted head cover on the east side was inspected. Only partial inspections of the cone portion used to be carried out. Since no abnormalities were observed during past inspections, accordingly the head cover was never dismantled to inspect the valve assembly and carry out a part-by-part inspection. TRCe-610 seat and housing were found damaged causing a restriction to plug movement and causing uncontrolled gas bypassing of 1101-C tube side leading to high temperature.

Overtightening of the valve:

TRCe-610 valve disc found slightly deformed along with slightly misaligned shaft due to overtightening efforts of the valve through hand jack to completely close the valve.

Thermal Stresses resulting from Frequent plant start-ups and shut downs:

The plug cone of the valve was found cracked and damaged indicating effects of prolonged operation under high thermal stresses (Creep phenomena) due to frequent startups during 2011 to 2020 owing to the natural gas crises in the country. These frequent start-ups / shut downs also resulted in increased bulging of transfer line liner leading to refractory loss from underneath.

Impact of High Temperature on HTS Catalyst at SOR

HTS Catalyst was replaced during TA-2023 after the completion of a useful service life of 8 years of the previous (Haldor Topsoe) HTAS catalyst charge. EOR CO slip, differential pressure (dP) and inlet temperatures were at 2.7 mol%, 7.3 psi (50.3 kPa) and 678 °F (359 °C), respectively. The new catalyst charge loaded is of the same type as that of the previous. TOPSOE SK-201-2 was loaded having the following specifications;

TOPSOE

SK – 201 – 2

Size	6 x 6 mm
Volume	1448 ft^3/ 41 m^3
Fe_2O_3/Fe_3O_4 %	89 %
Cr_2O_3 %	9 %
Carrier/Base	Al_2O_3
SOR	March 2023

10th April 2023

New Catalyst SOR CO slip, dP and Inlet Temperature were satisfactory at 1.9 mol%, 4.5 psi (32.0 kPa) and 625 °F (329.4 °C), respectively.

24th May to 12th June, 2023

HTS Catalyst was exposed to high temperature due to WHB Bypass valve passing which resulted in high HTS inlet temperature reaching to a maximum of 680 °F (360 °C) for 18 days where the fresh catalyst charge was exposed to the EOR temperatures at its SOR.

13th to 18th June, 2023

A six days outage was taken to address the root cause as explained earlier in this paper.

19th June, 2023

After plant start-up, the CO slip, dP and Inlet Temperature were 2.36 mol%, 4.7 psi (32.4 kPa) and 632 °F (333 °C), respectively.

1st and 2nd week of May, 2024

Before the 2024 Outage the CO slip, dP and Inlet Temperature were observed to be 2.33 mol%, 4.7 psi (32.4 kPa) and 647.7 °F (342 °C), respectively.

13th to 28th May, 2024

Planned Outage in which secondary reformer catalyst was replaced with a new charge. Refractory liner job and inspection was done as well.

June, 2024

Plant started after outage and the CO slip, dP and Inlet Temperatures at 2.51 mol%, 4.8 psi (33.1 kPa) and 643.6 °F (340 °C), respectively.

26th Aug, 2024, Catalyst Assessment

The CO slip, dP and Inlet Temperatures at 2.43 mol%, 4.8 psi (33.1 kPa) and 644 °F (340 °C), respectively. Inlet temperatures increased to maintain the CO slip at the outlet of the HTS.

Analysis and Evaluations

The reaction completion curves and temperature profiles of the new catalyst charge from its SOR to 1 year ahead of it are analyzed and evaluated indicating the catalyst activity deterioration. The reaction completion curves (Figure 20) and reactor bed temperatures (Figure 21) at various depths of the catalyst bed of the old catalyst EOR and new catalyst charge at SOR and 1-year post incident are compared as shown below in Figures 20 and 21, indicating a 17% drop in the catalyst bed activity of the catalyst from May 2023 until Aug 2024.

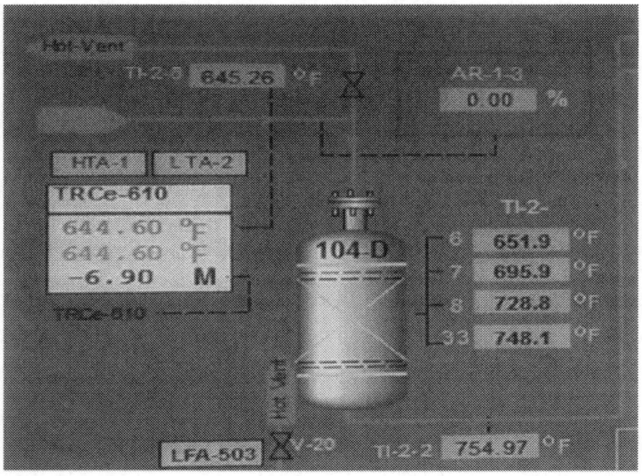

Figure 19: DCS display of the HTS reactor along with its Catalyst bed Tis

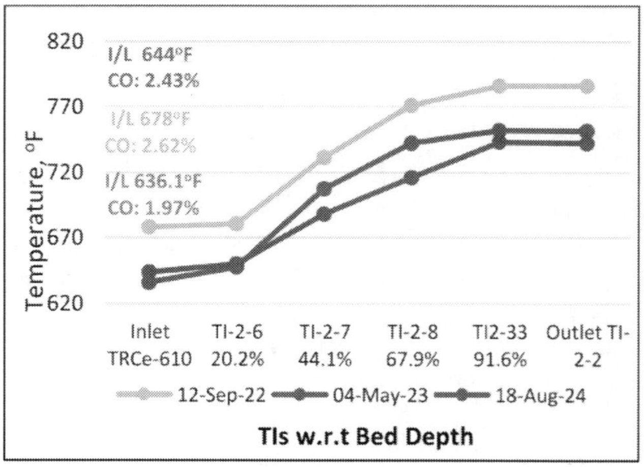

Figure 21: Catalyst bed temperature profiles before (04-May-2023) and after the incident (18-Aug-2024)

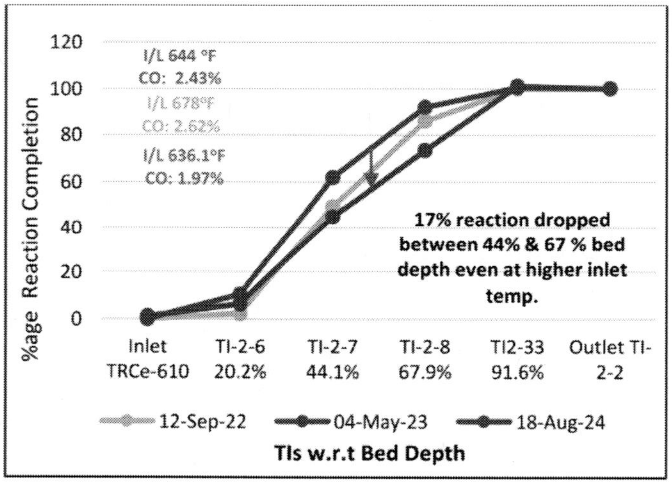

Figure 20: %age reaction completion curves before (04-May-2023) and after the incident (18-Aug-2024

New catalyst bed temperatures and delta Ts across the catalyst bed are also plotted for 1-year including the incident and post incident period as reflected in Figure 22 below, showing that the reaction between the 1st and 2nd bed has significantly reduced and shifted towards the middle and lower portion of the catalyst bed.

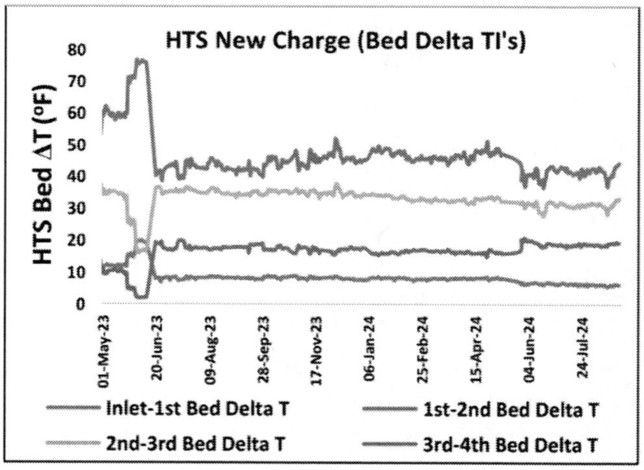

Figure 22: HTS bed change ΔT (°F) over time since the incident and 1 year onwards.

Catalyst Remaining Life Estimation:

HTS – High Temperature Shift Convertor

The catalyst should be operated at the minimum practical inlet temperature that affords the desired conversion of carbon monoxide, since this results in an increased life of the catalyst limiting sintering. Magnetite crystallites are not completely stabilized by the chromium oxide and, as a result, during its life the catalyst gradually loses activity as the surface area decreases due to thermal sintering. This is exacerbated at high temperatures, particularly above 750°F (399 °C). A consequence of this is that new catalyst will operate satisfactorily at a lower inlet temperature than older, slightly deactivated catalyst. As the catalyst ages it is necessary to increase the converter inlet temperature in order to maintain a satisfactory exit carbon monoxide level, and the temperature profile through the bed typically changes with time. Eventually the overall carbon monoxide conversion decreases, and the carbon monoxide at the exit runs to a level at which the catalyst has to be changed. During operation therefore, care should be taken not to raise inlet temperatures unnecessarily, since this may result in some loss of low-temperature activity and a somewhat shortened economic life of the charge [2]

Increase in the temperature profile causes the equilibrium to shift to the left, thereby CO slip at the exit is greater, but kinetically, the high temperature accelerates the reaction rate and approach conversion at equilibrium conditions. Elevated temperatures are problematic as they can harm the HTS catalyst, leading to reduced performance. Maintaining an optimal temperature is crucial for the efficient conversion of CO to CO_2, preventing issues such as catalyst deactivation through sintering. Sintering results in a decreased contact surface area, consequently diminishing catalyst activity and leading to higher pressure drop.

Detailed regression analysis was carried out with respect to CO slip and inlet temperature. Life of new the HTS catalyst charge is predicted on the basis of HTS inlet temperature and CO slip at HTS exit. Trends in Figure 23 and 24 show this data plotted for old and new catalyst charges. When this inlet temperature data for the new catalyst is extrapolated keeping in view the maximum EOR (End of Run) temperature 680 °F (360 °C), it can be forecasted that the catalyst life is expected to reduce by 1 to 2 years.

ATE (Approach to Equilibrium) of HTS increased from 20 °F (13.3 °C) to 60 °F (38.9 °C) from May 11, 2023 to Aug 6, 2024, respectively, which indicates that the reaction has moved further away from the equilibrium.

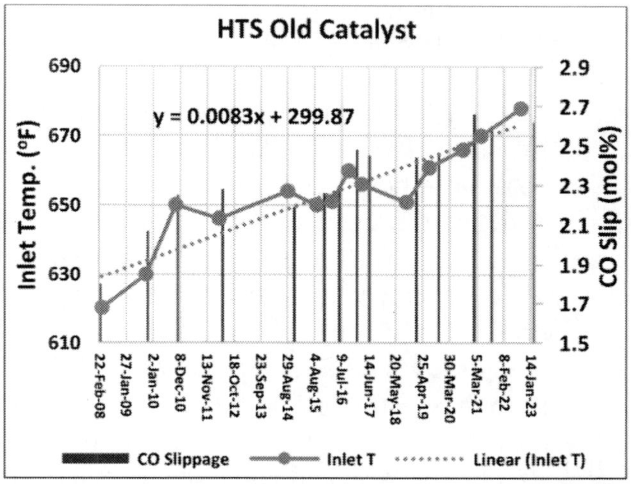

Figure 23: HTS Inlet Temperatures and CO slip of previous charge from its SOR till EOR.

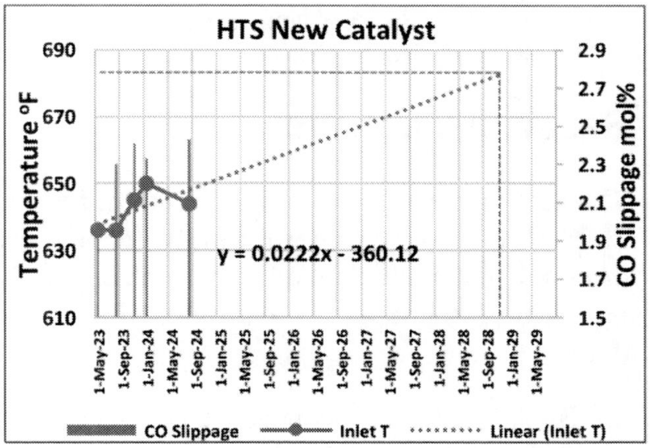

Figure 24 HTS Inlet Temperatures and CO slip of current charge

LTS – Low Temperature Shift Convertor

As the HTS inlet temperature increases, temperatures at the LTS inlet also increases due to limiting heat transfer margins between HTS and LTS. As a result of this higher CO slip is witnessed at the downstream of the LTS which limits plant load.

Corrective Actions

The plant was restarted after taking the following corrective action;

1. Loose refractory material was removed from bulged liner bottom, exchanger tubes and by-pass tube.
2. Welding of damaged cone and seat was carried out to ensure smooth operation of the bypass valve and avoid uncontrolled bypassing of hot gas.
3. A detailed procedure for the refractory job, including stringent post application checks were developed to prevent refractory failures.
4. A regime for detailed inspection of 1101-C, its bypass line, and valve, along with mechanical testing of its reducer (plus cone), shaft, valve disc and valve seat to detect any metal surface or weld failures beforehand is set to be done as a part of a PM plan every 3 years, during the major turnarounds.
5. Implementing stringent condition monitoring of the transfer line to detect loss of refractory in the transfer line small portion.

Recommendations

The seat disc crack was not repaired in the June 2023 Outage, but valve assembly (cone, seat, disc and shaft) are planned to be replaced in the next Turn Around as per approved work procedures.

I. Bypass valve seat concentric cone area to be replaced in next TA-2026.
II. Bypass valve seat plate to be replaced due to crack initiation.
III. Repair of Secondary Reformer outlet Transfer line protection liner to be carried out in the next Turnaround as per approved repair procedure.
IV. Casting of missing refractory beneath the liner will be carried out after liner repair as per OEM's recommended procedure, followed by a dry out procedure.
V. HTS inlet temperature to be maintained with respect to margin in LTS CO slip for optimal plant operation
VI. HTS process parameters monitoring to be continued as per routine practice and OEM feedback to be acquired on quarterly basis.

Conclusion

Refractory repair at any location upstream of the WHB is very critical based on higher velocity and temperature. Even minor refractory repair jobs should only be carried out based on recommended procedure followed by a dry out procedure as per OEM recommendations.

WHB bypass valve internals inspection is very critical with respect to its health and mechanical integrity. Accordingly, a detailed inspection of valve internals should be part of a PM regime.

HTS catalyst life is expected to reduce by 1 to 2 years as evaluated on the basis of HTS inlet temperatures and CO slip by using data from a previous catalyst charge and extrapolating onwards. HT shift catalyst should be operated at the minimum practical inlet temperature that affords the desired conversion of carbon monoxide, since this will result in an increased catalyst life by limiting sintering.

References

[1] AIChE Ammonia Technical Manual 2008 "A Waste Heat Boiler Replacement from bayonet to fire tube"

[2] Catalyst Handbook, second edition edited by Martyn V. Twigg

Problems Encountered in the HTS Reactor When Revamping to Higher Rates

Koch Fertilizer Ft. Dodge operates a Kellogg ammonia plant in Ft. Dodge, Iowa that was originally designed for 600 STPD. In 2022, the facility completed plant improvements that allowed increased ammonia production rates from 950 STPD to approximately 1,280 STPD. After the improvements, it was observed that the High Temperature Shift Reactor displayed an increasing trend of pressure drop. Upon investigation, it was found that the incremental higher rates (> +100% of original design rates) created a maldistribution in this reactor. CFD modeling also confirmed this observation. A new distributor was designed and installed to mitigate the issue in the future.

Jackson Lee
Koch Fertilizer

Karen Braden
Clariant

Dean Potter
MKEC Engineering

Introduction

Koch Fertilizer ammonia plant in Ft. Dodge, Iowa was originally designed for 600 STPD and now operates at a revamped operating capacity of 1280 STPD. In 2023, the Ft. Dodge facility high temperature shift (HTS) reactor experienced increased differential pressure (dP) due to waste heat boiler (WHB) leaks and high space velocities. The latter caused catalyst lifting and milling. This document provides an overview of the timeline, mitigation measures, and best practices.

Experiences at Koch Fertilizer Ft. Dodge

The Ft. Dodge ammonia production facility implemented a major revamp in August 2022 with upgrades on the desulfurization section, primary reformer, compression (air compressor and syngas compressor), and ammonia converter. During the plant construction outage, the HTS reactor catalyst was replaced. When the vessel was opened, a mound of catalyst was found piled up in the middle (see Figure 1). This finding was noted, and the catalyst was replaced with Clariant ShiftMax® 120 catalyst and had a raw SOR dP of 7 psi (0.5 bar).

Figure 1. August 2022 – Initial Entry

Starting in January 2023, the raw and normalized dP began to increase. In March 2023, a plant outage was taken and several leaking tubes in the 102-C WHB were plugged. The WHB leak was believed to be the source of increasing dP at the time, but the dP continued to climb ~1.5 psi (0.1 bar)/week (see Figure 2). In April 2023, the raw dP had increased to 20 psi (1.4 bar).

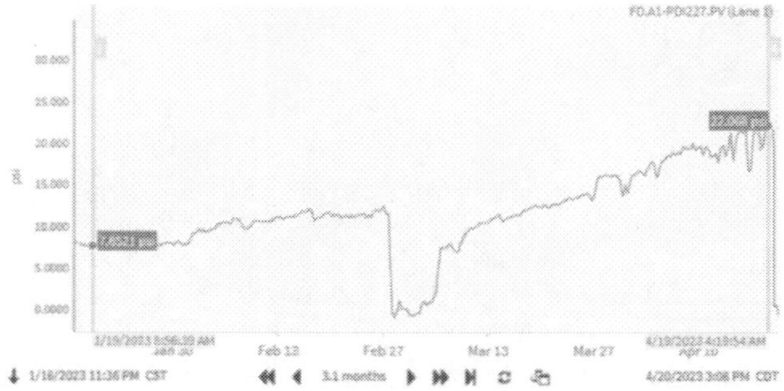

Figure 2. Pressure Drop Increase in early 2023.

HTS spent catalyst samples from the August 2022 outage were shipped to Clariant's testing facility for analysis. The sample results showed phosphorous, potassium, silica, sodium and magnesium supporting the theory that the known WHB leaks were causing the HTS reactor dP to increase. Clariant also tested the side crush strength to check for physical stability with results showing the catalyst was within the normal range of 20-50 N.

Figure 3. Spent Catalyst Samples

On April 24, 2023, the 102-C WHB developed a leak causing the dP rate of rise to increase from 1.5 psi (0.1 bar)/week to 3 psi (0.2 bar)/day (see Figure 4). The dP was reaching the crush strength limit of the catalyst and hurting production leading the site to prepare for a replacement of the top two to three ft (0.6 – 0.9m) of catalyst believed to be impacted by the WHB leak.

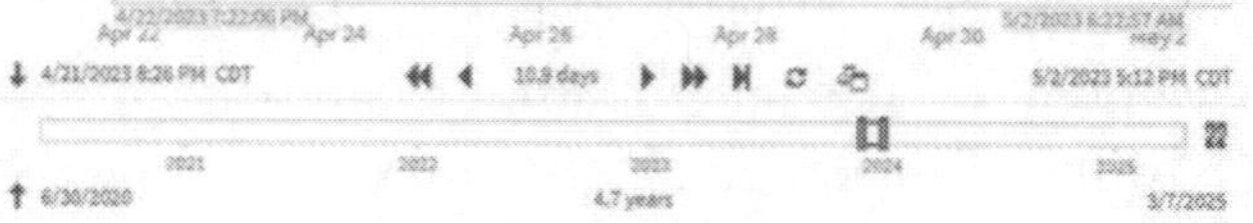

Figure 4: A trend showing the dP increase following the April 2023 outage.

During this time, a computation fluid dynamics (CFD) analysis was conducted due to the milling that was observed during the expansion outage. This CFD analysis showed high velocities from the inlet distributor were creating eddy currents causing lifting and milling of the catalyst (see Figure 5).

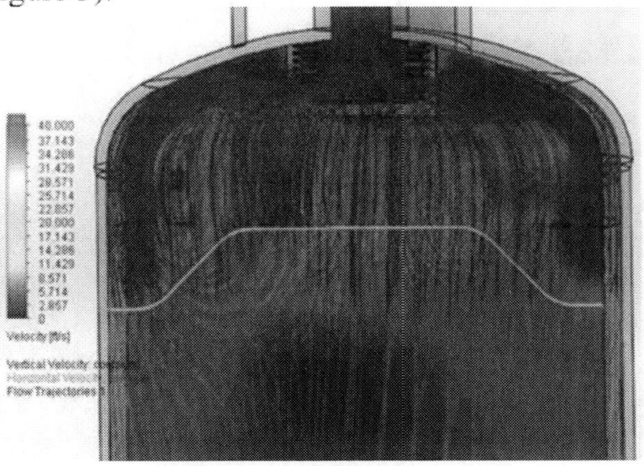

Figure 5. CFD results

The Ft. Dodge facility designed and fabricated a distributor to even out the velocity gradients, this design was based on the CFD simulation results (see Figure 10). The site also decided to add Clariant ShiftGuard® 100GG after consultation with Clariant support. ShiftGuard® 100 GG is an inert material designed to protect HTS catalyst from WHB leaks.

The unit was shut down on May 2, 2023. At this time, the CO outlet was 6% (previously at 2.8% due to the WHB leak acting as a quench which reduced the temperature to the HTS reactor.) The vessel was opened and entered the afternoon of May 3, 2023. There was a large pile of catalyst in the center of vessel, worse than was seen in August 2022 before the revamp (see Figure 6). In addition, catalyst particles had migrated onto the top of the screen from below the screen settling on top of Clariant's ShiftGuard® 100 GG inert material (see Figure 7). Clariant personnel were on site and recommended not to go below a 0.7 L/D ratio. The catalyst pile on the top was removed (4'-4" (1.32m)) and flattened out. Visual observation showed evidence of boiler feedwater (BFW) in the first two to three feet (0.6 – 0.9m) with visually normal catalyst below. (Four fresh bags of HTS catalyst leftover from the outage (1'-8" (0.3 m)) were added along with the inert guard material and hold down grating on May 4th. (Hold down grating was not previously installed in this converter.) On top of the guard material, a layer of 2" (~50 mm) inert balls were added to minimize movement at high plant rates. Total HTS catalyst volume was reduced from 1065 ft^3 (30 m^3) to 811 ft^3 (23 m^3). The distributor was not installed at this time due to safety concerns of doing the installation under supplied air in inert atmosphere.

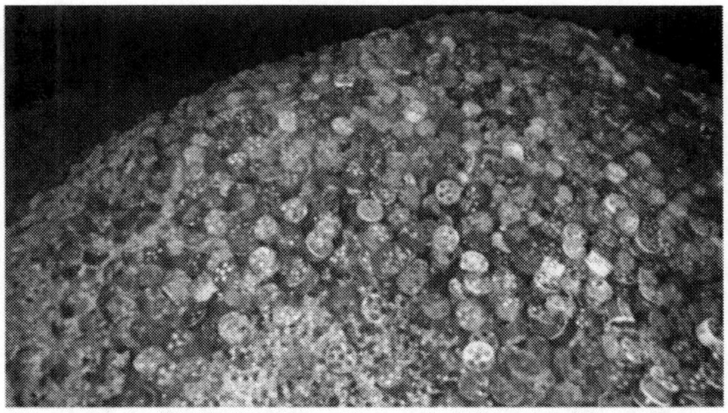

Figure 6. HTS Bed as Found, May 2023 – Large Mound

Figure 7. HTS Catalyst Dust Mixed with Inerts – Above Screen

Figure 8. HTS Bed and Screen as Found, May 2023

Clariant and Koch Fertilizer discussed start-up procedures including a hold step at dew point. dP data from May 12, 2023, showed an 8 psi (0.6 bar) dP.

Unfortunately, the site experienced a significant WHB leak in August 2023 on the 101-C WHB (not the 102-C which was the problem in January 2023) which resulted in water being sent to the HTS reactor. The catalyst was fully changed out in September 2023 while the 101-C exchanger was being repaired. Although the catalyst was performing acceptably, since the last change-out was a short load, it was decided to take the opportunity during the outage to replace the catalyst with a full charge to increase the probability that the HTS performance would be acceptable until the next turnaround. The new distributor was installed during this outage.

The new distributor was developed in collaboration with Dean Potter of MKEC Engineering, a specialist in catalyst optimization, and Scott Munhall, a Process Subject Matter Expert (SME) at Koch Fertilizer. The basic problem shown in the computational Fluid Dynamics (CFD) analysis revealed that the previous distributor (Figure 9) was undersized and positioned too close to the vessel inlet to effectively slow the incoming gas, a common issue with many path of least resistance or gas deflection distributor designs. The schedule for fixing these issues was accelerated. From the start of the CFD study 4/11/23 CFD results presented 4/18/23 Distributor Design complete 4/26/23 delivery of a fully customized 304SS distributor onsite delivered 5/3/23. 22 days from problem statement to delivered equipment.

Figure 9: The original HTS inlet gas distributor.

Instead of evenly distributing the gas across the catalyst surface, the old distributor functioned like a horizontal jet nozzle. This caused turbulent eddy currents, as shown in Figure 5, which impacted the outer edges of the catalyst bed with gas forces over 300% greater than those produced by a professional Stihl leaf blower. Additionally, the old distributor created a low-pressure zone beneath the distributor plate, contributing to the formation of a centralized catalyst pile approximately 2 feet (60 cm) tall, as well as catalyst grinding and milling. Drawing on prior experience optimizing similar systems, Dean chose to model a Stagnation Weir catalyst distributor (Figure 10), (US Patent 12,280,332 B2).

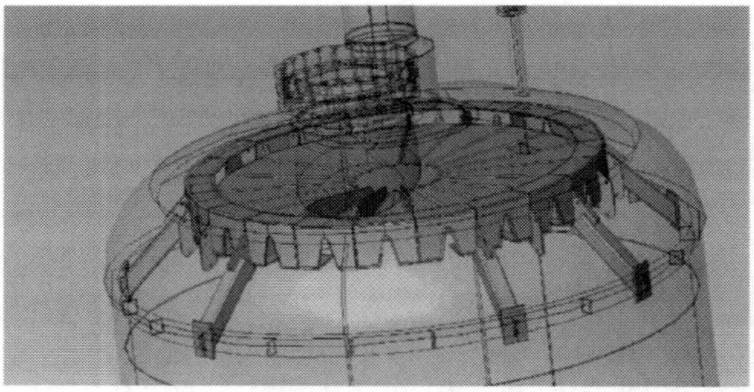

Figure 10: A 3D scan of the new distributor.

This distributor operates much further away from the top vessel inlet nozzle than the old style. This gap creates a space where the gas entering the vessel can decelerate. The gas then flows outward across the flat surface which further decelerates the gas. When the gas reaches the edges of the distributor some gas is directed back toward the center by baffles while other gas flows through the "weir" or slots and out to the outer edges of the catalyst bed. The key design feature is that this distributor is redirecting gas that has decelerated, in this case below 14mph (22.5kph). These two design features significantly slowed the incoming gas compared to the original distributor before the flow was redirected by the weir to evenly distribute across the entire catalyst top surface, see Figures 10 and 11.

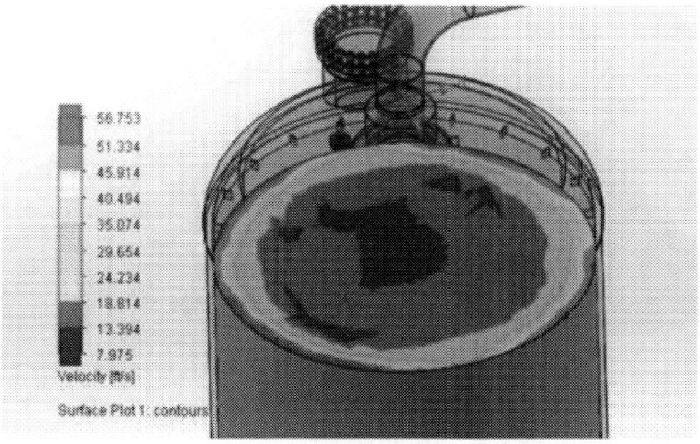

Figure 11: The velocity profile of the original distributor.

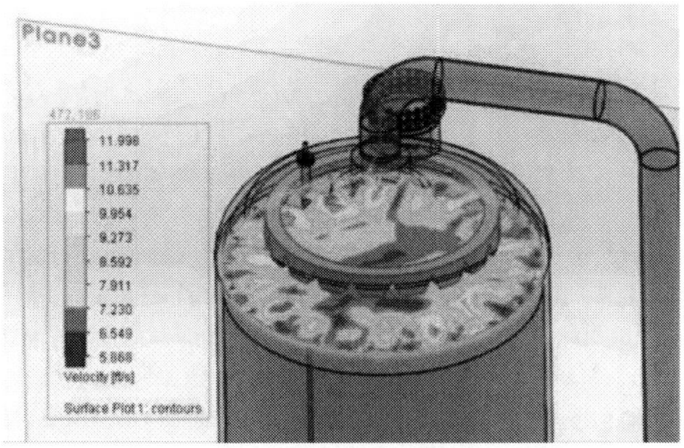

Figure 12: The velocity profile of the new distributor showing the top of reactor.

Figure 11 (old or original distributor) and Figure 12 (new distributor) show that the peak gas velocity at the catalyst bed surface has been both reduced and more evenly distributed. Peak velocities were lowered by approximately 490%, from 59 ft/s (18 m/s) with the old distributor to just 12 ft/s (3.6 m/s) with the new design.

The closer proximity of the distributor to the bed also helped to eliminate the possibility of eddy currents.

The final design used 21 pie shaped pieces that bolted together on underside seams, connected to a central circle piece to which each pie piece was anchored (figure 13).

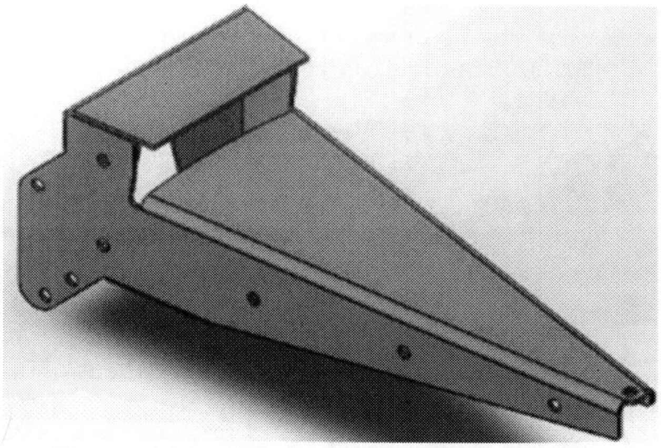

Figure 13: A 3D drawing of one of the pie pieces used to create the new weir shaped distributor.

Each piece was sized to fit through the manway opening and was light enough to be handled by two crew members inside the vessel. The new Stagnation Weir Distributor support design made use of a series of evenly spaced metal tabs along the inner circumference of the vessel, originally installed for catalyst bed leveling. The distributor's leg slots were designed to slide over these wall tabs, allowing the distributor to be securely held in place by gravity after installation (Figure 14).

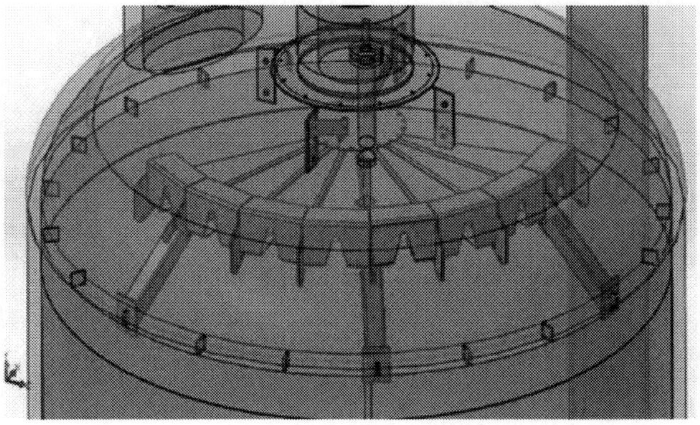

Figure 14: A 3D scan showing how the new distributor was affixed to the wall tabs.

The three pie-shaped sections located directly below the manway (Figure 15) were designed with their bolting seams positioned along the top rather than the bottom. This configuration enabled the final pieces to be slid into place and secured by a technician reaching in from outside the vessel. A reinforced hole was integrated into the distributor to accommodate the Axial Ribbon temperature indicator (TI).

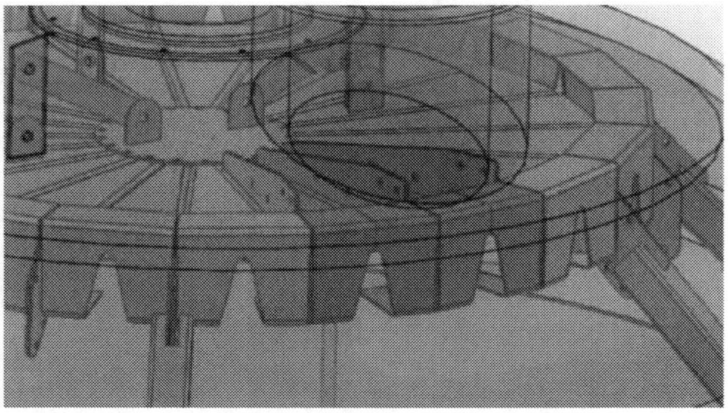

Figure 15: 3D scan showing how the pie pieces below the manway are built.

The distributor was designed for one-time installation, with three easily removable sections located below the manway future catalyst replacements. Although this configuration results in a slightly more confined workspace inside the vessel, the site has successfully completed catalyst changeout and inspection of the HTS bed with the distributor remaining in place. See Figure 16 for an as installed picture.

Figure 16: A picture from the manway showing the distributor as installed after an outage in March 2024 before the vessel was closed.

Figure 17 shows differential pressure (dP) trends for the HTS over time periods with no known WHB leaks. Data points prior to the installation of the new distributor are shown in orange, while those following the installation are shown in green. A steady increase in pressure losses across the bed is expected; however, the notable improvement resulting from the new distributor technology and catalyst design compared to previous installations is particularly significant.

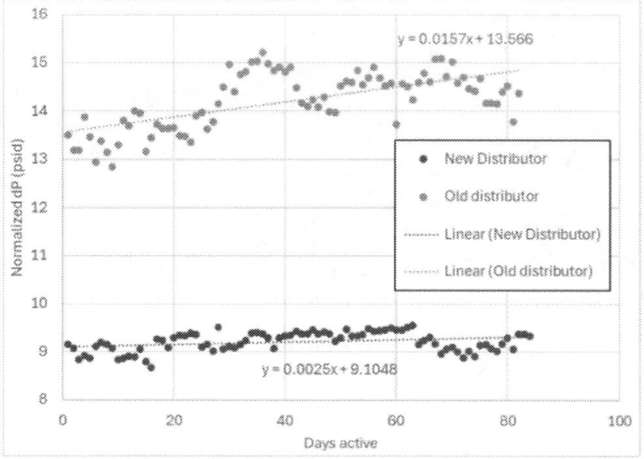

Figure 17 A trend comparing dP across the HTS with the old distributor in orange, and the new distributor plus support balls in blue.

Due to another WHB leak (101-C) in January 2024, the vessel was opened and inspected in March 2024. A full catalyst replacement was necessary but upon initial opening of the vessel, the catalyst was not a mound of catalyst but a flat level of supports with catalyst below (Figure 18). The site continued to add the additional layer of inert support balls on top of the catalyst to further aid in catalyst hold-down. In addition, when replacing the catalyst, visual indication from the severe WHB leak showed evidence of BFW much further down the bed and more evenly distributed than previous change-outs. These visual observations (better distribution from the WHB leak and therefore better distribution during normal operating conditions) signal that the new distributor is functioning better than the previous distributor.

Figure 18. HTS Bed as Found, March 2024.

Summary

Although several WHB leaks contributed to increased dP in the Koch Fertilizer Ft. Dodge HTS reactor, another underlying issue contributed to dP gain in the HTS reactor. After the plant expansion, increased rates led to increased velocities with a distributor that was not able to work effectively and contributed to milling and piling of the catalyst.

After installing a newly designed distributor, the problem was resolved helping to better distribute the feed into the reactor but also prevent catalyst piling leading to better HTS catalyst performance in the future.

Some best practices for HTS reactor loading post-expansion are:

1) Check that the distributor is designed at expected rates using CFD
2) Consider adding hold down material grating or a hold down grid between catalyst and inerts on top and bottom to keep catalyst in place
3) Consider adding additional large inert supports on top when there is room to further keep material in place

Safe, Reliable and Industrial Operation of SynCOR™ Enable Large-scale Low Carbon Ammonia Production

The Topsoe SynCOR™ technology based on advanced autothermal reforming operating at very low steam-to-carbon (S/C) ratio is the leading and proven technology for large-scale syngas production. Low Carbon Ammonia production with SynCOR™ can enable a carbon capture of more than 99%. The safe operation of SynCOR™ builds upon the vast experience gained in SynCOR™ plant operating at very large scale and the same high expectations apply in the ammonia industry.

The safe and reliable operation of the SynCOR™ technology is proven over the past 20+ years of industrial syngas production, equal to more than 6000 MTPD (6614 STPD) ammonia underline SynCOR™ technology as the unmatched reliable technology solution.

Thomas S. Christensen, Per Juul Dahl, Lari Bjerg Knudsen
Topsoe A/S

Introduction

Low Carbon Ammonia plants are foreseen to use autothermal reforming for syngas production as this technology enables large scale production and provides advantages in drastically reduced CO_2 emissions [1]. Application of autothermal reforming (ATR) in ammonia production was introduced by Topsoe many years ago [2, 3]. Today large-scale low-carbon ammonia production using autothermal reforming is generally accepted by the ammonia industry.

Ammonia was produced with stand-alone autothermal reformers already back in the 1950's and several plants with Topsoe technology were put in operation in 1960's and 1970's [2]. The stand-alone autothermal reforming technology was further developed to the Topsoe SynCOR™ technology for low steam-to-carbon operation [3] which has large advantages in compact design and enabling large-single line capacity plants [4, 5].

SynCOR™ technology

Topsoe SynCOR™ is a stand-alone autothermal technology for syngas production, in which the pre-reformed natural gas is sent directly to a SynCOR™ reformer and thereby eliminating the fired tubular reformer (SMR) normally used in the ammonia process. In SynCOR™, the hydrocarbons are combusted by oxygen and converted to syngas. Without a tubular reformer, steam addition can be reduced significantly, resulting in much smaller plant equipment due a lower mass flow.

Conventionally in traditional plants syngas for ammonia synthesis is produced by a primary reformer (SMR) followed by a secondary reformer. Air is added to the secondary reformer to combust residual methane from the primary reformer and to obtain the H_2/N_2 ratio required for ammonia synthesis (approximately 3.0) Alternatively, syngas for ammonia synthesis can be produced by oxygen-blown reforming and stand-alone autothermal reforming [2, 6]. The required N_2 is added in the purification section before the ammonia loop, thereby by-passing the entire front-end and further reducing the equipment sizes in syngas section.

In a stand-alone ATR reformer, the flame temperature, combustion intensity and combustion chamber temperatures are higher than in an air-blown secondary reformer because pure oxygen is used and the steam-to-carbon (S/C) ratio is lower in the ATR-based design. Therefore, robust design of ATR burner, refractory lining and catalysts are required.

Topsoe's SynCOR™ is an advanced Autothermal Reforming with several advantages:

- Operation at lower S/C ratio.
- Operation at higher temperature.
- Smaller equipment of ATR, waste-heat boiler and fired-heater for same capacity compared to conventional designs, and SynCOR™ enable larger single line production capacity.

Figure 1 and Table 1 show typical configuration and operating temperatures of various syngas reformer types.

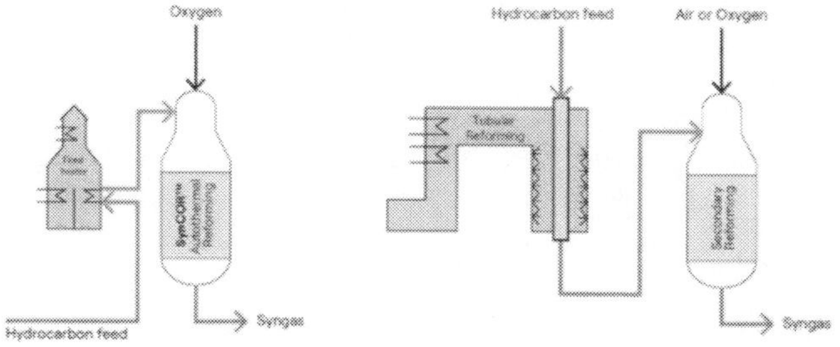

Figure 1. Lay-out comparison of autothermal reforming and air-fired secondary reforming.

The SynCOR™ – as shown in Figure 2 - combine adiabatic pre-reforming, fired heater with low carbon emission, pure oxygen feed with SynCOR™ autothermal reforming operating at cutting edge low steam-to-carbon ratios for efficient conversion of natural gas to syngas.

	Oxygen-blown secondary reforming	**ATR at high S/C**	**SynCOR™ ATR at low S/C**
Feed			
Methane in ATR feed	15-25 mole%	30-40 mole%	50-65 mole%
Higher hydrocarbons in ATR feed	0 mole%	0-2 mole%	0-3 mole%
Feed ratio			
Steam/Carbon	1.6	2.0	0.60
Oxygen/Carbon	0.28	0.64	0.55
Reactor temperatures			
Adiabatic flame temperature, °C (°F)	1485 (2705)	1635 (2975)	2060 (3740)
Combustion chamber temperature, °C (°F)	1250 (2282)	1210 (2210)	1300 (2372)
Catalyst exit temperature, °C (°F)	1025 (1877)	1015 (1859)	1050 (1922)

Table 1. Typical operating conditions of various oxygen-fired reformer types

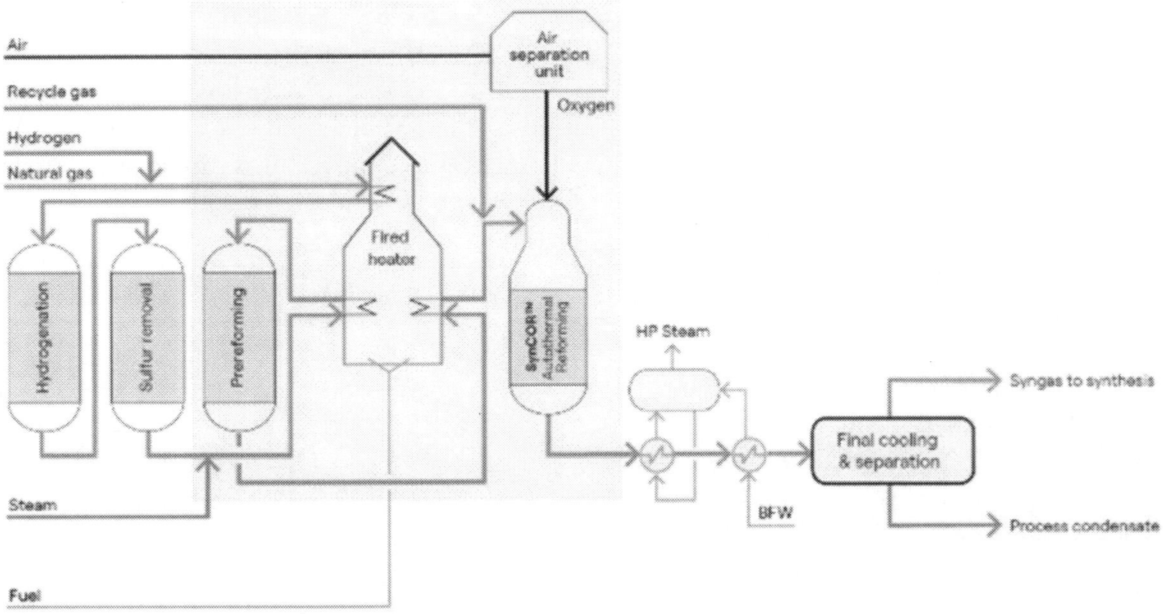

Figure 2. Schematic diagram for stand-alone Topsoe SynCOR^{TM}

Advantage of low S/C operation

The effect of S/C on the syngas production and the overall ammonia plant performance has been studied in a sensitivity case study covering stand-alone autothermal reforming with S/C ratio in the range from 0.6 to 1.8.

For 99% carbon capture Low Carbon Ammonia production, it shows that low S/C ratio reduces plant size and equipment size by 20% in syngas section resulting in lower overall CAPEX. Additionally, low S/C enable reduced oxygen consumption by 7% resulting in smaller ASU and reduced CAPEX and energy consumption.

The catalyst system in the autothermal reformer converts the syngas into the chemical equilibrium gas composition resulting from methane steam reforming and water gas shift reactions. With highly efficient Ni catalyst in SynCOR^{TM} operating at high temperature, the approach to equilibrium will be very small.
Syngas composition and methane leakage are dependent on several parameters (operating conditions):

- Pressure. Low pressure favors low methane leakage
- Catalyst temperature. High operating temperature favors low methane leakage and conversion
- S/C ratio. Low oxygen consumption is favored by low S/C ratio and methane leakage increase with low S/C ratio. Higher steam content requires more oxygen feed to heat feedgas, reaction mixture and syngas product up to fixed reactor outlet temperature.

Figure 3 shows variation of the methane leakage with S/C-ratio for a constant pressure and catalyst outlet temperature. The variation of methane leakage is small and only span from 0.2 to 1.0 mole% over a very large S/C range of 0.6 to 1.8. A high S/C ratio of 1.8 represents a very mild autothermal reformer operation and low S/C ratio of 0.6 is the highly attractive S/C ratio of SynCOR^{TM}.

In Low Carbon Ammonia schemes the methane content in the syngas, which is the equilibrium content at the catalyst exit temperature, can be recovered in downstream purification section and, in the high 99% carbon capture processes, the carbon slip from methane, CO and CO_2 can be recycled to the reforming system. For such layout, the high carbon capture degree of 99% can be achieved independent of the methane slip.

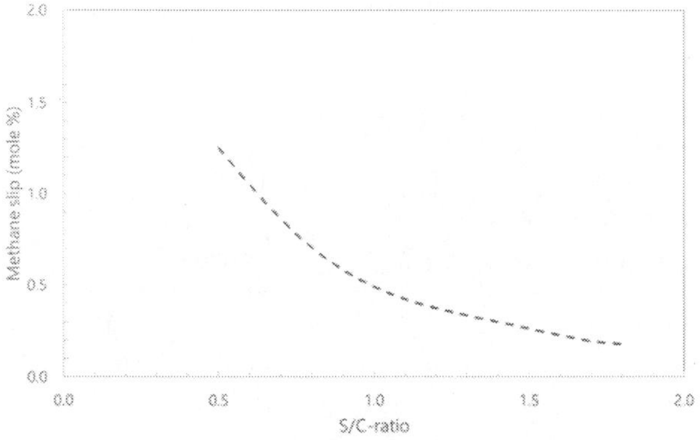

Figure 3. Effect of S/C ratio on methane leakage for constant pressure and temperature at ATR outlet [32 barg(464 psig) and 1050°C(1922°F)]

Significant benefits in ammonia production scale have been identified with Topsoe SynCOR™ technology: (1) large single line production capacity (2) production of syngas with high CO/CO_2 ratios, (3) low steam requirements, which lead to lower CAPEX and OPEX in large ammonia plants, and (4) low steam-to-carbon enables larger single-line capacity [4].

SynCOR™ technology has produced syngas in single line industrial units, with capacities greater than 6000 MTPD (6614 STPD) ammonia.

SynCOR™ reactor design

The reactor design for SynCOR™ consists of a burner, a combustion chamber, target tiles, a fixed catalyst bed, a catalyst bed support, a refractory lining, and a reactor pressure shell as illustrated in Figure 4.

The fuel-rich combustion of hydrocarbons with pure O_2 at high flame temperatures requires robust equipment design of the internals, and high focus on safety, due to the handling of pure oxygen. Topsoe has solved these challenges and SynCOR™ uses the proprietary CTS™ (Cool Tip Swirler) burner equipment, which has a short, centralized flame resulting in low temperature on the refractory lining ensuring safe reactor operation. The CTS™ burner has proven to be safe and reliable in large-scale industrial operations for several decades and the robustness of the equipment is highlighted by its continuous operation for more than 4 years [9], please also see Figures 6 and 8.

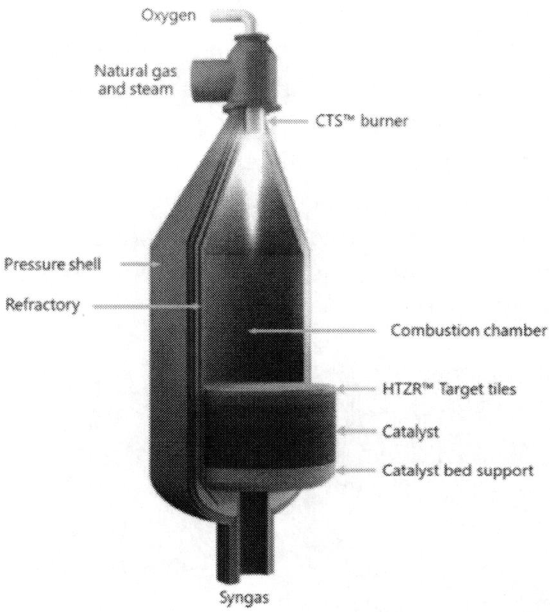

Figure 4. Schematic of the Topsoe SynCOR™

A good performance of these individual elements is critical for an overall safe performance of the Topsoe SynCOR™ technology and reformer. In the following sections, the critical internals are described together with safety considerations.

The continuous improvements of the critical elements in Topsoe SynCOR™ reformers have now all been demonstrated in industry for more than 4 years, thereby contributing to reliable operation for our customers and shorter turn-around time, and longer time between turnarounds.

CTS™ Burner

The flame in the ATR is a diffusion flame where hydrocarbon and oxygen are fed as separate feed streams and mixed by the burner nozzles. The sub-stochiometric combustion of hydrocarbons to CO and hydrogen has a very high flame temperature.

The CTS™ burner is designed to provide efficient mixing of the feed streams which is important for clean combustion without soot formation, and at the same time essential to keep flame core away from burner nozzle. It is also essential to keep the hot gas away from the refractory wall by the central flame and the circulation of colder gas along the walls.

The CTS™ burner was first introduced to industrial operation in stand-alone ATR reformers for ammonia production in 1992 [6] and has since then been applied in both ATR reformers and in oxygen-blown secondary reformers. The 2nd generation CTS™ burner was developed for low steam-to-carbon SynCOR™ and in operation since 2002 [9].

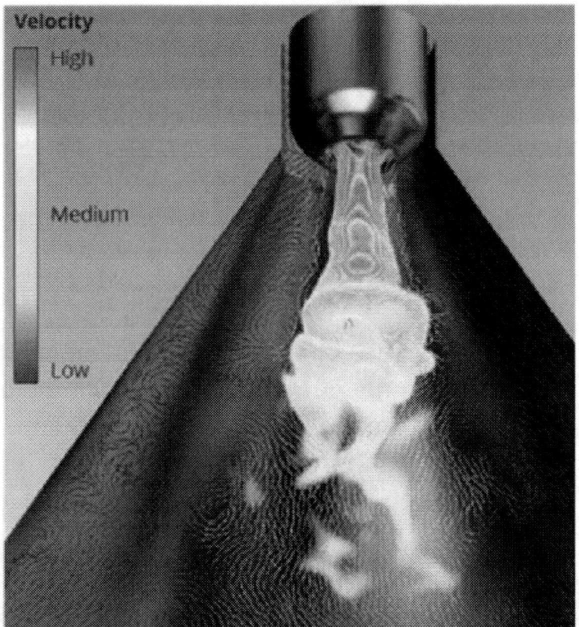

Figure 5. CFD simulation of flow pattern in flame combustion and combustion chamber

Continuous development of materials and fluid flow dynamics has ensured improved robustness and reliability of the CTS Burner. The 3rd generation CTS™ burner was introduced 15 years ago in 2009 with improved fluid flow and burner cooling, please see Figures 5 and 6. Further material upgrade took place in 2013 to specifications with increased robustness towards operation variations and today 3rd generation CTS™ burner have proven more than 4 years of continuous operation.

Figure 6. CTS™ burner (3rd generation) after the first continuous use of 3 years.

Catalyst bed

The SynCOR™ internals also include an innovative catalyst bed configuration with Topsoe wear resistant HTZR™ zirconia target tiles and a graded catalyst bed loading with RKA-10 and RKS-2 catalysts, as shown in Figure 7. The catalyst bed design also ensures more than 4 years continuous operation.

Topsoe catalyst and target tiles have a proven industrial track record. The HTZR™ target tiles show outstanding chemical stability and wear resistance. The RKA-10 noble metal-promoted catalyst and optimized catalyst loading has now been in industrial operation for more than four years and has been installed in several autothermal reformers [5].

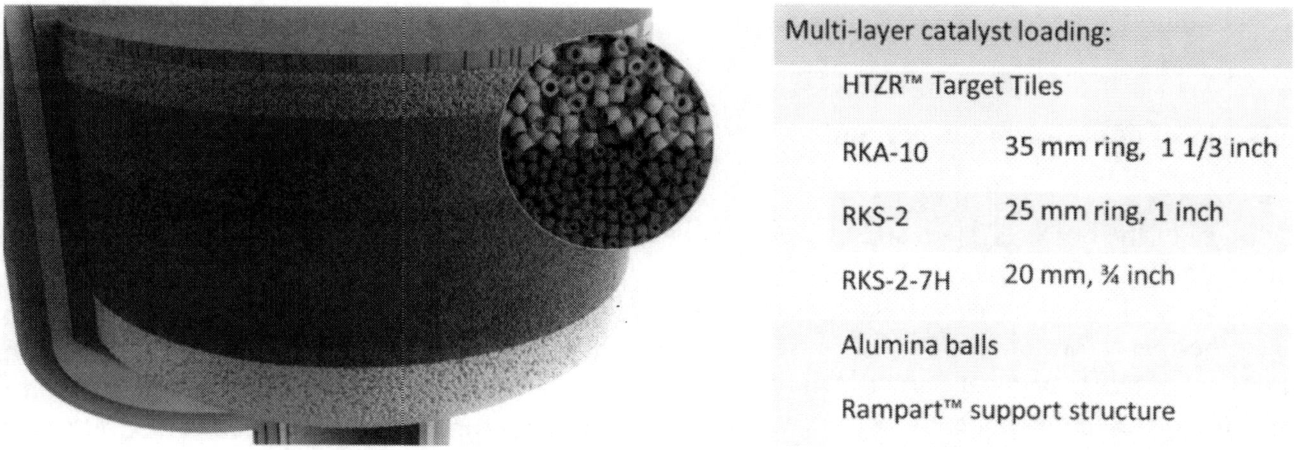

Figure 7. Catalyst bed configuration

Safety and protection features

Stand-alone autothermal reforming includes many safety and protection design features, which have been developed, implemented and matured over several decades.

Safety aspects of oxygen system

In oxygen-blown reforming systems, such as two-step reforming, autothermal reforming and SynCOR™ technologies, the oxidant is pure oxygen produced by cryogenic air separation in a separate ASU unit. The oxygen purity is typically above 98% and in most modern ASU systems above 99.5%. The remaining components are argon and nitrogen, with argon being the predominant component for the high purity systems with >99.5% oxygen purity. For a low inert free ammonia loop, it is an advantage to operate with as low argon as possible.

For very pure oxygen systems with oxygen purity above 98%, more stringent design is required. With high oxygen purity systems, follow additional safety considerations. Good guidelines for safe design have been established by EIGA [7].

Important aspects of safe design include:

- ***Material selection.*** Two design principles enhance the level of safe design: 1) Use of nickel-copper alloy, such as Monel-400 as material of construction in parts of the oxygen system with high temperature and high velocity regions. 2) Use of stainless steel and high Ni-Cr alloys in parts of the oxygen system and burner system with high temperature and when oxygen is reduced by inerts or by steam addition.
- ***Velocity.*** The velocity in oxygen piping should be kept low and below 10 m/sec [32.8 ft/sec] at typical oxygen feed pressure of 40 bar [580 psi] to eliminate the risk of ignition in the piping system [7].
- ***Cleanness of system.*** The pressurized oxygen system must be always kept clean. Good design practice includes installing particle filters before oxygen is heated to remove potential substances which could act as an ignition source. Good maintenance practice includes maintaining a separate workshop area for equipment and instrument components used in the oxygen system to always ensure cleanness of valves and instruments. Good maintenance is also to use special lubricant approved for oxygen service. Proce-

dures for line cleaning before commissioning and start-up is of utmost importance.

Safety aspect of design of burner and combustion chamber

The design of the burner system and of the combustion chamber with refractory lining and pressure vessel are crucial aspects of safe design of the autothermal reformer reactor.
The crucial design aspects of burner include

- Keep the flame centered in the combustion zone and avoid flame impingement with gas temperatures at the refractory walls above critical material values
- Provide even supply of feed streams from the connecting piping system
- Ensure refractory temperatures below critical values by design with recirculation of combusted gas and high resistance time in post combustion zone.

In burner design and development, Topsoe has tested various burner designs and extensively used Computational Fluid Dynamics (CFD) modelling to correlate experimental combustion observations and plant data simulations and to validate models [3]. In the following various burner designs are briefly described.

Jet burners
Jet burners with high velocity oxidant injection is a well-known burner principle, which has been widely used in fuel firing for heating purposes. The high velocity jet entrains the surrounding fuel gas along the length of the flame and recirculation zones are created. Even though the jet flame is simple it requires a very precise and even gas distribution around the flame jet to avoid a skew flame and impingement of high temperature on the cylindrical combustion chamber walls, which can be a safety risk. The skew flame is a higher risk at low flow scenarios, such as turn-down operation.

Jet flames are generally longer than other types of flames and require a long combustion chamber to obtain a homogeneous gas mixture and low post combustion zone temperatures. Jet-flames are subject to soot formation in high methane-rich zones where radiation from the flame bring hydrocarbon-rich zones into soot formation temperatures before they are burnt.

So, all in all, jet flames require operation at high S/C ratio and high flow velocities to provide stable performance and low refractory temperatures. The operation of jet-flames is less flexible to feed variation and turn-down and the flame type is not the optimal choice.

Multi nozzle burner.
Multi nozzle burners have been widely used in air-blown service, but in pure oxygen service with much higher exotherm heat release by combustion reactions, the temperature gradients increase and will result in uneven mechanical stress on individual burner nozzles and differences between warm and cold nozzles. Further, in autothermal reforming with methane rich feed gas, a high degree of soot formation has been observed and steam additions for inhibition of soot chain reaction is not sufficient. So multi nozzle burners should generally be avoided.

Swirl burner
Swirl burners provide efficient mixing of feed streams and short flames, and high velocity swirl burners are superior to flexible operation in load (incl turn-down) and feedstock composition variations. Swirl burners are more complex than jet burners and requires a good fluid dynamic understanding and extensive experimental basis. In burner testing, Topsoe found that swirl burners are superior to suppress soot formation and enable low S/C operation.

Topsoe CTS™ Burner is an example of a successful swirl burner. Topsoe SynCOR™ design with CTS™ burner have been in reliable industrial operation at low S/C ratio since 2002 and the industrial reference plants have accumulated more than 150 years of safe operation time without catastrophic failures. Turn-down with robust and stable flame in the range 30-110% has been proven by the swirl CTS™ burner, and this a much broader operating range than jet-flames where some are limited to 70% turn-down to avoid safety issues, such as unwanted critical high refractory temperatures.

Double swirl burner

In double swirl burners, both oxygen and hydrocarbon feed are rotated to enhance mixing, but such flame can result in flame impingement where the flame core is directed out toward the combustion chamber walls. In conical shaped combustion chambers, which should promote an external circulation of colder combustion gases, flame impingement has been observed. The following case story shows an example of safety risk with double swirl burners, which should be avoided.

Burner case story. There have been a few examples in chemical industry of burner failures. A case story was reported from a hybrid oxygen-blown autothermal reformer in 1994 [8], where a failure of the burner system and misdirected flame increased refractory temperature above critical values and resulted in a loss of containment with a hole in the pressure vessel. A detailed check of the design of the burner nozzle revealed that the flame shape resulted in high refractory temperatures. The mechanical nozzle construction and flame shape together with metal dusting in the feed supply systems contributed to the failure [6, 8].

The particular burner system was later replaced with the Topsoe CTS™ burner and has been in operation for more than 25 years [9]. The central and short flame position of the CTS™ burner ensures a centered flame with homogenous temperatures of the refractory lining in the combustion chamber [5, 6].

Safety aspect of ATR vessel design and operation

The refractory lining is an integrated protection feature of the reactor design to protect the pressure vessel from excessive temperatures. Advanced autothermal reforming requires a multilayer refractory lining, and in Topsoe SynCOR™ several layers are applied including; high purity alumina brick as inner layer towards the process, high insulating alumina bricks as the first temperature insulation, and then dense castable layers towards the pressures shell. These layers form the major thermal insulation. Topsoe has applied such refractory systems with high quality materials and installation for many years and it has proven to provide safe operation The refractory lining has been untouched for over 20 years in reference plants [9].

The pressure shell is typically below 200°C (392°F) and exposed to the surrounding air and the natural draft air circulation along the warm surface, gives a natural cooling of the shell even in hot climate.

Alternative shell design selected by some licensors includes an additional water-cooling jacket outside of the pressure shell to cool and provide a thermostat in case refractory protection have limitations. A drawback of a water jacket is that one can only detect refractory problem but not determine the location of refractory issue or determine the magnitude of the refractory problem.

The advantage of the bare wall design is that better monitoring is possible, and several possibilities for temperature monitoring is available. The vessel is painted with thermo-indicative paint and shell surface temperature can be measured and shown in the DCS. With IR techniques, the full shell monitoring by fixed mounted IR cameras or batch-wise operator monitoring of surface temperatures by 2D IR portable cameras is possible as routine measurements. It is Topsoe's experience that robustness and reliability of refractory lining in SynCOR™ reactors are very high, and only routine measurement with 2D IR portable cameras is recommended to ensure safe operation or the ATR reactor.

Safety aspect of Fired heater

The fired heater in stand-alone autothermal reforming provides a significantly simpler furnace system than tubular fired steam methane reformers (SMR). The fired heaters are typically bottom-fired cylindrical furnaces for smaller plants and box-type furnaces for larger plants. A common feature of such fired heaters are only few burners for the fuel combustion. Because the burners are few, they use flame detection system with flame scanners and pilot burners as safety measures.

Safety aspect of process control and instrumentation

Topsoe SynCOR™ is designed with an automated LTM (Load and Temperature Management) system. The system prevents operator mistakes during startup and normal operation. The system ensures smooth load changes by safe adjustment of feed, steam, oxygen, fuel and combustion air within acceptable limits. The system can be designed to cover changes in feed and fuel composition and to cover special trip scenarios initiated by downstream units or from incoming supply of feedstocks or utilities. The LTM is an addition to the standard DCS and ESD systems.

In operation, Advanced SynCOR™ autothermal reforming processes are protected by 3 levels of monitoring and control:

- Alarms when process and equipment of selected parameters approach abnormal values
- Cut-back of typical heat input or flow to avoid too low or high critical values and maintain production
- Trip and shutdown in case of critical parameters cross critical values and before equipment damage occurs. Typical parameters for trip values are low Steam/Feed ratio, high Oxygen/Feed ratio, high temperature protection of the burner system, combustion chamber and catalyst bed, high fired heater coil temperatures, fuel and combustion air pressure.

Advanced instrumentation and control provide more reliable operation, which leads to fewer stops and starts and hence longer lifetime of all equipment. Fewer trips also mean reduction of CO_2 emissions from fewer start-ups and flaring during transient operation.

Safe and reliable industrial operation of SynCOR™ technology

The Topsoe CTS™ burners [6] were first installed in several revamp projects during the 1990s, both in stand-alone ATR reformers and in oxygen-blown secondary reformers. A 2500 MTPD (2756 STPD) methanol plant has been operating with a CTS™ burner for more than 25 years as part of the Topsoe two-step oxygen-blown reforming process. For the same period, CTS™ burners have been operating in a large Gas to Liquid (GTL) plant with three parallel syngas units.

Industrial plants using Topsoe SynCOR™ at low S/C ratios have been in operation in European chemical plants since 2002 and at a large syngas hub in a chemical complex since 2004 [9].

Fully integrated GTL facilities producing synthetic fuels (i.e. diesel and kerosene) utilize Topsoe SynCOR™ technology for very large-scale production of syngas. The first large-scale SynCOR™ GTL plant was commissioned in 2006 [9], and later similar plants were constructed and started in 2014 and 2022. The plants produce syngas equivalent corresponding to more than 6000 MTPD (6614 STPD) ammonia in single train, please see Figure 8. The 3rd generation CTS™ burner [5] is applied for large-scale operation and Topsoe CTS™ burners have a proven record of providing safe operation and a long service life at the advanced conditions using pure oxygen as oxidant, please see Figure 6.

The operating periods between planned maintenance are gradually increasing, and 4 years is the norm of many operators today. Reliable operation and technology solution are pre-requisites for the strategy of long operating periods between turnarounds. The proven operation experience of SynCOR™ technology and the CTS™ burner from the past 15 years support the plant maintenance strategies of four years or more between planned maintenance turn-around.

Figure 8. Industrial SynCOR™ plant producing syngas equivalent to >6000 MTPD (6614 STPD) Ammonia.

Testimonial

A client who has been operating a Topsoe SynCOR™ Syngas plant for the past 5 years made the following statements:

- *Operating a SynCOR™ ATR is much easier and safe for our personnel*
- *We have not had a single operational issue with the ATR, which resulted in a unit trip. The ATR is very reliable and safe*

Low Carbon Ammonia production with SynCOR™

Low Carbon Ammonia production with SynCOR™ can enable a carbon capture more than 99% (scope 1) through integration of SynCOR™ with shift section, CO_2 removal technologies and proven synthesis gas purification technologies such as nitrogen wash unit, PSA and/or membrane solution. This technology combined with the Topsoe shift catalyst, SK-501 Flex, can fully utilize the low steam carbon operation of the reforming section and a syngas purification section to provide "inert free" make-up gas for the ammonia synthesis loop operation and an optimal large scale ammonia production configuration. CO_2 is captured from the process and CO_2 capture from flue-gas of fired heater is not required because low-carbon fuel is burnt Reference to typical process lay-out is shown in Figure 9.

The carbon capture reduction by 99% by SynCOR™ is classified as reduction of Scope 1 emission, which is the direct emission from the ammonia plant production cycle. Scope 1 emissions is defined as the release when a company manufactures a product or delivers a service. Scope 2 emission is indirect emission from upstream the ammonia production

Optimized Low Carbon Ammonia lay-out with SynCOR™ for 4000 MTPD (4409 STPD)

SynCOR™ is recognized to be the preferred layout for Low Carbon Ammonia, and it provides attractive performance numbers. SynCOR™ Ammonia technology allows a very high carbon recovery due to a relatively low fuel consumption because of low S/C ratio low internal oxygen combustion in ATR, and further CO_2 capture from the syngas and addition of nitrogen to correct H_2/N_2 ratio in the syngas purification section just upstream the ammonia loop. The SynCOR™ blue Ammonia technology can meet a Scope 1 carbon capture of more than 99%.

Autothermal reforming also permits high hydrogen production because of the high outlet temperature of typical 1050°C (1922°F) from the ATR. The syngas leaving the ATR is in equilibrium with respect to the steam reforming and water shift reactions favoring low methane slip as well as high CO content.

The CO in the syngas is converted by steam in the shift section to form more hydrogen and carbon dioxide. Following the CO_2 in the process gas is captured in the CO_2 removal section (pre-combustion carbon capture) and the hydrogen rich stream further purified in the hydrogen purification section. The necessary amount of nitrogen is introduced to obtain a H_2/N_2 ratio of 3.0 in the inert free make-up gas which after compression is sent to the ammonia synthesis loop.

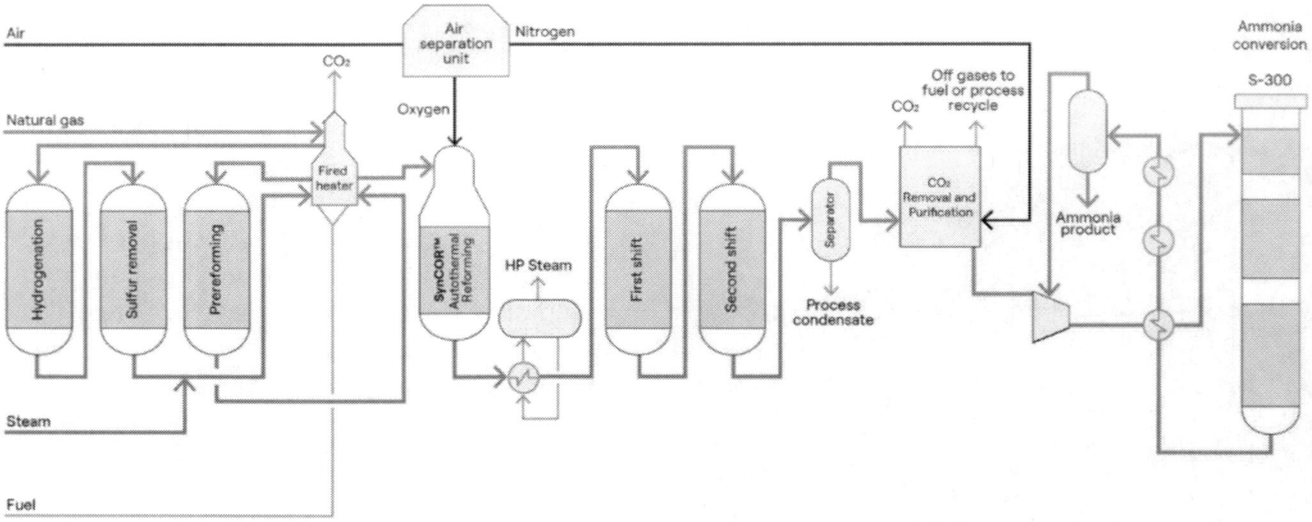

Figure 9. Typical process lay-out of Topsoe SynCOR Blue Ammonia™

Carbon capture efficiency

The process can be arranged with carbon capture efficiencies between 85 and > 99% [1]. 85% carbon recovery is inherently obtained with the process scheme. Further reduction of CO_2 emission is easily obtained by using off gas and purified synthesis gas from the purification section as fuel in the fired heater and auxiliary boiler. Alternatively, a carbon dioxide recovery unit (CDR) can be installed in the flue gas section for reducing flue gas CO_2 emission. The stripper reboiler duty for the post-combustion carbon capture unit can be provided by LP steam generated in the ammonia plant. Further increase in carbon capture efficiency can be obtained by increasing the hydrogen content in the fuel stream. 99% carbon recovery (scope 1) and reduction of CO_2 emission is possible by optimizing the off-gas utilization, by recycling part the off gas from the purification section to the ATR to be able to obtain the high carbon recovery.

Summary

Advanced autothermal reforming in form of the Topsoe SynCOR™ technology operating at low steam-to-carbon ratio has been in industrial operation for more than 20 years. The operation experience shows safe and reliable, long-time industrial operation at very large scale of syngas production. The Topsoe SynCOR™ technology enable Low Carbon Ammonia production for very large production plant capacities today corresponding to more than 6000 MTPD (6614 STPD) ammonia in single train.

References

1. A. Kokoti and P. J. Dahl, "Decarbonize with blue ammonia", Ammonia Plant Saf., 62 (2022)
2. "Topsoe-SBA Autothermal Process" Nitrogen (May 62), 35
3. T. S. Christensen and I. I. Primdahl, "Improve syngas production using autothermal reforming", Hydrcar. Proc. (March1994), 39.
4. P. J. Dahl, C. Speth, A. K. Jensen, M. Symreng, M. K. Hoffman, P. A Han, S. E. Nielsen, "New ATR based ammonia process", Ammonia Plant Saf., 58, (2017), 161
5. T. S. Christensen, J. R. Larsen, L. J. Shah and M. Stenseng," Market-leading development for oxygen-blown secondary and autothermal reformers", Proc. Nitrogen+Syngas 2014 conference, London, February 2017
6. T. S. Christensen, I. Dybkjaer, L. Hansen and I. Primdahl, "Design and performance of secondary and autothermal reforming burners", Ammonia Plant Saf., 35, (1994), 205
7. Oxygen Pipeline and Piping Systems. EIGA Doc 13/20
8. G. Shaw, H. de Wet and F. Hohmann, "Commissioning of World's Largest Oxygen Blown Secondary Reformers, Ammonia Plant Saf., 35, (1994), 316
9. "Process burners for syngas production", Nitrogen+Syngas, 320, Nov-Dec 2012, 38

Sichuan Lutianhua: A Case History of Operational Excellence and Safety - 20 Years of Proven Reliability with Autothermal Reforming Technology

Hydrogen and synthesis gas are the major intermediates in the conversion of hydrocarbons (e.g., natural gas) into many important chemicals such as ammonia, methanol, Fischer-Tropsch liquid fuels, alcohols, and aldehydes from oxo-synthesis. The main alternative to Steam Reforming, especially for large capacity plants, is Autothermal Reforming (ATR), where hydrocarbon conversion is achieved through the use of oxygen. Oxygen-blown ATR reactors can also be used in combination with SMR (two-step reforming), where the ATR is placed downstream of a steam reformer operating with high methane slip and/or with the addition of natural gas to the synthesis gas stream between the steam reformer and the ATR. In the ATR, a catalytic bed is installed within a refractory lined pressure vessel to bring the methane steam reforming reaction as close to equilibrium as possible.

Since 2005, Sichuan Lutianhua Lvyuan Alcohol Industry Co., Ltd. has been operating a 1,350 MTD methanol plant with two-step reforming featuring a Casale ATR. This ATR has proven reliable and flexible, handling variable loads due to market fluctuations without shutdowns. In this paper, we present a case study of a two-step reforming process featuring a Casale ATR in operation since 2005 in a 1,350 MTD methanol plant. We highlight the reactor's long-term reliability, safety, and flexibility under variable loads, emphasizing the critical role of burner design in ensuring stable operation and catalyst longevity.

Marco Mazzamuto Carlucci, Francesco Baratto
Casale SA

Sun Guang
Sichuan Lutianhua Lvyuan Alcohol Industry Co., Ltd.

Introduction

Hydrogen and synthesis gas are key intermediates in the chemical industry, enabling the production of ammonia, methanol, Fischer-Tropsch fuels, alcohols, and aldehydes. Steam Methane Reforming (SMR) has traditionally been the dominant technology for syngas production. However, for large-capacity plants, Autothermal Reforming (ATR) presents a compelling alternative due to its ability to achieve high conversion efficiency using oxygen. ATR is also employed in two-step reforming configurations, where it complements SMR to optimize methane conversion and overall process efficiency.

Since 2004, Sichuan Lutianhua Lvyuan Alcohol Industry Co., Ltd. has operated a 1,350 MTD methanol plant, located in Luzhou (China), featuring Casale's ATR technology in a two-step reforming configuration.

This paper presents the operating experience of this plant and highlights the design features that contribute to the ATR reactor's reliability, safety, and high performance.

Lutianhua group background

Lutianhua Group Inc. is a large industrial complex engaged in fertilizer production, fine chemicals manufacturing, research and development, construction, installation and logistics. Headquartered in Luzhou City, Sichuan Province, the company holds total assets of 21 billion yuan.

Established in 1959 as Sichuan Luzhou Natural Gas Chemical Plant, Lutianhua Group is recognized as the birthplace of China's modern urea industry. The company has achieved significant milestones while also navigating challenges stemming from industrial restructuring and market fluctuations.

Lutianhua's production capacities include 1.0 million tons per year of synthetic ammonia, 2.1 million tons per year of urea, 750,000 tons per year of methanol, and several other chemical products, totaling more than 20 product varieties.

CASALE revamping in Luzhou

In 2004, the ammonia plant owned by Lutianhua Group in Luzhou, originally licensed by Toyo, was retrofitted to methanol production by Casale implementing Autothermal Reforming technology.

Toyo plants employ a two-step reforming process for ammonia production, consisting of an SMR followed by a Secondary Reformer powered by air to complete natural gas conversion and introduce nitrogen for ammonia synthesis.

Casale revamping consisted in the replacement of the original Secondary Reformer burner with Casale ATR burner, modifying the existing pressure vessel in view of the new technology.

Casale ATR burners are normally installed in a low alloy steel pressure vessel with two-layer refractory protection. Vessel design temperature is lower than 300°C (572°F) and it is painted with thermochromatic paint to easily identify any possible refractory breach.

In this case, to reduce the overall project investment cost, the existing water-jacketed pressure vessel of Secondary Reformer was also retrofitted; the top section of the vessel was removed and replaced with a new one featuring an extended neck. Additionally, the transfer line from the SMR was modified and upgraded to accommodate the new height of the ATR gas inlet nozzle.

A new ASU (Air Separation Unit) was installed to feed the required amount of oxygen to the new ATR.

Additional plant modifications to achieve the new revamping scheme included:

- Steam superheater coils replacement
- NG feed heating coil replacement
- New combustion air preheater

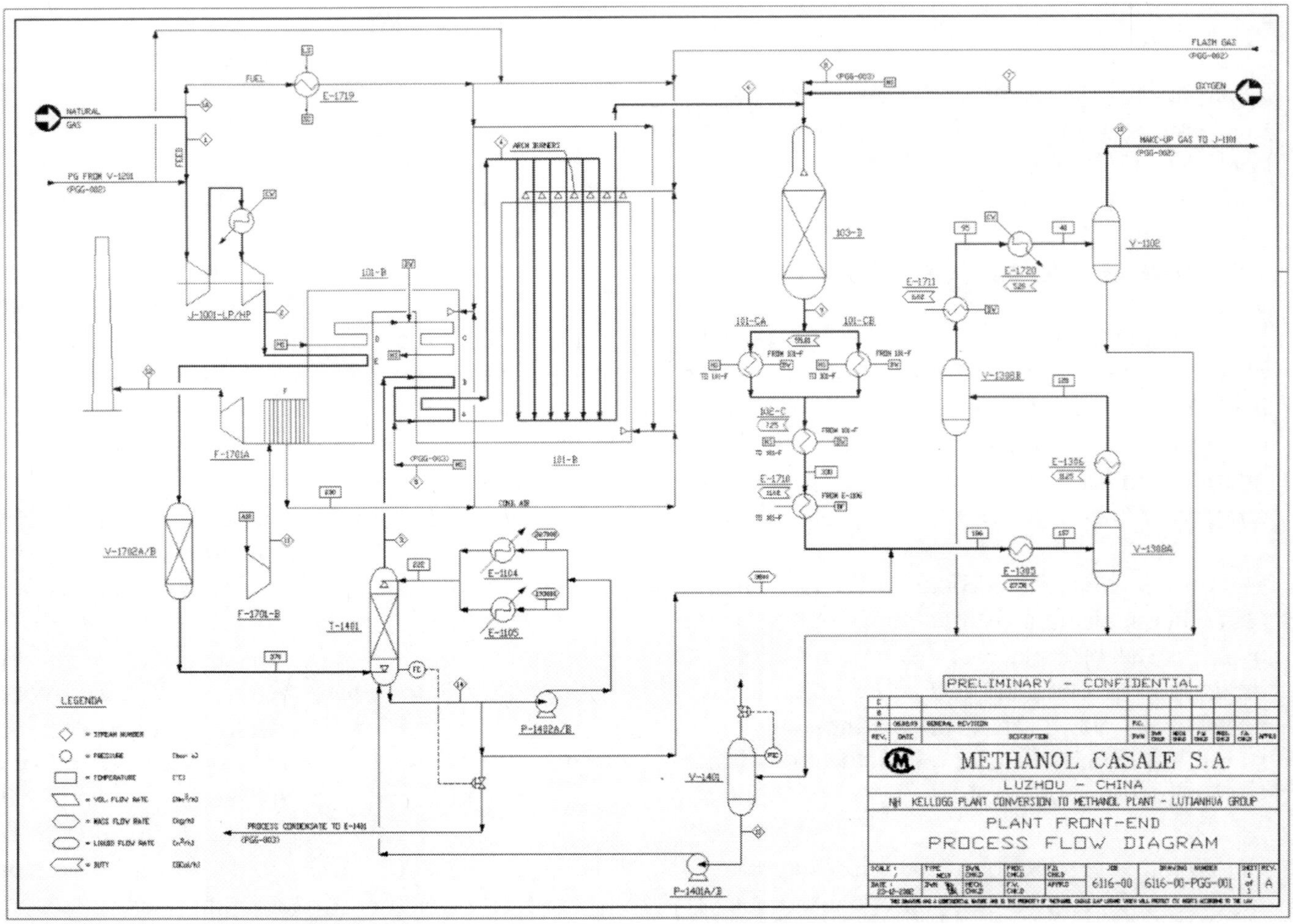

Figure 1. Process Flow Diagram of Retrofitted Luzhou Plant

Post retrofit, plant capacity of 1350 MTD was unchanged, but thanks to the superior process performances of ATR with respect to secondary reforming, it has been possible to decrease the operating temperatures of SMR as shown in Table 1.

The decreased operating temperatures and heat flux across catalytic tubes led to a reliability enhancement and longer operating life.

The ATR burner is the most critical component in this application as it significantly impacts both the process and the lifecycle of the unit. It controls outlet composition through fluid dynamics and mixing, and its optimized design ensures long catalyst life and stable pressure drop by maintaining a uniform inlet temperature (refer to Table 1).

Additionally, an optimized burner design allows for soot-free operation across various plant loads and during startup or shutdown phases, preventing soot buildup on catalysts and pipelines.

Casale ATR Technology

Casale ATR's main feature is a proprietary water-cooled burner at the top for mixing reactants and generating the flame for optimal performance in terms of minimum temperature approach and pressure drops.

		Ammonia	Methanol
Production	MTD (STD)	1'350 (1,488)	1'350 (1,488)
Tubes material		HK 40	
Tubes number dimensions	N° mm	504 114.4 x 21.6 x 9600 (4.5" x ¾" x 31'6")	
Tubes inlet T	° C (°F)	510 (950)	405 (761)
Tubes outlet T	° C (°F)	854 (1,569)	712 (1,314)
Max metal tube T	° C (°F)	936 (1,717)	< 880 (< 1,616)
I.S. Tubes heat flux	$kcal/m^2h$ (Btu/ft^2h)	> 70'000 (25,807)	< 55'000 (20,277)

Table 1. Operating Parameters after Luzhou Toyo Plant Retrofit

The catalytic bed is supported by inert material for final hydrocarbon conversion.

Casale has designed ATRs for 17 plants since 2001, with capacities from 10,000 Nm^3/h to 740,000 Nm^3/h of equivalent hydrogen production.

11 of these syngas generation sections have been put in operation with capacity ranging from 30,000 Nm^3/h to 630,000 Nm^3/h of equivalent hydrogen production which is currently the largest single train syngas generation operating worldwide.

Casale ATR Operating Principles

Oxygen is introduced at high velocity at the top of the combustion chamber, mixing with partially reformed gas introduced perpendicularly. This enables perfect mixing and results in a uniform temperature and composition at the catalytic bed entrance, enhancing catalyst performance.

Figure 2. Casale ATR for 475 kNm3/h eq. H2 production

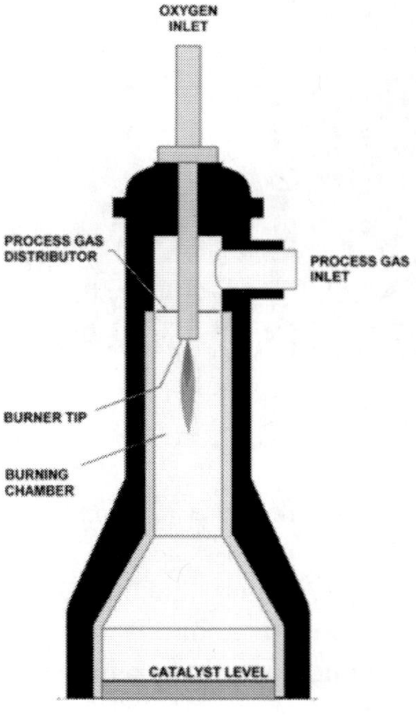

Figure 3. Casale ATR sketch

The combination between the kinetics and the fluid dynamics principles embedded in Casale's proprietary model for burner design ensures a stable flame with reduced length, with benefits in terms of reactor size, catalyst lifetime and performances. A shorter and controlled flame avoids catalyst impingement by the flame while achieving an almost uniform distribution of temperature and gas composition at the chamber outlet before the vessel cone.

The fluid dynamic flow field inside the combustion chamber is designed to protect the refractory lining from the high temperature core of the flame, preventing hot spots on the lining surface.

Figure 4. Casale Water-Cooled ATR 3D model

The gas recirculation that builds up inside the combustion chamber shields the refractory lining from the flame's hot core, which can reach temperatures as high as 2700°C (4,892 °F). This keeps its surfaces well under the refractory maximum operating temperature.

Casale's ATR operation has proven to be soot-free in industrial applications.

Autothermal Reformer Burner Tip Cooling

Casale ATR burners use a water-cooling system similar to those in POx (Partial Oxydation) applications, to ensure a long operating life.

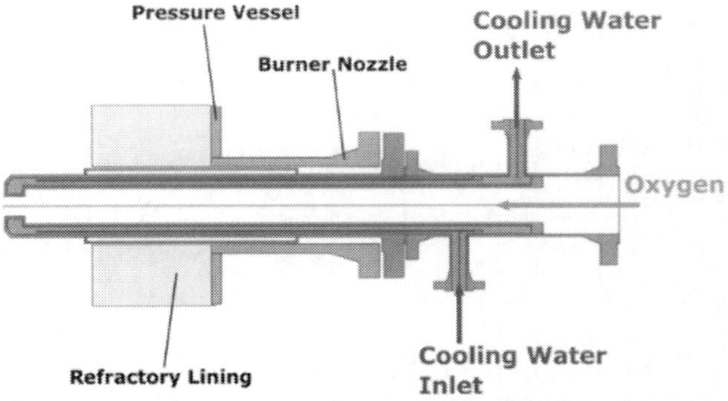

Figure 5. Casale Water-Cooled ATR Burner working principle

This system maximizes cooling efficiency on all surfaces exposed to the flame and the entire burner body, enhancing safety and fault tolerance under unexpected conditions.

The first Casale ATR burner with water-cooling has been operating since 2001 in Daqing (China – 350 MTD/385 STD), delivering solid performance, safe operation and a lifespan of more than 10 years. The cooling system and mechanical strength allow the burners to withstand severe conditions while maintaining integrity, performance and safety.

Figure 6. Casale Proprietary Water-Cooled ATR Burner

The last upgrade of Casale ATR technology is represented by the 3D printing with Inconel alloys of burner tip, by means of an additive manufacturing process. The 3D printed tip enables scale-up at high capacities and pressure (hydro-tested up to 350 bar, over 5070 psi) while providing effective cooling of all surfaces exposed to flame.

Operational Experience in Luzhou

The operational experience of Casale ATR in Lutianhua Group plant (Luzhou, Cina) can be summarized as follows:

- Plant Capacity:	1350 MTD
- ATR type:	Single burner (1 main + 1 spare)
- Started-up:	2004
- Status:	Currently in Operation
- Operational range:	40 to 120% of its design
- Pressure drop (O2 side):	~ 2 bar (29 psi)
- Methane slip:	< 0.5% dry basis

During the first 4 years of ATR operation (2004-2008) Casale monitored the burner status when possible, taking advantage of a plant turn-around caused by NG shortage.

Figure 7 shows the burner tip condition at different time intervals over 4 years of operation.

To be underscored, the tip is the part of the burner which operates at the most extreme conditions, therefore it generally requires more maintenance and attention with respect to the rest of the burner.

As can be noted from Figure 6, after 4 years of operation the burner tip is essentially intact. This testifies to the robustness and reliability of the burner tip.

After 2008, the main burner was replaced with the spare burner, which operated until 2017, when it was replaced by the refurbished main burner. Since 2017, the plant has been running with the main burner, which remains in operation to this day.

It should be noted that between 2008 and 2017, the plant underwent multiple shutdowns and start-ups due to natural gas shortages. It should be noted that repeated thermal cycling is one of the most severe stress conditions for ATR burners. Despite this, the spare burner consistently demonstrated excellent performance and reliability.

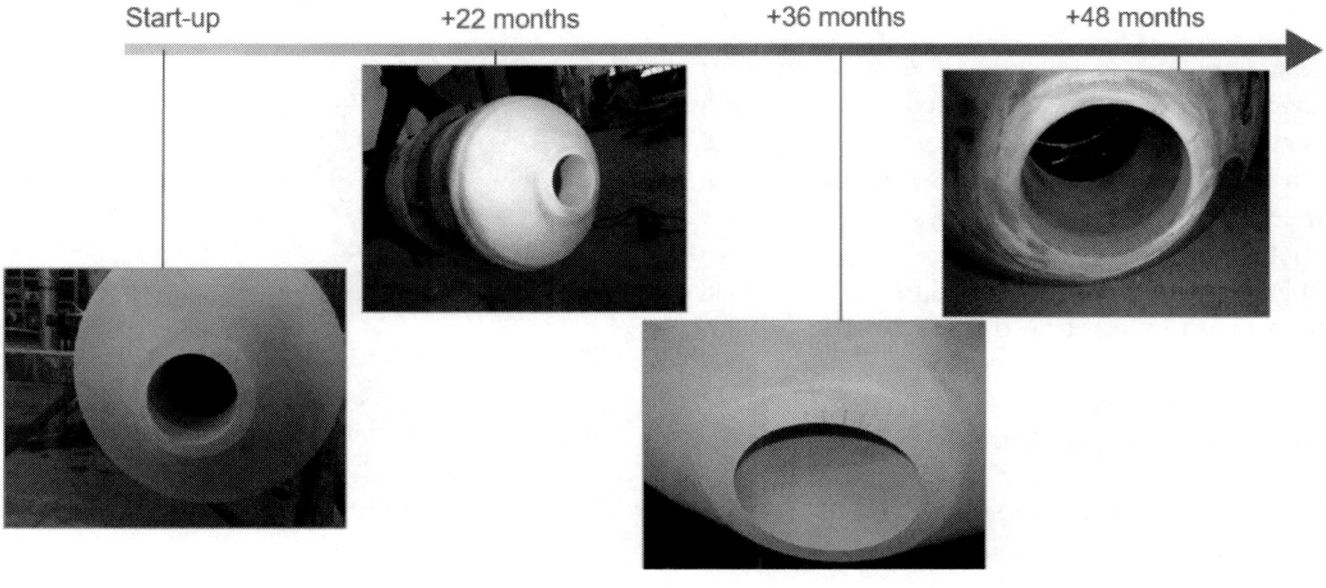

Figure 7. Luzhou Casale ATR Burner Tip Conditions over 4 years

Considering the shutdown periods, it can be concluded that both the main and spare burners have effectively accumulated approximately 4 to 5

years of continuous operation.

In any case, it is recommended to inspect the burner at least every 4–5 years. If cracks or other issues are detected, the burner can be promptly replaced with the spare unit, allowing maintenance to be carried out on the main burner without impacting plant operation.

Feedback from Luzhou Plant Operators

Luzhou plant operators and managers have been interviewed regarding their experience with Casale ATR technology. Overall, their feedback has been extremely positive.

Operators have highlighted that the ATR burner has never shown any reliability issues: the cooling circuit operates as designed, and no unplanned shutdowns have ever been caused by burner malfunctions.

The water-cooled design, particularly, is regarded as highly robust and reliable, allowing operators to focus on managing other critical events without the need to worry about the burner during plant upsets.

Operators also emphasized the advantages of Casale's patented water-cooled design, which ensures intrinsic reliability across a wide range of plant loads. Since the burner is equipped with an independent cooling water system (i.e. burner cooling is not dependent on process gas flowrate) operators feel confident in adjusting plant load according to market needs without risking damage to the ATR.

Additionally, during inspections of the ATR vessel, no soot deposits have ever been found on the catalyst, confirming the burner's clean and efficient operation.

The expected methane slip, below 0.5% on a dry basis, is consistently achieved at full plant load. The temperature distribution within the catalytic bed has been reported as uniform, which is a key factor in prolonging catalyst life and maintaining high conversion efficiency.

After about 20 years of operation, the general feedback from operators and managers is that the Casale ATR burner has always been able to meet required process operating conditions. Furthermore, it is considered reliable and safe regardless of plant load.

Conclusions

The operational experience at Sichuan Lutianhua Lvyuan Alcohol Industry Co., in Luzhou (China) highlights the effectiveness of Casale's ATR in maintaining reliability and high performances under variable load conditions.

Since its commissioning in 2004, the ATR unit in Luzhou plant has demonstrated:

- **Zero** unscheduled shutdowns despite fluctuating market demand.
- **No** maintenance issues or failures, reflecting the robustness of its design.
- **Consistent** syngas composition, essential for stable methanol synthesis.

The latest state-of-the-art Casale ATR burner design uses 3D printing of the burner tip with additive manufacturing, making it much more robust and effective in tip cooling. This combination is making it a preferred choice for large-scale syngas production.

The first ATR with a 3D printed tip has been started up in February 2025 in Sichuan Zhengdakai plant, for ethylene glycol production.

Process Gas Backflow Causes LTS Runaway Reaction

In 2024, Koch Fertilizer's Fort Dodge ammonia production facility experienced a potentially destructive runaway reaction in their "short shift" reactor. The incident occurred during startup after replacing the bed but before the reduction process began. Soon after natural gas was introduced to the front-end, a rapid increase in both temperature and pressure of the short shift was observed, despite it being isolated from the process gas flow path. This led to the maximum allowable working temperature of the vessel being exceeded. Inspections following the event determined no equipment damage had occurred, though the catalyst in the bed experienced significantly shortened life. The subsequent investigation identified reverse flow through a pilot PSV as the source of gas entering the vessel, which had not previously been identified as a risk. The investigation led to several recommendations to mitigate this risk, and the learnings from this incident were shared with other Koch Fertilizer facilities.

Jackson Lee
Koch Fertilizer

Introduction

On March 18, 2024, the Koch Fertilizer owned ammonia production facility located in Fort Dodge Iowa experienced a potentially destructive runaway reaction in a low temperature shift (LTS) converter. This paper will cover the sequence of events, subsequent investigation, and adopted solutions to prevent similar events from occurring in Koch's fleet of fertilizer plants.

Background

The Fort Dodge ammonia production facility was originally constructed in the 1960s with a nameplate rate of 600 STPD (544 MTPD). In the original configuration, the plant featured an LTS guard bed, called the short shift, located before the LTS. As the name "guard" implies, the purpose of this configuration was to protect the LTS from poisons or the effects of water leaks, while also converting incoming carbon monoxide (CO) to carbon dioxide (CO_2). In the 1990s, the plant went through an expansion which brought nameplate production up to 900 STPD (816 MTPD). This expansion included modifications to the piping in the shift converter section that changed the flow pattern from series to parallel on the LTS and short shift, allowing for lower pressure drop to accommodate the increase in flow. The short shift catalyst bed contains approximately one half of the catalyst volume of the LTS. This lower volume gives it a shorter run time with more frequent catalyst changes, compared to the LTS.

Plant conditions

In early March of 2024, the Fort Dodge facility experienced a shut down. Leading up to the shutdown, gas samples showed a deterioration in short shift performance, leading to high CO slip. The convenient timing and presence of replacement catalyst onsite led the process team to recommend a short shift catalyst change to site leadership, which was accepted.

During the outage, a third-party contractor removed the spent catalyst and installed the new material. The work was completed without issue. After replacement, the site decided to proceed with startup, achieve ammonia production, and then perform the reduction on the new short shift catalyst with the assistance of the catalyst vendor.

Incident

As the plant startup began, the short shift, which contained unreduced catalyst, was isolated from the operating process system. This configuration is shown in Figure 1.

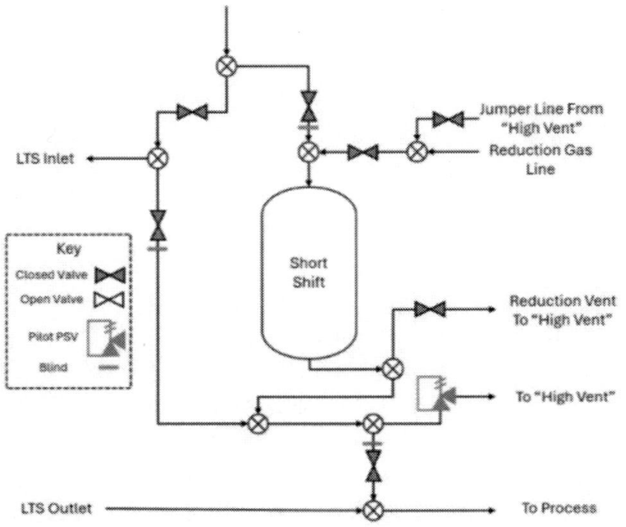

Figure 1: Configuration of the short shift system when startup was attempted.

At approximately 10:30 PM on March 17, steam was added to the process as a part of the startup procedure and was vented out of PV-102 into the "high vent" process vent. The high vent is also the discharge location of other process vents and pressure safety valves (PSVs), including the PSV and reduction vent line for the short shift. PV-102 is located downstream of the High Temperature Shift converter (HTS) and thus the gas being vented contained a high concentration of hydrogen gas.

Shortly after steam began to flow through PV-102 and out the high vent, the pressure in the short shift vessel began to increase, as shown in Figure 2.

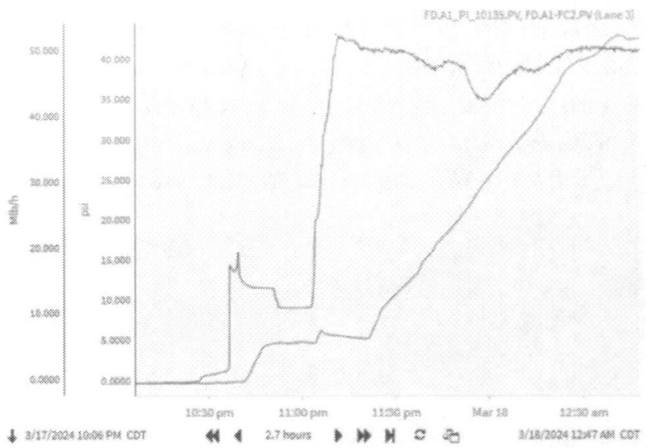

Figure 2: Trend showing steam flowrate into the process in brown, and short shift vessel pressure in pink.

The pressure indication on the short shift vessel was only transmitted into the site's PI data historian, and therefore did not appear on the distributed control system (DCS), used by operators to monitor and control the plant.

Shortly before 3:00 AM on March 18, operators began the process gas flow to the plant as a part of normal startup (reference the first time marker on Figure 3). Approximately twenty-five minutes later, the value of several in-bed temperature indicators (TIs) began to climb:

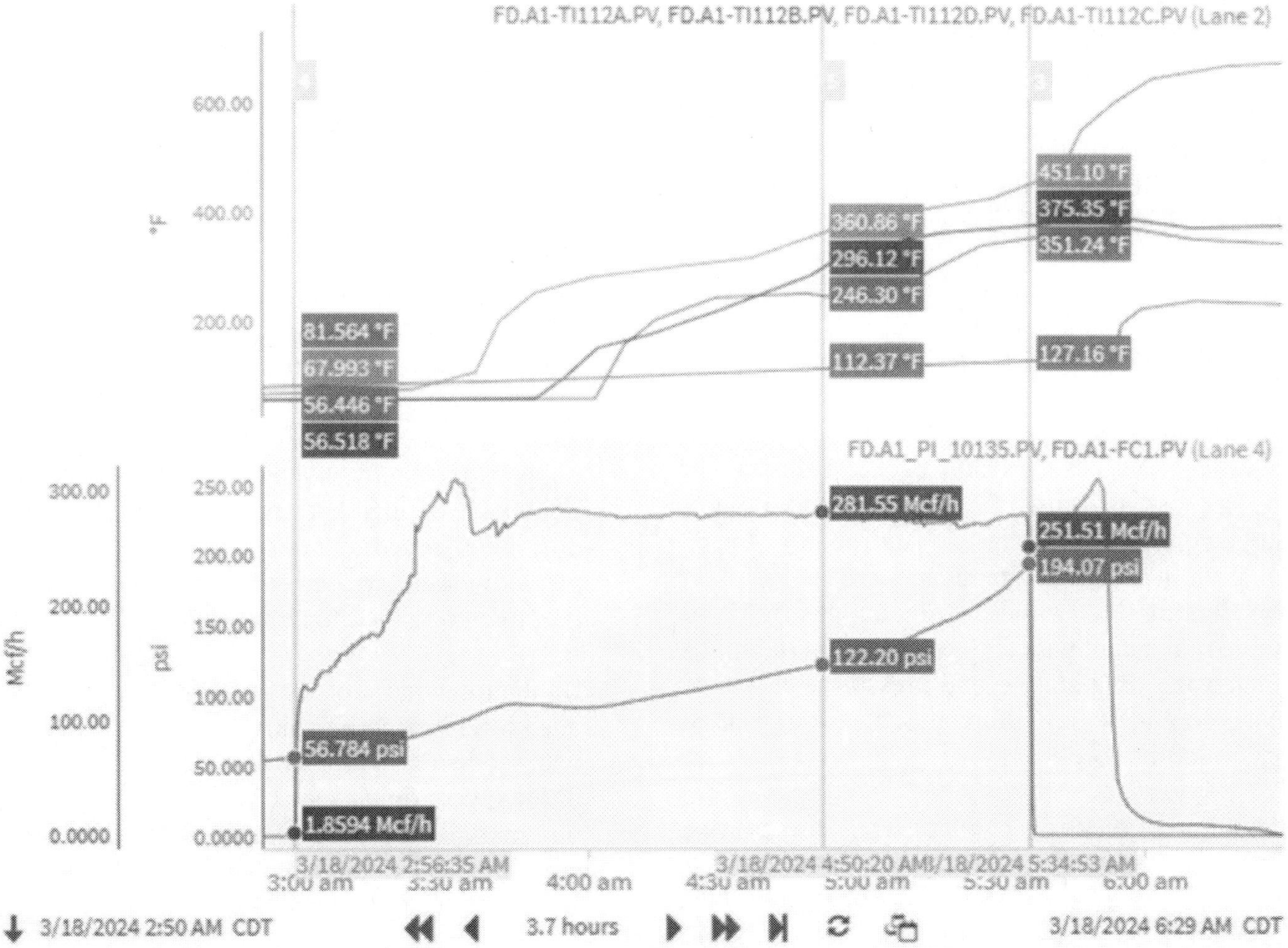

Figure 3: Trend of short shift TIs on the top lane, and gas flow (blue) and short shift pressure (pink) on the bottom lane.

Approximately thirty minutes after process gas feed began, the site experienced an upset in the high-pressure steam header. This diverted operator attention to other areas of the plant as they attempted to stabilize the unit. Around 4:50 AM, operations had stabilized the rest of the plant and noticed the short shift bed TIs had climbed from 50 – 80°F (10 – 27°C) to a maximum temperature of 360°F (182°C). At this time, operations called the site's process engineer and informed him of the situation (noted by the second marker on Figure 3). After confirming with operations that the vessel's inlet and outlet were blocked and blinded, the process engineer contacted another, more experienced, engineer for assistance.

The process team identified this situation as a potentially dangerous runaway reaction. LTS catalyst is normally loaded with the active copper sites in the oxidized state. To activate the catalyst, a small amount of hydrogen gas is mixed with a carrier gas, usually natural gas at the Fort Dodge facility, and fed to the vessel at a temperature sufficient to reduce the oxidized copper, according to the following reaction (1):

$$CuO(s) + H_2(g) \rightarrow Cu(s) + H_2O(g) + Heat$$

This reduction process is highly exothermic and controlled very carefully according to a procedure with the help of an experienced process engineer and a representative from the catalyst company. A typical reduction of the Fort Dodge short shift bed can take 36 – 48 h under normal conditions.

If excess hydrogen gas is added, temperatures in the bed can exceed 500°F (260°C) at which point copper oxide (CuO) can be reduced directly by methane present in the carrier gas according to the reaction (1) (2):

$$4CuO(s) + CH_4(g) \rightarrow 4Cu(s) + 2H_2O(g) + CO_2(g) + Heat$$

This reaction produces more heat than the hydrogen reduction which further speeds up the process, causing a snowball effect as catalyst temperature rises, exacerbated by the fact that the carrier gas is primarily methane, providing a large volume of reactant gas. Additionally, the reduction of CuO with methane increases pressure by creating three moles of product gas from one mole of reactant gas. This combination of rapidly increasing temperature and pressure can lead to catastrophic vessel failure and release of process gas if left unchecked.

The initial recommendation from the process team was to use a nitrogen truck that was onsite from the outage to push nitrogen through the bed, cooling it down and diluting the concentrations of methane and hydrogen. However, the contractors who controlled the truck were not onsite at the time as no nitrogen use was expected, so they were called and asked to return to the site as soon as possible. Meanwhile, catalyst temperatures and vessel pressure continued to increase dramatically, at rates reaching 15°F/min (8°C/min) and 4 psi/min (28 kPa/min).

When the highest bed temperature reached 450°F (232°C) at 5:35 AM, the process engineers advised operations to initiate an emergency shutdown, as pulling all feeds from the plant would theoretically remove the source of hydrogen and methane from the vessel (refer to the third marker on Figure 3). The point of 450 °F (232 °C) matches the point at which the site's procedure requires stopping hydrogen flow to the vessel during a planned reduction. However, even after gas flow was stopped, vessel temperature and pressure continued to climb.

By the time that process gas was removed (reference the first marker on Figure 4), the reaction was occurring so rapidly that the "runaway" conditions were quickly reached (reference the second marker on Figure 4). As indicated by the increasing slope of the blue line, the rate of temperature rise began increasing.

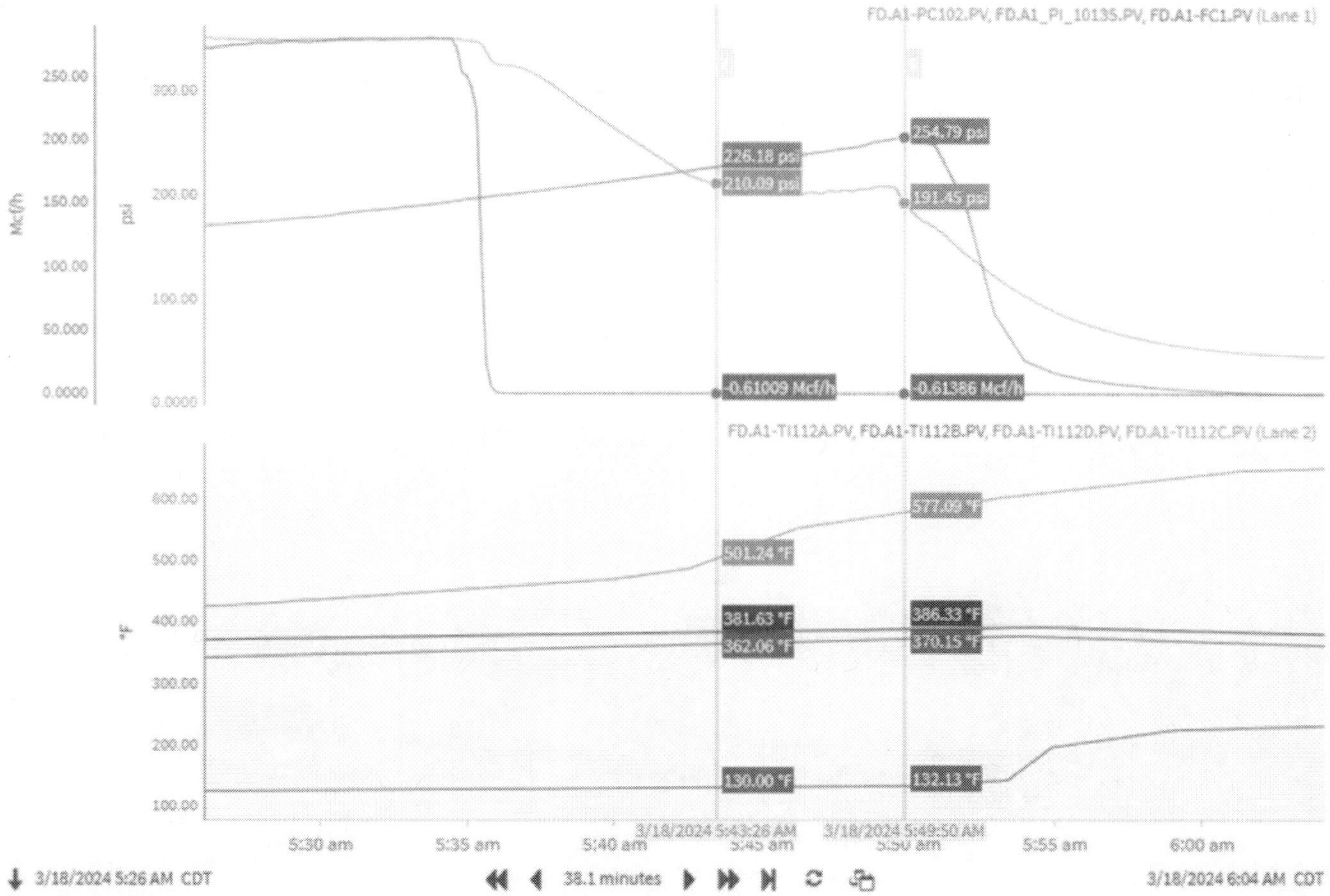

Figure 4: Trend showing short shift temperatures on the bottom lane, and vessel pressure, vent pressure (green) and gas flow on the top lane around the time the emergency shutdown occurred.

In addition to the temperature continuing to rise, removal of feeds from the front end of the plant failed to reduce the vessel's pressure. This is shown in Figure 4 as the pressure (pink line) continued to increase even after the gas flow (blue line) was stopped. Seeing this, the process engineers recommended that operations open the manual "reduction vent" on the bottom of the vessel shown in Figure 1. Operations was able to open the vent at approximately 5:50 AM and vessel pressure began to drop (reference the third marker on Figure 4). Additionally, the process engineers advised operations to push cold nitrogen through the catalyst bed and out the vent, after the nitrogen truck operators arrived onsite.

Before nitrogen flow began, the highest indicated catalyst bed temperature reached 919°F (493°C). This temperature exceeded the maximum allowable working temperature (MAWT) of the vessel which is set at 491°F (255°C). It should be noted that while the MAWT was exceeded, this setpoint is based on an assumed maximum allowable working pressure (MAWP). MAWP and MAWT have an inverse relationship, so as MAWP increases, MAWT will decrease. At no point during the incident did the pressure exceed the MAWP of the vessel, set at 500 psig (3,500 kPa), reducing the risk of high temperature damage to the pressure vessel.

Following the incident, it was believed that the cause of the incident was the reduction vent line shown on Figure 1. The valve on this line was closed, but the process blind had been removed in preparation for usage during the reduction. The consensus was that the block valve had leaked, allowing higher pressure gas in the high vent system to enter the short shift vessel, which was operating near atmospheric pressure. Members of the site's fixed equipment team conducted an inspection of the short shift vessel. The inspection did not reveal findings of concern, and the team determined that the short duration of the high temperature, along with the fact that high temperatures had been experienced at pressure much lower than the MAWP, meant that the risk for vessel damage was low.

The site decided to pursue startup again after inserting a blind into the reduction vent line, thinking this would stop the backflow of process gas into the vessel. At 3:35 PM on March 18, steam was added into the process and began to flow out PV-102. However, at about 7:00 PM, Operations again contacted the process engineers and informed them that the bed temperatures had begun to increase. (Figure 5)

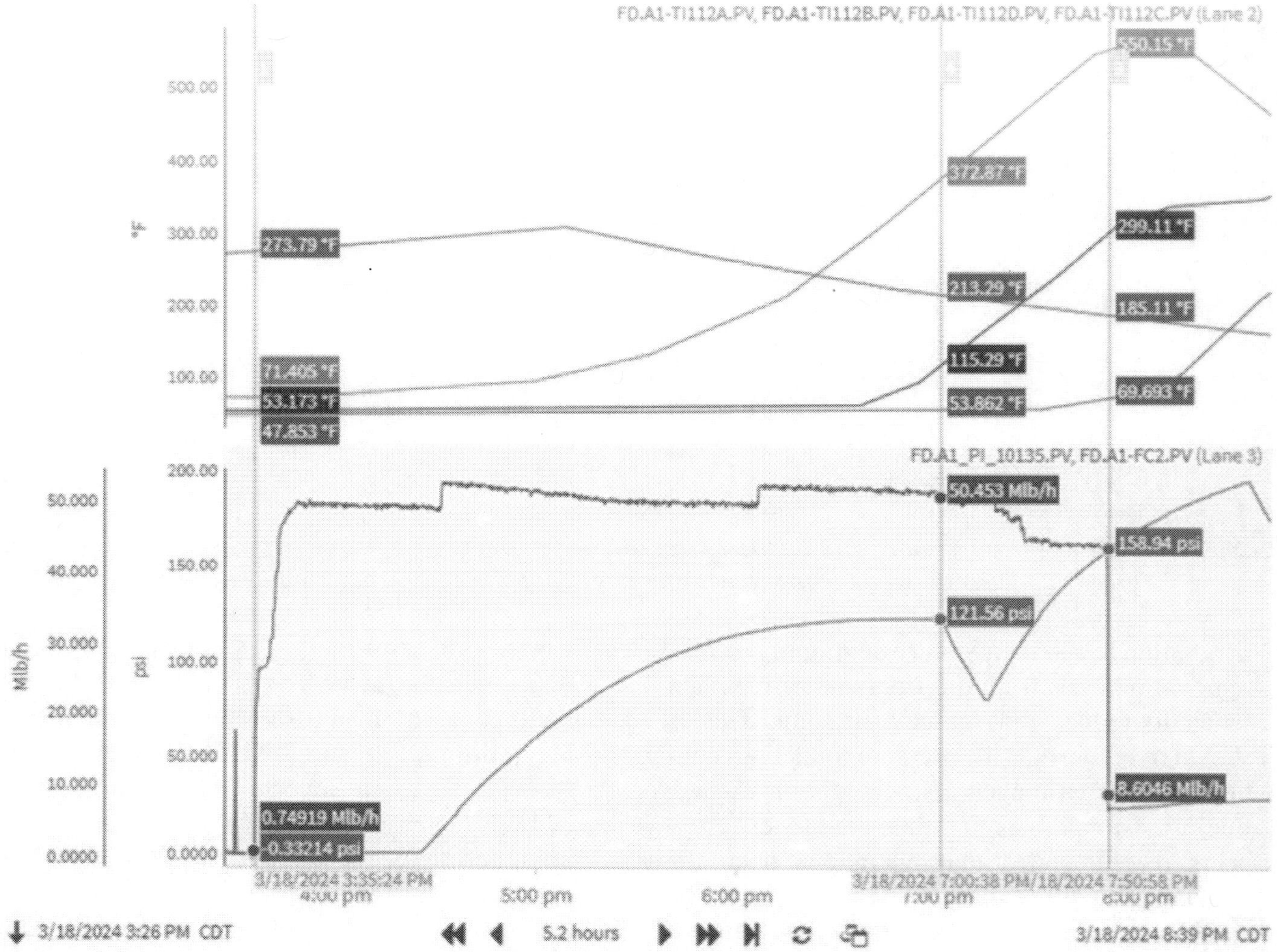

Figure 5: Trend showing short shift temperatures on the top lane, and vessel pressure and steam flow on the bottom during the second attempted startup.

Process engineering immediately recommended opening a small bleed on the vessel to reduce pressure and begin nitrogen flow again, which was done at approximately 7:15 PM. However, the temperature continued to climb and at 7:50 PM the decision to once again initiate an emergency shutdown was made. The short shift vessel was again purged with cool nitrogen to lower the catalyst bed temperatures.

The site team determined that this incident was not a runaway reaction, as no hydrogen was being produced in the plant at that time. After consulting the catalyst vendor, the site concluded that the temperature increase was from the chemisorption of steam onto catalyst that had been reduced during the runaway reaction. This is normally seen when a reduced bed is placed in service the first time. The steam in the process gas causes a temperature wave to pass through the catalyst bed as steam contacts the fresh copper.

Investigation

Following the event, an investigation was conducted to determine the cause of the incidents and identify potential solutions.

The inlet and outlet valves for the short shift were closed and blinded, eliminating these locations as potential entry points for process gas and steam. The hydrogen source was determined to be process gas that was being vented through PV-102 and into the high vent. This gas comes from the outlet of the HTS and contains approximately 43 %(vol) hydrogen, which is much higher than the hydrogen content used in reduction gas, which is typically 2 %(vol) or less. The higher pressure in the high vent system, created by backpressure from the vent and piping, would have allowed this hydrogen rich gas stream to enter the short shift vessel, which was near atmospheric pressure.

There are several ways for gas from the high vent to enter the short shift including:

- *The high vent jumper line on the reduction gas line at the inlet of the vessel shown in orange on Figure 6.*
- *The reduction vent line on the outlet of the vessel, shown in purple on Figure 6.*
- *The pilot PSV on the outlet of the vessel, shown in green on Figure 6.*

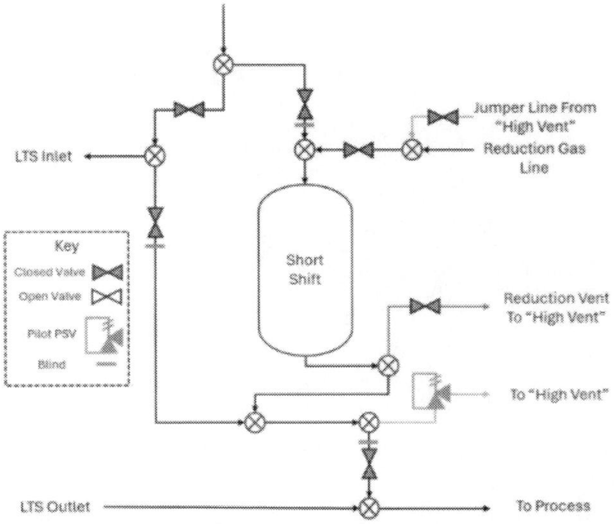

Figure 6: A diagram showing the possible sources for high vent gas to enter the short shift.

The temperature indicators in the catalyst bed indicated that the short shift catalyst warmed from the bottom up. This, along with the multiple closed block valves, led the team to discredit the theory that the orange high vent jumper that feeds into the reduction gas line was the source of hydrogen.

As for the two remaining sources, it was noted that the purple reduction vent line had been blinded before the second startup, indicating it could not have been the flow path for both incidents. These facts led the investigation team to conclude that the pilot PSV had leaked backwards at least once, leading to steam entering the vessel and causing the chemisorption-based wave of increased catalyst temperatures seen on the second attempted startup.

Unlike conventional PSVs, which are held tightly closed by a spring, a pilot PSV uses process pressure to maintain a tight seal. A cross-section of a typical pilot operated PSV is shown in Figure 7.

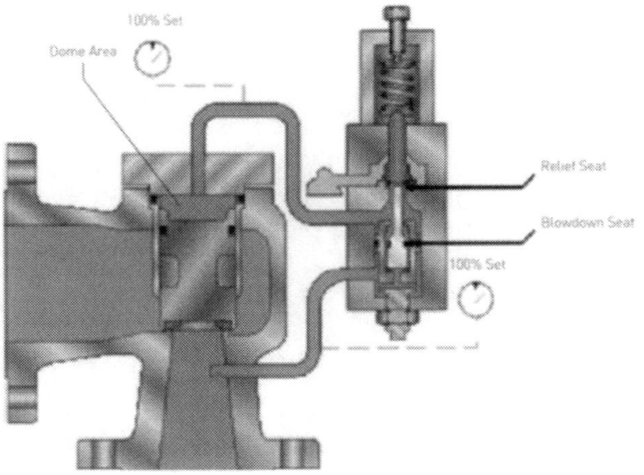

Figure 7: Diagram showing a pilot PSV (3).

Pilot PSVs are named such because they contain smaller conventional PSV that "pilots" the larger PSV. When process fluid (shown in orange on Figure 7) pressure is below the set pressure of the small pilot PSV, it flows to the dome area of the PSV and provides a seating force on the PSV. The dome area is larger than the process area (shown as a small green disc in Figure 7), so the seating force is higher than the process lifting force as both sides experience the same pressure. However, if the process pressure exceeds the set pressure of the pilot PSV, it will relieve, removing pressure from the dome area. The lifting force of the PSV then allows the process pressure to be relieved through the main body of the valve.

This mode of operation means that pilot PSVs with no process pressure do not have a positive seating force. This is usually mitigated through the installation of a backflow preventer seen below in Figure 8:

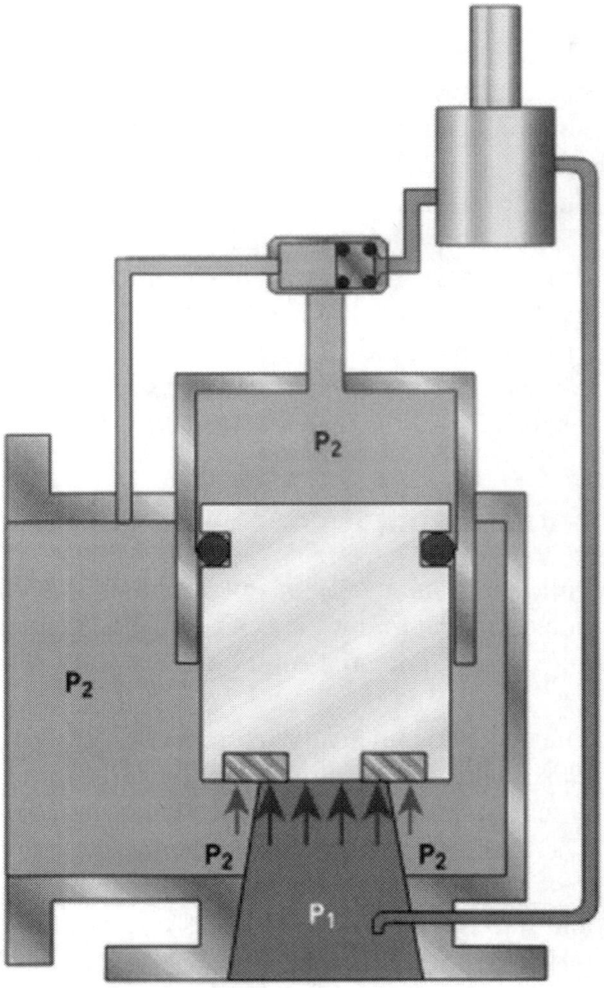

Figure 8: Diagram showing a pilot PSV with a backflow preventer as the small purple square (3).

The backflow preventer is a slider that changes the pressure source on the dome based on whether back pressure (P2) or process pressure (P1) is higher. If P2 is higher than P1, P2 should act as the seating force of the dome.

The pilot PSV on the outlet of the short shift had a backflow preventer installed, but the investigation concluded it must have become stuck during the second event as this was the only possible flow path for steam to enter the vessel. The lack of evidence and process indications prevented the team from determining definitively that backflow through the pilot PSV was the source of hydrogen during the runaway reaction, but it seems to be the most likely explanation, given the experience of the second event.

The behavior of the pressure in the vessel during the first incident also gives credibility to the theory that the PSV allowed backward flow from the high vent to the short shift vessel. Once the emergency shutdown was initiated, pressure in the short shift vessel did not decrease even as high vent flow and pressure did. If the flow path had been an open line to the high vent, pressure should have decreased as soon as the pressure in the high vent was below vessel pressure. However, reverse flow through a pilot PSV would allow pressure in the vessel to increase. Then, when the high vent pressure began to decrease during the shutdown, the pressure of gas in the short shift would provide a positive seating pressure to the dome, closing the PSV and trapping the gas inside. This very behavior is reflected by the process data shown in Figure 4.

This was a novel flow path that had not been considered as possible at the site before the incident. It should be noted that it is unusual for the short shift bed at Fort Dodge to be in the conditions that allowed this incident to occur. Normally, the short shift catalyst is replaced and reduced while the plant is running. During these times, the plant would cut rate and operate with the LTS only. In these cases, there should be no flow through the high vent, and thus no opportunity for reverse flow into the short shift via the pathways identified above, unless an upset caused the plant to trip and introduce gas to the high vent.

The catalyst in the LTS reactor, which operates in parallel with the short shift, is normally replaced during turnarounds. Because of this, the plant will start up with oxidized catalyst in the LTS vessel. The LTS also has a pilot PSV with a backflow preventer that discharges to the high vent, and is located within the isolation blinds for the vessel. This indicates that backflow incident the site experienced in March 2024 was a possibility during every plant startup when new LTS catalyst was installed, as hydrogen rich process gas will be vented at PV-102 and out the high vent during every startup. It seems that the site had just been fortunate in the past that the backflow preventer on the LTS PSV never failed.

Solutions and Results

The site identified several solutions and actions to prevent similar incidents from occurring in the future:

1. Modify Standard Operating Procedure (SOP) to note that oxidized LTS and short shift beds should be pressurized with nitrogen from a truck during plant startup and until the reduction begins. This should keep the pilot PSV seated tightly and the atmosphere inside the vessel inert.
2. Add block valves under the short shift and LTS PSVs so they can be isolated as a source of process gas during catalyst loading or startup if needed.
3. Investigate adding conditional temperature alarms to the short shift and LTS beds to help catch runaway conditions, as the normal high temperature alarms are set too high to be of use in case of a runaway.

Following the second incident, the site pressurized the short shift vessel with nitrogen to seat the PSV and brought the plant online without issue.

Shortly after attaining production, the planned reduction of the short shift was conducted. The reduction was completed in record time (about 16 h) at the site due to a large portion of the bed having been reduced during the runaway reaction.

After the plant was in operation, catalyst sampling showed that the new short shift bed was experiencing more rapid loss of performance than the previous bed. (Figure 9).

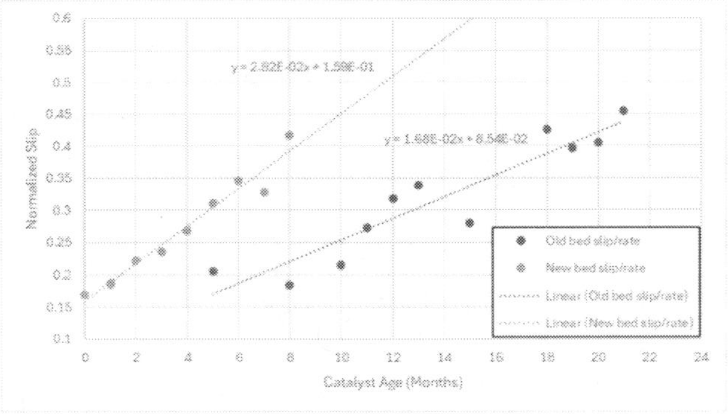

Figure 9: Trend showing the normalized dry CO slip for the "new" bed, which experienced the runaway, and the "old" or previous bed that did not, plotted against catalyst age in months.

Figure 9 shows that the CO slip from the short shift worsened more rapidly for the bed that experienced the runaway reaction, with the slope being nearly 75% steeper than that of the previous catalyst installation. The catalyst vendor who provided this charge confirmed that the high temperatures seen in the runaway reaction likely damaged the affected catalyst via thermal sintering. This mechanism essentially kills the active sites on the catalyst, providing less surface area for the reaction to occur and leading to higher CO slip.

As a result of poor performance and increasing catalyst pressure drop from a water leak, the short shift bed was again replaced in December 2024. A thorough Potential Problem Analysis (PPA) was conducted beforehand to minimize the risk of a repeat event. Nitrogen was used to pressurize the beds during startup and this succeeded in preventing another runaway scenario.

Conclusion

The runaway reaction at Fort Dodge was determined to be caused by a pilot PSV leaking backwards which had not been identified as a possibility before. Most previous short shift bed swaps had been carried out in conditions that made this flow path less likely. However, the parallel LTS bed was found to be at risk for a similar event during bed replacements, but the site had been fortunate and avoided that occurrence.

Recommendations from the investigation were shared within the Koch Fertilizer fleet, and subsequent short shift changeouts included a PPA before beginning work.

References

1. Clariant. *Operating Manual SHIFTMAX ® 200 SERIES* , CLARIANT INTERNATIONAL LTD CATALYSTS , Muttenz, Switzerland, 2013.
2. "Reduction of Copper(II) Oxide with Methane." *Royal Society of Chemistry*, 2009, www.rsc.org/cpd/teachers/content/filerepository/CMP/00/000/804/CFNS Experiment 45 - Reduction of copper(II) oxide with methane.pdf?v=1542030143498.
3. *Pressure Relief Valve Engineering Handbook*. Technical publication no. Tp-V300, Emerson, 2012, *emerson.com*, https://www.emerson.com/documents/automation/pressure-relief-valve-engineering-handbook-en-gb-4244934.pdf, Accessed 14 Feb. 2025.

Prevention of Process Safety Incidents by Establishing Strong Process Safety Leadership and Culture

As part of its continuous improvement journey in process safety, QAFCO conducted an analysis of process safety incidents (Tier 1) over the past four years. This analysis aimed to identify gaps in process safety leadership and cultural practices, against organization's Core Principles of Process Safety. This paper sets out practical experience of preventing process safety incidents by establishing strong process safety leadership and culture. By addressing these gaps, QAFCO seeks to enhance its process safety performance.

Kamleshkumar Modi
Qatar Fertiliser Company (QAFCO)

Introduction

To prevent process safety incidents in the industries, it is essential to establish strong process safety leadership and culture. QAFCO conducted an analysis of two Tier 1(definition as per API RP 754) process safety incidents to identify gaps with the aim to improve leadership and culture.

Incident 1

A carbamate solution leak was reported from the Reactor down comer drain line, due to a rupture of a 2-inch drain line, (18 m floor area) at the Urea plant.

Immediately, the plant was stopped. The release of carbamate solution, ammonia, urea solution and CO2 gas (to atmosphere and surface drain) was mitigated by adequate water curtains by the Emergency Response Team.

The system was depressurized through the drain valves and the leaky portion of the pipe. No severe consequences (i.e. Injury Fire/Explosion) occurred.

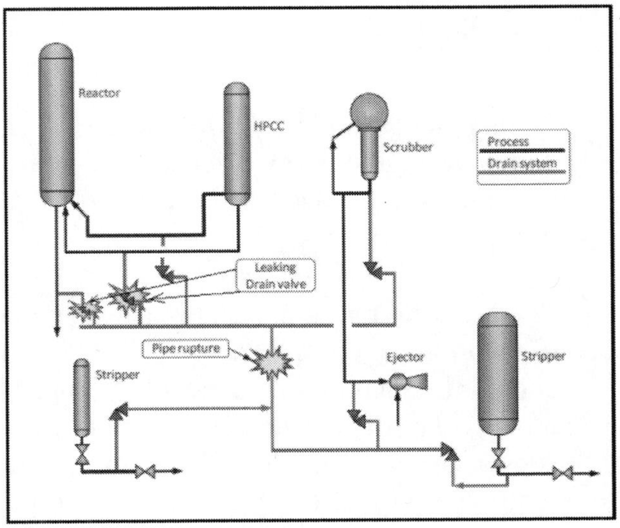

Figure 1. Carbamate solution leak diagram

The following causal factors were identified as part of the root cause analysis:

1. Understanding the Hazard/Risk:
   - Both the initial HAZOP and subsequent Re-HAZOP analyses failed to adequately cover the drain system.
2. Ineffective Root Cause Analysis (RCA) and Maintenance Practices:
   - Maintenance issues such as pump flushing maloperation, isolation valve passing, and discharge Pressure Control Valve (PCV self-regulating) malfunction were not properly addressed in the investigation.
   - Recurrent issues related to these maintenance activities were also not properly addressed.
3. Normalization of Deviation:
   - The flushing routine frequency, a critical operation for the Urea Synthesis drain system, was altered without documented justification or clarity at the time of the change.

Incident 2

During the startup of the synthesis gas compressor, a fire took place at the synthesis gas interchanger due to the release of high-pressure synthesis gas.

Synthesis gas mixture contains a high amount of Hydrogen (approximately 72%~74% vol.) and minimum ignition energy of hydrogen in air is one of the lowest among known substances at 0.02 mJ. The hydrogen got ignition energy from the friction of higher velocity gas with nearby steel structures and resulted in fire.

The Emergency Response Team provided a water curtain to nearby equipment and piping which enabled field operators to reach the high-pressure vent valves. Once the vent valves were opened and the loop was depressurized, the fire situation came under control and later fully extinguished.

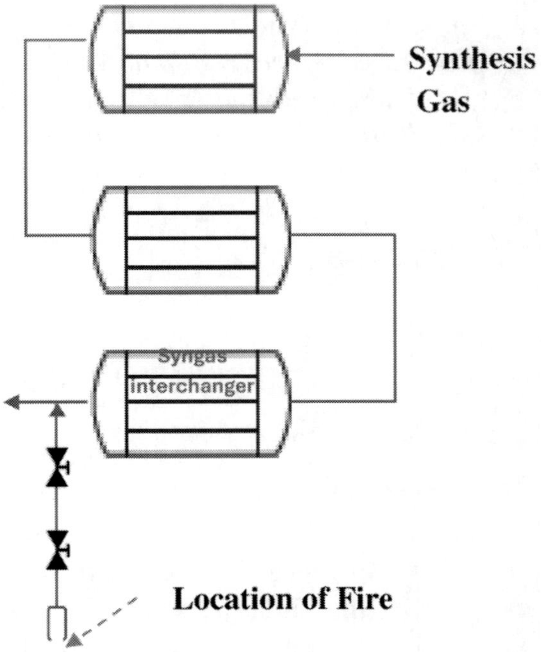

Figure 2. Synthesis gas fire diagram

The following causal factors were identified as part of the root cause analysis:

1. Understanding the Hazard/Risk:
   - Ineffective Readiness Checks: The operating procedure lacked a proper description of leak tests at higher pressures.
   - Insufficient Secondary Protection: No secondary protection (i.e., NRV) in place to mitigate the risk of high-pressure gas release in case both primary valves failed.

2. Human Factor:
    - Valves associated with high-pressure service were not properly closed.

Process Safety Gap Assessment

With the goal of eliminating process safety incidents and achieving an incident-free workplace, QAFCO conducted an independent third-party assessment based on the 10 Process Safety Core Principles:

1. Establish an Imperative for Process Safety
2. Production not possible without process safety. Provide Strong Leadership

   Leaders inspire others to process safety excellence and Walk the Talk. Leaders demonstrate tangible commitment and influence the same commitment in their subordinates.
3. Foster Mutual Trust

   Everyone does what they say and says what they mean. In a trusting relationship, people believe that their supervisors, peers, and reports will make the right decisions, act honestly and if necessary resolve conflict.
4. Ensure Open and Frank Communication

   Communication channels are open and encouraged and messenger not blamed. Accurate communication avoids confusion, wastage of time and incorrect decision making.
5. Maintain a Sense of Vulnerability

   Healthy level of respect for hazards and risk of facility. It is opposite of complacency, naturally follows a serious incident or near-miss.
6. Empower Individuals to Successfully fulfill their safety responsibilities

   Workers have authority and resources to performed assigned process safety roles.
7. Defer to Expertise

   Technical knowledge related to process safety valued and technical opinions accepted.
8. Combat the Normalization of Deviation

   Deviation from approved rules and standards never tolerated.
9. Understand and Act Upon Hazards/Risks

   Hazards and risks analyzed, controlled with appropriate safeguards, and managed.
10. Learn to Assess & Advance the Culture

    Lessons-learned sought internally and externally. Learnings used to maintain and enhance culture.

The key findings from the assessment was to improve:

- Hazard Identification/Risk assessment process
- Proactive incident/Near Miss Reporting
- Human Factors in plant operation
- Workforce Empowerment

The following key trainings were conducted by an external trainer to improve process safety leadership and culture:

- Process safety leadership workshop for all level of employees.
- Other training covering process safety core principles such as employee empowerment, stop work authority, chronic unease, effective communication, questioning attitude, blame free culture provided to all employees.
- Specific competency training which includes Layer of Protection Analysis (LOPA), Management of Change, Process Hazard Analysis, Pre startup safety review, and Safety in plant design, etc. The training conducted for Head of Sections, Engineers, Supervisors, Operators, and technicians.

Best Practices adopted at QAFCO

The following best practices were adopted at QAFCO:

- Leaders regularly discuss and resolve process safety risks.
- Leaders maintain high visibility on site.
- Conducted key process safety competency training for all employees (as per training matrix) to ensure competent resources are always available.
- Encouraged near-miss and incident reporting without fear of retaliation. Thus, improving near-miss reporting (30% increase) and its investigation.
- Promoted the sharing of process safety information across the organization including lessons learned from other industries.
- Empowered workforce with authority to question unsafe practices.
- Thorough incident investigations to identify root causes comprehensively.
- Enhanced communication of risks and associated controls.
- Empowered workforce to exercise stop-work authority.
- Launched Walk the Line (WTL) Program. The objective of the program was to address process safety incidents that occur when line-up changes are made to a process, and most of these changes occur when preparing equipment for maintenance and returning equipment to service. QAFCO has developed checklist which are used by plant area operators as part of WTL program. This has resulted in significant reduction in human errors, fires, and leakages.

As a result, QAFCO achieved Zero Process Safety Total Incident Rate as of 1st June 2024 (12 months rolling) and sustained this achievement as of date.

Conclusion

To prevent process safety incidents in industries:

- Strong leadership and commitment are key to enhancing a process safety culture.
- Site leaders lead by example in improving organization's culture by what they do and what they pay attention to.

- Consistent and timely informed decision making improves process safety performance.

References

[1] Lessons learned from process safety incidents at QAFCO - Internal online QAFCO portal.

[2] Essential Practices for Creating, Strengthening and Sustaining Process Safety Culture (Published by Center for Chemical Process Safety of the American Institute of Chemical Engineers)

[3] Process Safety Performance Indicators for the Refining and Petrochemical Industries. ANSI/API Recommended Practice -754.

Major accident hazard (MAH)! Are You Prepared to Avoid and Mitigate the Consequences of an Unlikely Event?

In compliance with OSHA's Process Safety Management (PSM) requirements, a Process Hazard Analysis (PHA) was conducted for all seven ammonia plants and five urea plants. This paper focused on the elaboration of the Major Accident Hazards (MAH) for a typical ammonia and urea production, using the information gathered during the PHA studies.

The study revealed considerable variability in the identified MAH across the plants, reflecting differences in design, operations, and site-specific conditions. To address these discrepancies, the risks were systematically adjusted and benchmarked against known major incidents in the ammonia industry. A critical element of the study was the identification of ammonia release scenarios, safety barriers, evaluation of emergency response measures, and the development of additional safeguards.

Additionally, the study emphasizes the need for the use of PHA as a dynamic risk management tool, and the importance of enhanced training and updates to the emergency plan.
The study culminates with lessons learned from the PHA studies and discussions on some recommended MAH scenarios for ammonia plant safety.

This framework integrates proactive risk management strategies, emergency preparedness, and mitigation measures, offering significant value to the ammonia production industry in preventing catastrophic incidents.

Cristiano Azevedo, Fertiglobe, Abu Dhabi, UAE

Faritha Banu Fertiglobe, Abu Dhabi, UAE

Importance of safety in Ammonia and Urea plants.

Ammonia and Urea production poses significant risks due to the hazardous nature of their processes and materials. It involves handling hazardous substances like ammonia, which is toxic and can cause severe health issues if inhaled. The production processes involve high pressures and temperatures, which can lead to explosions or fires if not properly managed. Accidental releases of ammonia can have significant environmental impacts, including water and soil contamination. Proper safety measures are essential to prevent acci- dental releases, ensuring equipment integrity and following strict safety protocols can mitigate these risks.

Objective

The objective of this MAH review study is to review the MAH scenarios, standardize the consequence and risk rankings, identify barriers with safety critical devices and safety critical tasks and ensure competency assurance of the team to manage MAH in plants.

Historical Incidents

Historical incidents involving ammonia and urea have highlighted significant safety concerns in the chemical industry. One of the most severe ammonia related accidents occurred at ammonia handling facility in Dakar, Senegal, in 1992, resulting in numerous fatalities and emphasizing the dangers of pressurized anhydrous ammonia. Additionally, ammonia and urea plants have faced issues like atmospheric corrosion, leading to equipment failures and safety hazards. These incidents have driven the industry to adopt stricter safety regulations and continuous improvements in process safety.

Company's Risk Assessment and Management System

"Risk Assessment and Management System" outlines the process of risk assessment and management in an organization. It focuses on the systematic approach to identifying, assessing, and managing risks through various stages from project initiation to operation phases using both qualitative and quantitative methods. The process flows from hazard identification to risk screening, then to management systems or quantitative assessments depending on the level of risk identified.

The initial step is to identify potential hazards through various methods like HAZID (Hazard Identification), HAZOP (Hazard & Operability Study), Design Reviews, MOC (Management of Change), Incident learnings, Audits etc., and the identified hazards go through initial screening of risks to determine their severity and potential impact.

The next step is the Risk Assessment, which encompasses the overall evaluation of risks within the organization, using qualitative methods to assess and prioritize them based on their potential impact. Non-Major Accident Hazards will be considered in Management Systems, which include the Work Management System (WMS), Procedures, and Job Hazard Analysis. These administrative processes are designed to organize work processes to ensure safety and guide employees in performing tasks safely and effectively.

The Major Accident Hazards (MAH) will be assessed through more detailed and numerical assessments with the help of studies like QRA (Quantitative Risk Assessment), FERA (Fire Explosion Risk Analysis), EERA (Escape, Evacuation, and Rescue Analysis), Bow Tie Analysis, SBRA (Scenario Barrier Risk Assessment), LOPA Assessments (Layers of Protection Analysis), Periodic Review & Audits. The output from the detailed assessments will be captured in the HSE Risk Register with defined owner and approval levels to conduct a periodic review and manage the MAH.

Overall, the Risk Assessment and Management System aims to convey the critical components of a comprehensive risk assessment and management system to manage Major Accident Hazards in Ammonia and Urea plants, including safety culture, training, competence, regular audits, inspections, and emergency response drills to maintain the integrity of safety barriers and ensure preparedness for potential incidents

Regulatory Framework

The regulatory framework including safety standards, inspection protocols, and emergency response procedures are crucial for mitigating major accident hazards in ammonia and urea plants, ensuring the safety of workers, the public, and the environment

OSHA Process Safety Management (PSM): The Occupational Safety and Health Administration (OSHA) PSM standard (29 CFR 1910.119) focuses on preventing the release of hazardous chemicals that could lead to catastrophic events.

EU Seveso Directive: The Seveso III Directive (2012/18/EU) aims to prevent major accidents involving hazardous substances and limit their impact on human health and the environment.

API Standards: The American Petroleum Institute (API) provides numerous standards and recommended practices related to safety, environmental protection, and mechanical integrity for ensuring safe operations and safety guidelines for handling hazardous chemicals.

CCPS Guidelines: The Center for Chemical Process Safety (CCPS), part of the American Institute of Chemical Engineers (AIChE), promotes process safety management excellence through its Risk-Based Process Safety (RBPS) framework.

Major Accident Hazard (MAH)

A major accident is an 'Occurrence' such as an emission, fire or explosion resulting from uncontrolled developments during operation leading to major, catastrophic or disastrous consequences irrespective of their likelihood and serious consequences affecting people, assets, the environment and/or company reputation. The consequences may be immediate or delayed.

The company risk matrix includes specific categories derived from the PHA studies to classify it as MAH.

Case Study:

MAH Review Methodology

The potential sources of ammonia release like storage tanks, pipelines, refrigeration systems and factors that increase the risk of release, such as equipment failure and human error, were identified during the PHA study. The study was initiated with a brainstorming exercise to understand and identify the MAH based on the Ammonia and Urea industry events database. Later, review of different studies, PHA (Process hazard Analysis), Bow - Tie, SG (Safeguarding), IOW (Integrity Operating Window), ER (Emergence Response) documents carried out to confirm identified MAH scenarios are captured in those studies. Figure 1 presents a summary of the brainstorming session, highlighting the extent to which each scenario was considered (Yes, No and Partially yes) in the referenced studies.

S.No	Process	Risk Title	Risk detail	MAH	PHA	SG	IOW	ER	Industry events , in the same technology and equipment
1	Ammonia	LOPC of Natural Gas	Natural gas leak from piping, flange joints	Yes	Yes	Yes	No	Partially Yes	Many events every year
2	Ammonia	Reformer (Top/Bottom) fire	Natural gas leak in the Primary/Syn gas leak from Secondary reformer tubes	Yes	Yes	Partially Yes	No	No	One major and other small incidents reported
3	Utilities	Boiler explosion	Boiler explosion	Yes	Yes	Partially Yes	No	No	Few events reported in the last 20 years
4	Ammonia	LOPC of Process Gas	Process gas due to piping, flange joints	Partially Yes	Yes	No	No	Partially Yes	Many events every year
5	Ammonia	Fire box explosion	Fire box explosion	Yes	Yes	Yes	Partially Yes	Partially Yes	One major and other small incidents reported
6	Ammonia	Runaway reaction in Methanator	Runaway reaction in Methanator	Yes	Yes	Yes	Yes	Partially Yes	Few events reported in the last 10 years
7	Ammonia	LOPC from Waste heat Boiler	Waste heat Boiler leak	No	Yes	Yes	Yes	No	Many events every year
8	Ammonia	LOPC from Syngas compressor	Syngas leak from compressor piping, seals flanges	Yes	Yes	No	Yes	No	Few events reported every 5 years
9	Ammonia	Compressor house explosion	Compressor house explosion	Yes	No	No	No	Partially Yes	One event reported in the last 30 years
10	Ammonia	LOPC from start-up heater	Start-up heater leak	No	Yes	Yes	Yes	No	Few events reported in the last 10 years
11	Ammonia	LOPC of liquid ammonia from a flare/stack	Liquid ammonia carry over to flare, Open drains and vessels	No	No	No	No	No	Three major events reported in the last 10 years
12	Ammonia	LOPC from Ammonia transfer pipeline	Ammonia leak from transfer pipeline	Yes	Yes	No	No	Partially Yes	Three major events reported in the last 10 years
13	Ammonia	LOPC from ammonia storage and/or Ammonia transfer	Ammonia leak in storage vessels and/or Ammonia transfer (tanks to vessel and trucks)	No	Yes	No	No	Partially Yes	Many incidents reported every 5 years
14	Ammonia	LOPC from Ammonia refrigeration	Ammonia refrigeration systems leaks	Yes	Yes	No	Yes	No	Many events every year
15	Urea	PSV open/lift in urea reactor	PSV open/lift in urea reactor	No	No	No	No	Partially Yes	Few events reported in the last 10 years
16	Urea	PSV open (ammonia system)	PSV open (ammonia system)	No	No	No	No	Partially Yes	Few events reported in the last 10 years
17	Urea	Urea HP Ammonia & carbamate pump system leak	Urea HP pump system leak	Yes	Yes	Yes	Yes	Partially Yes	Few events reported in the last 10 years
18	Urea	Urea Stripper tube failure	Stripper tube failure	Yes	Yes	Yes	Yes	Partially Yes	Few events reported in the last 10 years
19	Urea	Urea synthesis leaks flanges connections	Leaks in high-pressure flanges connections in Urea synthesis section	Yes	Yes	Partially Yes	Yes	Partially Yes	Many incidents reported every 5 years
20	Urea	Radiation source exposure	Radiation source exposure	No	Yes	No	NA	No	Few events reported in the last 20 years
21	Flare	Vent/flare header contains liquid / blockage		No	Yes	No	Yes	No	Few events reported every 5 years

Figure 1. Brainstorming on MAH

In the Safeguarding benchmarking study, gaps were observed against the MAH scenarios since safety critical systems are not built with the same features due to different technologies and age of the plants. For example, from all 5 urea plants, only 2 have automated urea synthesis high pressure trips as per the original design. Also, Integrity Operating Window limits and Emergency response documents do not include all the MAH scenarios.

The scenarios considered during the study had different risk rankings due to difference in risk perception, teams experience & expertise and the risk matrix, used in each facility, as shown in Figure 2.

S:No	Hazardous scenarios	Site A	Site B	Site C
1	Loss of containment of natural gas due to piping leaks, etc .	4B	4B	4B
2	Top/ bottom reformer fire	5E	4B	4B
3	Fire box explosion	6F	4C	3C
4	Methanation runaway reaction	5C	3B	2C
5	Loss of containment at syngas compressor piping, seals flanges	4F	4H	4H
6	Compressor house explosion	3D	4B	1C
7	Start-up heater fire/explosion/LOPC	5D	4C	3B
8	Loss of containment of ammonia from a flare, open drains or vessels	3F	2B	3C
9	Loss of containment Ammonia transfer pipeline	6C	4B	4A
10	Loss of containment in ammonia storage and/or Ammonia transfer (tanks to vessel and trucks)	6E	4B	4A
11	Loss of containment Ammonia refrigeration	4D	4C	4B
12	PSV open/lift in urea reactor	INA	4C	2C
13	Urea HP pump system leak	6E	4C	4B
14	Stripper tube failure	5F	4C	3C
15	Radiation source exposure	4B	No	No
16	Flare header contains liquid / blockage leading to excessive back pressure	INA	INA	INA
17	Ammonia converter catastrophic failure (insurance case)	INA	INA	INA

Figure 2. Initial list of MAH Scenarios

65% of the MAH scenarios are considered as a high risk in PHA study by Site A, however Site B and C have considered it in low risk and, detailing the associated consequences and effects of MAH were different. This is primarily due to the level of awareness at the sites. During the comprehensive review of the PHA study, it was observed that some credible scenarios involving asset damage, such as the ammonia converter failure and Liquid ammonia ingress to the flare header were not considered. Additionally, linking barriers to these scenarios was deemed inappropriate (e.g. Ammonia tank pressure safety valve considered for liquid overflow protection).

Several factors influence the effectiveness of HAZOP risk rating:

Culture that Design is Correct: There is often a prevailing belief that the initial design is flawless, which can hinder the identification of potential hazards. Encouraging a culture of continuous improvement and critical evaluation of designs is essential.

Experience: The experience level of the team conducting the HAZOP can significantly impact the accuracy of risk ratings. Ensuring that the team comprises individuals with diverse and extensive experience in process safety is vital.

Right People Around the Table: Having the right mix of expertise and perspectives during HAZOP sessions is crucial. This includes representatives from design, operations, maintenance, and safety.

Risk Awareness and Appetite: Understanding the organization's risk tolerance and ensuring that it is reflected in the HAZOP process is important. This involves setting clear guidelines on acceptable risk levels, a risk matrix to be used during risk determination and ensuring that all team members are aware of these guidelines

Categorization and definitions of typical MAHs scenarios

The MAH list was categorized into three main categories: Toxic Release, Fire & Explosion and Mechanical Integrity, as summarized in Figure 3. These scenarios highlight critical areas of concern in industrial safety, emphasizing the importance of maintaining containment, preventing leaks, and ensuring the integrity of mechanical systems to avoid catastrophic failures.

Event	Scenario
Toxic release	Loss of Primary Containment of Ammonia from storage vessels
	Loss of Primary Containment of Ammonia during loading operation
	Loss of Primary Containment of Ammonia from transfer pipelines
	Ammonia refrigeration systems leaks
	Loss of Primary Containment of liquid ammonia from open vent/flare
	High pressure in Urea HP scrubber due to high hydrogen
	Urea Stripper tube failure
	Urea HP pump system leak
Fire and Explosion	Loss of Primary Containment of Syngas from compressor piping, seals flanges
	Loss of Primary Containment of Natural gas in the Primary/Secondary reformer tubes
	Run-away reaction in Methanator
	Loss of Primary Containment from flange connections due to Urea High-Pressure Synthesis Loop Failure
	Loss of Primary Containment of Natural gas from piping, flange joints failures
	Loss of Primary Containment of Process gas from piping, flange joints
	Fire box explosion
Mechanical integrity	Catastrophic failure of Ammonia converter
	Compressor house explosion, Boiler Explosion
	Catastrophic failure of compressor/ turbine (Syn gas)

Figure 3. Final list of MAH scenarios

Loss of Primary Containment (LOPC) events involving ammonia and urea present significant major accident hazards due to their potential for toxic releases, vapor cloud explosions, and jet fires. These events can occur from various sources, such as storage vessels, transfer pipelines, refrigeration systems, and high-pressure equipment.

Carbamate leaks, particularly in urea plants, pose significant risks due to the highly corrosive nature of ammonium carbamate. When leaks occur, they can lead to the rapid deterioration of carbon steel equipment, causing severe structural damage. The release of carbamate can result in toxic ammonia exposure, which is hazardous to personnel and the environment. Additionally, the high pressure and temperature conditions in urea plants aggravate the risks, potentially leading to jet fires, vapor cloud explosions, and significant business impacts.

Toxic Release

Scenarios considered for Toxic Release include, Hot ammonia vapor entering storage tanks, which can cause overpressure, leading to tank failure and/or ammonia release, posing risks of toxic exposure and offsite impacts. Similarly, the presence of oxygen air infiltration during loading operations can lead to stress corrosion cracking and tank failure. Corrosion, gasket failures, and mechanical defects in transfer pipelines and refrigeration systems can result in ammonia leaks, causing fires and toxic releases. Figure 4 summarizes the details of the Toxic release scenarios in Ammonia plant.

Risk Title	Risk detail	Potential Causes	Consequence description
Toxic release	LOPC of Ammonia from storage vessels	Hot ammonia vapour from ISBL enters the storage tank,	Potential risk of overpressure resulting in tank failure or mechanical failure of tanks resulting in release of ammonia. Potential personnel exposure, offsite impact, significant business impact
	LOPC of Ammonia during loading operation (L.Ammonia @ 22 bar/319 psi, -33 deg C)	Oxygen infiltration into the ammonia storage tanks leads to stress corrosion cracking and tank failure	Potential risk of overpressure resulting in tank failure or mechanical failure of tanks resulting in release of ammonia. Potential personnel exposure, offsite impact, significant business impact
	LOPC of Ammonia from transfer pipeline (L. Ammonia @ 22 bar/319 psi, -33 deg C)	Corrosion, gasket failure, impact to pipe, valve open or leaking, railcar movement	Liquid ammonia leak from transfer pipeline/ unloading area. Potential risk of personnel exposure, fire, offsite impact
	LOPC of vapour Ammonia from refrigeration system (V. Ammonia @ 13.5 bar/196 psi)	Corrosion, overpressure, surging, mechanical failure, weld defect	Ammonia (g) leak from refrigeration compressor system piping, seals, flanges. Potential risk of fire and associated equipment damage, business impact.
	LOPC of liquid ammonia from open vent/flare	Malfunction of the level control system on the KO drum.	Potential risk of ammonia liquid carry over to flare/stack due to possible flame out. Risk of personnel toxic exposure to liquids released from the flare and significant business impact.

Figure 4. MAH toxic category (Ammonia)

High hydrogen levels in urea HP scrubbers, urea stripper tube failures, and leaks in urea HP pump systems can also lead to severe toxic release incidents. These scenarios highlight the critical need for robust safety measures, regular maintenance, and emergency response plans to mitigate the risks and protect personnel, assets, and the environment. Figure 5 summarizes the details of the Toxic release scenarios in the Urea plant.

Risk Title	Risk detail	Potential Causes	Consequence description
Toxic Release	High pressure in Urea HP scrubber due to high hydrogen	High pressure in scrubber due to high H_2 (increased carryover from upstream NH_3 plant or H_2 removal), insufficient O_2 injection	Potential increased H_2 in system and presence of NH_3. Potential failure at HP scrubber or LP absorber and release of flammable gas. Potential risk of fire, equipment damage and significant business impact.
	Urea Stripper tube failure (Carbamate vapor @ 145 bar/2102 psi)	Mechanical failure, failure of controllers, crystallization	Stripper tube failure resulting in release of urea/ammonia solution
	Urea HP pump system leak (L. Ammonia @ 162 bar/2351 psi)	Blocked discharge, corrosion, flange leaks	Urea HP pump system leak resulting in release of carbamate. Potential risk of fire and toxic atmosphere and associated equipment damage, business impact

Figure 5. MAH toxic category (Urea)

Fire & Explosion

The scenarios involving syngas, natural gas, and urea highlight significant fire and explosion risks. A syngas leak from compressor piping and seals at high pressure, corrosion, overpressure, surging, mechanical failure, or weld defects, leading to jet fires, with potential consequences of injuries, equipment damage, and substantial business impacts. Natural gas leaks in primary or secondary reformer tubes can occur due to overheating, fatigue, overpressure, mechanical failure, or weld defects, posing risks of internal fires and equipment damage. Figure 6 summarizes the details of the fire scenarios in Ammonia/Urea plant.

Risk Title	Risk detail	Potential Causes	Consequence description
Jet Fire	Syngas leak from compressor piping, seals flanges (Syngas @ 195 bar/2827 psi)	Corrosion, overpressure, surging, mechanical failure, weld defect	Syngas leak from compressor piping, seals flanges. Potential risk of fire and associated equipment damage, business impact
	Natural gas leak in the Primary/Secondary reformer tubes (Natural Gas @ 5 bar/73 psi)	Overheating, fatigue, overpressure, mechanical failure, weld defect	Natural gas leak in the Primary/Syngas leak in Secondary reformer tubes. Potential risk of fire inside the box. Also, risk of equipment damage and business impact
	Run-away reaction in Methanator (Syngas @ 28 bar/406 psi)	Excessive exothermic reaction due to incorrect composition from upstream upset	Rapid increase of temperature in the Methanator resulting in run-away reaction. Potential risk of vessel weak joint failure and release of flammable gas possible fire and associated equipment damage and business impact.
	Leaks in high-pressure flanges connections in Urea synthesis section (Ammonia/Carbamate @ 140 bar/2031 psi)	Corrosion, overpressure, mechanical failure, weld defect, controller malfunction, operator error	Leaks in high-pressure flanges connections in Urea synthesis section resulting in release of carbamate. Potential risk of fire, toxic atmosphere and associated equipment damage, business impact

Figure 6. MAH Fire category

Run-away reactions in Methanator, triggered by excessive exothermic reactions due to incorrect upstream gas composition, can lead to rapid temperature increases, vessel failures, and the release of flammable gases, resulting in jet fires and vapor cloud explosions. Figure 7 summarizes the details of the Explosion scenarios in Ammonia/Urea plant

Risk Title	Risk detail	Potential Causes	Consequence description
Explosion	Natural gas leak from piping, flange joints	Corrosion, impact, gasket leak, valve passing or left open, thermal over-pressure, mechanical failure, weld defect.	Natural gas leak from piping/ flange joints. Potential risk of fire / explosion, asset damage and business impact
	Process gas due to piping, flange joints	Corrosion, gasket leak, mechanical failure, weld defect, vibration, Flow control malfunction, Tubing leak / failure in reboilers	Process gas leak due to piping, flange joints and/or tubing failures in Shift converter & CO_2 removal section. Potential risk of fire, asset damage and business impact
	Fire box explosion	Flameout (during startups), operator error	Possible flameout leading to accumulation of gas inside the box and potential delayed ignition causing vapor cloud explosion and damage to reformer. Also, risk of fire propagate outside of the reformer wall or outside the bottom. Risk of multiple injuries, equipment damage and significant business impact.

Figure 7. MAH Explosion category

Leaks in high-pressure flange connections in the urea synthesis section, handling ammonia/carbamate at high pressure can be triggered by corrosion, overpressure, mechanical failure, weld defects, controller malfunctions, or operator errors, leading to the release of toxic carbamate, fires, and toxic clouds. These events can result in injuries to personnel, asset damage, production loss, environmental impact, and reputational damage.

Mechanical failure/degradation

From a mechanical failure or degradation perspective, several major accident hazards can arise in industrial settings. In the ammonia plant, a syngas leak from compressor piping and seals at high pressure (approximately 195 bar/2827 psi) can result from corrosion, overpressure, surging, mechanical failure, or welding defects. This can result in potential compressor house explosions due to a vapor cloud explosion (VCE), causing fires, equipment damage, and injuries due to projectiles. Similarly, a boiler explosion can occur due to blocked outlets or flameouts, leading to gas accumulation and delayed ignition, resulting in VCEs and injuries from projectiles. Figure 8 summarizes the details of the Mechanical failure scenarios in Ammonia plant.

Risk Title	Risk detail	Potential Causes	Consequence description
Mechanical failure/ Projectiles	Compressor house explosion (Syngas @ 195 bar/2827 psi)	Corrosion, overpressure, surging, mechanical failure, weld defect	Syngas leak from compressor piping, seals flanges. Potential risk of fire & explosion, injuries due to projectiles, associated equipment damage, business impact
	Boiler explosion	Blocked outlet, flameout	Potential flame out resulting in accumulation of gases. Potential risk of delayed ignition resulting in VCE and subsequent injuries due to projectiles, equipment damage and significant business impact.
	Compressor/turbine catastrophic failure		The release of high-pressure syn gas, which is a mixture of hydrogen, nitrogen, and other gases, can create an explosive atmosphere. Any ignition source can lead to a potential catastrophic explosion or fire. Potential risk of injuries due to projectiles, and possible ammonia and other toxic gases release, associated equipment damage, business impact.
	Ammonia Converter catastrophic failure		The ammonia converter operates under high pressure and temperature conditions, a failure can lead to the release of flammable gases, creating an explosive atmosphere. Potential risk of injuries due to projectiles, and possible ammonia and other toxic gases release, associated equipment damage, business impact.

Figure 8. MAH Mechanical failure category

Compressor or turbine catastrophic failures, often due to the release of high-pressure syngas, can create explosive atmospheres. Any ignition source can trigger a catastrophic explosion or fire, releasing ammonia and other toxic gases, causing significant equipment damage and business impact. Additionally, ammonia converter failures, operating under high pressure and temperature, can release flammable gases, leading to VCEs and injuries from projectiles.

These scenarios highlight the critical need for robust maintenance, regular inspections, and effective safety measures to mitigate the risks associated with mechanical failures and degradation in industrial environments.

Outcome of the study

Following the completion of the PHA studies, a Process safety program was adjusted, with focus on,

- Identify the typical Major Accident Hazards (MAH) of all the ammonia and urea plants
- Define critical safety barriers for each scenario (ongoing)
- Establish the basic requirements for each site's Major Accident Hazard- Prevention Plan (MAHPP)

The definition of Major Accident Hazards (MAH) varies significantly across different sites, leading to inconsistencies in risk management.

In one assumption the MAH is considered only the disastrous events. For example, public or more than 10 fatalities within the organization. In other approach, one fatality is already critical and considered a MAH.

Each site has a different risk matrix (Figure 9) which resulted in varying levels of preparedness and response to potential hazards.

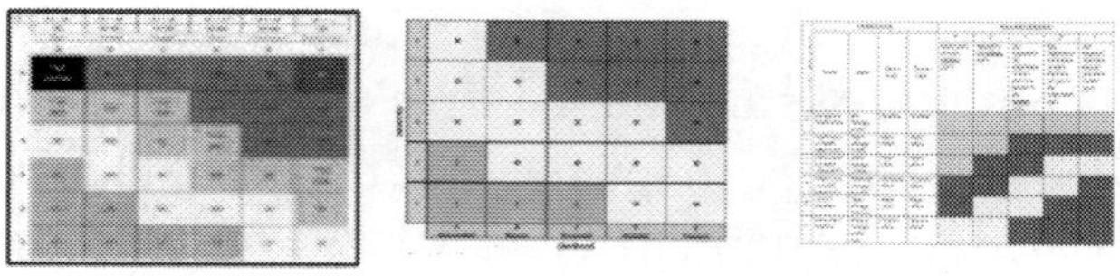

Figure 9: Risk Matrix used in the PHA reviews

To ensure a unified understanding, the following definition has been adopted for MAH.

A major accident is an 'Occurrence' such as an emission, fire or explosion resulting from uncontrolled developments during operation leading to major, catastrophic or disastrous consequences irrespective of their likelihood and serious consequences affecting people, assets, the environment and/or company reputation. The consequences may be immediate or delayed. Consequences affecting people includes, includes any fatality or permanent impairment.

Standardizing the risk matrix across all sites is crucial to ensure a uniform approach to identifying and mitigating MAHs. The new Risk matrix is demonstrated in the Figure 10, includes in the marked box the events classified as MAH.

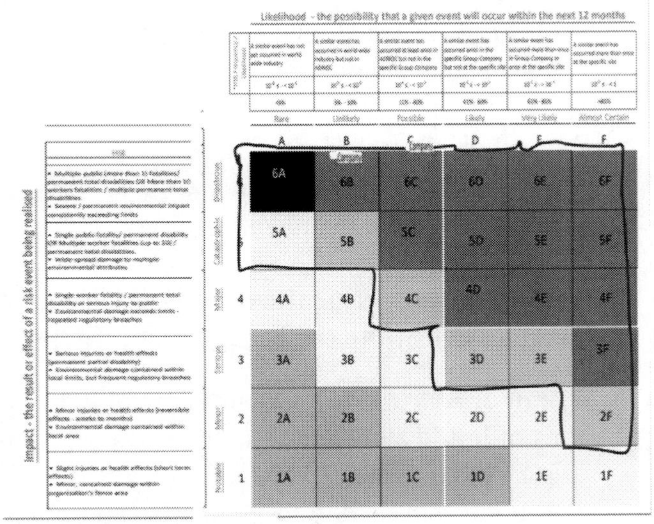

Figure 10: Unified Risk Matrix

During the MAH workshop, a review was conducted of ammonia incident papers presented at previous Ammonia Safety Symposiums. The objective was to assess MAH-related incidents within the ammonia and urea industry. Based on this review, the scenarios listed in Table 1 represent the minimum set to be considered across all production facilities.

Also, in line with the MAH definition, which includes fatalities and permanent impairments, the scope of MAH has been expanded to incorporate the 10 Life-Saving Rules and health hazards, as outlined in Table 2.

Category	Scenarios (Process related)
Fire & Explosions from Natural Gas, high rich hydrogen process streams.	Piping (leaks/Fire): Flammable gas leak/fire from flange joints and other fittings,
	Reformer (leaks/Fire): Process gas tube; Top flange, pigtails, transfer line and other fittings
	Reformer (explosion): Fire box internal explosion during start-up/ shutdown
	Waste heat Boiler (Leak/Fire)
	Syngas compressor (Leak/fire): Piping, seals flanges, oil and flammable materials
	Compressor house (Explosion)
	Start-up heater process gas (Leak /Fire- Explosion)
	Runaway reaction in Methanator (Leak/Fire)
	Synthesis loop, Major leaks,
	Boiler (Explosion)
Mechanical integrity failure	Turbine/ compressor catastrophic mechanical failure
	High pressure steam pipeline leak
Toxic release from Ammonia plant	Ammonia tank (overflow) due to high level
	Flare (overflow): Liquid Ammonia carry over
	Pipeline (Leak): Release in the public areas
	Vessel/jetty ammonia (Leak)
	Truck loading/ transportation (Leak)
	Ammonia refrigeration systems (Leak)
	Release from PSV (overpressure): Ammonia supply system
Ammonia/ ammonia carbamate from Urea plant	Release from Urea reactor PSV (overpressure)
	Urea HP Ammonia & carbamate pump system (leak)
	Urea Stripper (Leak): tube failure
	Synthesis high-pressures (Leaks): Flanges, Equipment connections
	Flare header/stack leaks/ carryover of liquid / blockage

Table 1: List of Process related MAH

Category	Scenario (Non-Process)
Operations/ Occupational safety	Working at height (Fall)
	Driving (Collisions)
	Work authorization (Injuries)
	Confined space (Intoxication)
	Energy isolation (Intoxication, Burns, Fires, Mechanical)
	By passing safety controls (Injuries, Leaks, Damage)
	Mechanical lift (Injuries)
	Toxic gas (Intoxications)
	Line of fire (Injuries)
	Hot work and ignition sources (Injuries, Fires, Explosions)
Health	Handling radiations sources (Injuries, Long term health issues)
	Heart attack, Suicide, Heat stress

Table 2: List of Non-Process MAH

Elaboration of the Major Accident Hazard Prevention Plan (MAHPP)

Process	Guide word	Deviation	Cause	Conse-quence	Safeguards	Recom-mendations
Scenario				Severity	Safety Barrier/Preventive measures	Risk
Major Accident Hazard				Fatalities, Permanent impairment and High Severity events	Operational controls Mechanical integrity Emergency response	Risk-based improvements

Figure 11- From PHA to MAH- PP

MAHPP for each scenario uses the extract of the HAZOP/ PHA, as demonstrated in figure 11. The development of the MAHPP could be summarized into these elements:

Hazard Identification - Conduct a thorough risk assessment to identify potential hazards. Use techniques like Hazard and Operability Study (HAZOP), Failure Mode and Effects Analysis (FMEA), and What-If Analysis.

Risk Assessment - Evaluate the likelihood and potential impact of identified hazards. Prioritize risks based on their severity and probability.

Preventive Measures/ Safety barriers - Implement engineering controls such as safety valves, pressure relief systems, and containment structures. Establish administrative controls including standard operating procedures (SOPs), regular maintenance schedules, and safety training programs.

Monitoring and Detection - Install monitoring systems to detect leaks, pressure changes, and other anomalies. Use alarms and automated shutdown systems to respond quickly to potential incidents.

Emergency Response Planning - Develop and document emergency response plans. Conduct regular drills and training sessions for employees to ensure preparedness.

Communication and Training - Ensure all employees are aware of the hazards and trained in emergency procedures. Foster a safety culture where employees feel responsible for reporting hazards and near-misses.

Review and Update - Regularly review and update risk assessments and safety measures. Learn from past incidents and industry reports, including those shared during ammonia safety symposiums, to avoid repeating the same mistakes.

Regulatory Compliance - Ensure compliance with local, national, and international safety regulations and standards.

Scenario: Ammonia Leak – Toxic release

One of the most critical scenarios is a large Ammonia release, is from the large Storage Tanks"

The atmospheric storage tank containing several tons of liquid ammonia, stored at –33 deg C ./-27.4 deg F. In case of release, ammonia rapidly vaporizes upon release due to its low boiling point, creating a dense, cold vapor cloud. The vapor cloud can disperse downwind, influenced by the prevailing weather conditions. Ammonia is highly toxic. Exposure to high concentrations can cause serious health effects (Acute Exposure Guideline Levels, AEGL 2 is 160 ppm for an hour), and can be life-threatening (Immediately Dangerous to Life or Health, IDLH is 300 ppm)

Modes of releases could be continuous release (leaks from pressurized & atmospheric tanks, pipes, hose), instantaneous release (catastrophic tank, pipe, hose rupture), short duration and time-varying release

Existing barriers are basic process controls, interlocks, pressure relief devices, gas detectors around the facility trigger alarms, escape mask, firefighting systems and emergency response.

During the barrier review, gaps identified related to operation control procedures, tank bund walls, emergency response, crisis management and shelter-in place. Figure 12 presents a summary of the key barriers to ammonia release from storage tanks. The barriers highlighted in red indicate areas that require improvement.

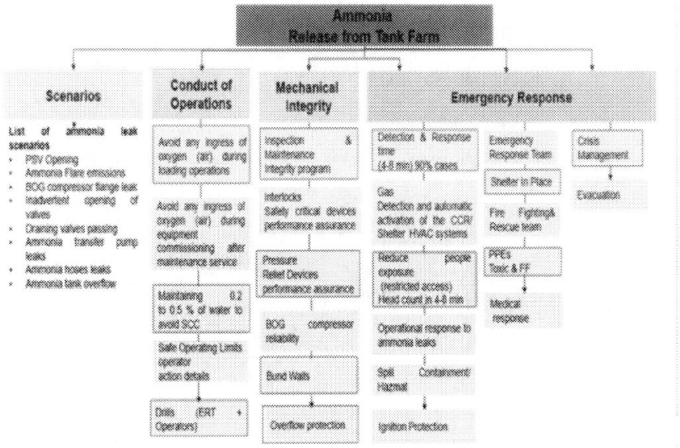

Figure 12. Key Barriers for Ammonia release from storage Tanks

A barrier assessment checklist has been developed to do a complete assessment across sites for the LOPC (Loss of Primary Containment) of Ammonia from Storage Tank scenario and plan to be developed to address the key barrier gaps with prioritization. Figure 13 shows the Barrier Assessment checklist.

The major outcome of the study is the identification of Major Accident Hazards identified for the company, with clear definition by reviewing the existing and external information and detailing effective barriers.

The next step will be to do barrier assessment and implementation of gaps with a definite plan to ensure safety and effectiveness of the organization.

MAH - Ammonia storage tank barrier assessment

S:No	Barriers	Implemented /Available	Relevant Details	Not Available	Relevant information	Remarks
1	**Avoid any ingress of oxygen (air) in the ammonia storage tank**					
i	Is the relevant procedure includes the requirement?					
ii	Is this hazard is addressed in the HAZOP study?					
2	**Avoid any ingress of oxygen (air) in the ammonia storage tank during equipment commissioning after maintenance service**					
i	Is the relevant procedure includes the requirement?					
3	**Safe Operating Limits**					
a	**Maintaining 0.2 to 0.5 % of water in the anhydrous ammonia**					
i	Is the relevant procedure includes the requirement?					
ii	Are all the storage tanks are maintained as per the above defined limits?					
b	**Are all the ammonia storage tanks have the operating limits for Level control**					
i	Is the relevant procedure includes the requirement? Share details					
c	**Are all the ammonia storage tanks have the operating limits for Pressure control**					
i	Is the relevant procedure includes the requirement? Share details					
4	**Drills (ERT + Operators)**					
i	Is there a document plan for leak from ammonia storage tank?					
ii	Are operators trained on the emergency Response for this scenario?					
5	**HSECES Performance Assurance**					
i	Are the storage tanks associated instrumentation/equipments listed below are included in HSECES?					
ii	Is the performance Assurance performed for listed below HSECES?					
	1. Pressure Relief Devices					
	2. BOG compressor					
	3. Gas Detection and automatic activation of the CCR/ Shelter HVAC systems					
	4. Interlocks (Pressure, level)					
	5. Pressure Control Valve					
	7. ESD Push button					
6	**Bund Walls**					
i	Does the storage tanks have bund walls?					
ii	Bund walls preventing the liquid ammonia from spreading during					
7	**Overflow protection**					
i	Do the storage tanks have overflow protection?					
8	**Detection & Response time**					
i	Is there ammonia sensors installed near to the tank?					
ii	Is the restricted access practiced near tank area?					
9	**Spill Containment/ Hazmat**					
i	Is there a spill containment methodology/process defined?					
ii	Are trained people available to contain ammonia leaks?					
iii	What is the disposal process defined? Is this practiced at site?					
10	**Ignition Protection**					
i	What are the ignition protection available? (e.g, lightning protection, static electricity control, flame arrestors, ignition source control etc..)					
11	**Emergency Response**					
i	Does the Emergency Response procedure includes Ammonia storage tank scenario?					
ii	Are operators trained on the emergency Response for this scenario?					
12	**Shelter In Place**					
i	Does the site have Shelter In Place?					
ii	Are the gas detectors installed at upstream and downstream of fresh air intake?					
iii	Is there a activation of red/green light to indicate the status? Manual or automatic?					
iv	Does the auto cut-off of ventilation system exists?					
v	Are the doors are adequately sealed to maintain positive pressure and part of routine inspection program?					
vi	Is the Breathing Air bank /supply available?					
vii	Are there alarm & communication arrangements available?					
viii	Is the communication and Emergency lighting has power back up?					
13	**PPEs for Toxic & FF**					
i	What are PPEs in place to handle ammonia leaks/during exposure?					
14	**Medical response**					
i	Are the Medical team trained to handle ammonia leak scenario?					
15	**Crisis Management**					
i	Are table top exercise conducted for ammonia tank leak scenario?					
16	**Evacuation**					
i	Is the Evacuation process defined in the procedure?					

Figure 13. Barrier Assessment checklist

Conclusion

Defining the MAH scenarios is a key activity to ensure emergency preparedness for overcoming unlikely events.

The participation of the employees in the development of the PHA and its integration into the company management system is a key element to ensure the success of any company approach towards prevention of MAH.

The study of the ammonia release from the storage tank has been identified as one of the critical scenarios, and the detailed study has provided critical insights into the potential hazards and the effectiveness of the existing safety measures. The release scenario demonstrated the significant risk posed by the failure of containment systems, leading to the rapid dispersion of ammonia vapor.

The study underscored the importance of operational control to avoid air ingress and maintaining 0.2 to 0.5% of water content to avoid stress corrosion cracking (SCC). Also, regular maintenance and inspection of storage tanks and associated valves to detect and address potential failures early.

Emergency response measures, including detection, alarm systems, and containment strategies, were evaluated. While effective in mitigating immediate risks, areas for improvement were identified, particularly in the speed of response and communication protocols.

Enhanced training for personnel on emergency response procedures was also recommended.

In conclusion, the effectiveness of each safety barrier and fundamental to avoid the development of the identified scenarios. While the designed safety measures provide a foundational level of protection, the implementation of the recommended actions will significantly enhance the facility's ability to prevent and respond to major hazard scenarios, thereby safeguarding both human health and the environment.

References

1) Occupational Safety and Health Administration (OSHA) PSM standard (29 CFR 1910.119)
2) DIRECTIVE 2012/18/EU (Seveso-III Directive) OF THE EUROPEAN PARLIAMENT AND OF THE COUNCIL
3) Management of Safety Critical Elements as a Base for Risk Management of Major Accident Hazards, Mariana Bahadian Bardy, Flavio L B Diniz, Joo Paulo Oliveira, Paula Silveira, AIChE Spring Meeting and Global Congress on Process Safety
4) Compendium of chemical hazards: Ammonia EGL, UK HAS
5) US Environmental Protection Agency (EPA) Acute Exposure Guideline Levels' (viewed 27 September 2024)
6) National Institute for Occupational Safety and Health
7) A comprehensive review of hazards identification and risk assessment in ammonia plants, Dan Cojocaru, Fertilizer Industrial Services Ltd, United Kingdom, Venkat Pattabathula, SVP Chemical Plant Services, Australia
8) The Lessons Learned from the First 100 Safety Hazards in Urea Plants, Prem Baboo, National Fertilizers Limited, Mark Brouwer, UreaKnowHow.com B.V., Jo Eijkenboom, UreaKnowHow.com B.V., Majid Mohammadian, OCI N.V.

Analyzing Ammonia Incidents Through the Lens of Process Safety

This paper describes AIChE's first ammonia process safety book that uses the risk-based process safety approach developed by the AIChE Center for Chemical Process Safety (CCPS). Selected ammonia incidents, presented at previous AIChE ammonia safety symposia, are analyzed using the method described in the book "CCPS Guidelines for 'Risk Based Process Safety' (RBPS)." The ammonia incident process safety book is scheduled for a 4Q 2025 release.

Bruce K. Vaughen
AIChE, CCPS, USA

Venkat Pattabathula
SVP Chemical Plant Services, Australia

Jim Richardson
Catalyst and Chemicals, USA

Introduction

It has been 70 years since AIChE started the Ammonia Safety Symposium in 1956, which is now the longest-running symposium in the chemical industry's history. With almost 1,000 ammonia plants producing nearly 200 million tons per year in 2025, ammonia is one of the most widely produced chemicals in the world. Ammonia is critical for food production; without it, the world could not provide the food needed for its ever-growing population.

Unfortunately, the global ammonia industry has experienced incidents that have caused fatalities, injuries, environmental harm, property damage, and business interruption. In addition, the history of the ammonia industry is peppered with many incidents that have been repeated after a lapse of a few years. People move on, the experiences move with them, and the lessons are forgotten. To help keep the memories alive for people new to the ammonia industry, selected ammonia incident-related papers presented at AIChE Ammonia Safety Symposia are analyzed.

This ammonia incident process safety book is the first to compile and present ammonia production and storage incident case studies. The incidents occurred during every ammonia-producing stage, including production, conversion, separation, heating, reacting, handling, and storing. Incidents with specific equipment, such as the fired heaters, steam boilers, and rotating equipment, are also discussed. A simplified process flow diagram for the stages in an ammonia process, from introducing the raw materials to storing the ammonia, is shown in Figure 1. These ammonia case studies provide the reader with invaluable learnings that, when applied and shared, can help reduce the likelihood of similar incidents occurring again at their facilities.

The process safety book aims to show what went wrong, describe the engineering solution at the time of the incident, and discuss the management systems that, had they been in place or rigorously applied, could have helped reduce the likelihood and consequences of the incident. Although many of the ammonia incident case studies did not result in severe consequences, such as fatalities, serious injuries, grave environmental harm, or acute property damage, they may have had the potential to do so.

Process safety management systems

The American Institute of Chemical Engineers (AIChE) started its Ammonia Safety Symposium in 1956, making it the oldest symposium in AIChE's history. In the early 1990s, Process Safety Management Systems (PSMS) were developed industry-wide and have evolved from minimum regulatory compliance systems to today's risk-based management systems. This book describes how the twenty Elements in the Center for Chemical Process Safety (CCPS) "Risk Based Process Safety" (RBPS) management system can be used to describe the root, systemic causes of these incidents. [1]. By developing and applying the CCPS RBPS management principles, a company can establish a consistent, structured management approach for safely designing, operating, maintaining, and sustaining its processes. The twenty Elements are systems developed to help prevent incidents or reduce their likelihood. Each case study concludes with a new systemic, root cause evaluation using the twenty Elements of the RBPS management system (Table 1), [1].

Although these incidents may involve more than one of the Elements, there often is not much additional information from which to work. Thus, the RBPS System review is based on the most likely Elements that, had they existed at the time of the incident, could have helped to prevent the incident, or reduced its likelihood.

Root cause analyses and the RBPS elements

CCPS defines a "Root Cause" as:

> *A fundamental, underlying, **system-related** reason why an incident occurred that identifies a correctable failure in a **management system**. There is typically more than one root cause for every process safety incident [2].* [Words **bolded** for this paper]

The "root cause" as it was first introduced has evolved to "root systemic causes" reflecting the better understanding of how inadequate management systems (or the complete lack of them) is the focus of a formal Root Cause Analysis (RCA).

CCPS defines the RCA as:

> *A formal investigation method that attempts to identify and address the **management system failures** that led to an incident. ... [An RCA] identifies the underlying reasons the event occurred such that workable corrective actions can be selected and implemented to help prevent recurrence of the event (or occurrence of a similar event).* [2] [Words **bolded** for this paper]

Often, especially before the CCPS RBPS guidelines were issued in 2007, the "root causes" of these ammonia incidents were specific engineering and administrative solutions to help prevent an identical incident from occurring again.

Unfortunately, as noted earlier, people move on, the experiences move with them, and the lessons are forgotten. If the systemic root causes are not determined during an incident investigation, new people working with the same, unchanged system will make the same error, achieving the same results—an incident.

The ammonia incidents

The proceedings from the yearly Ammonia Symposium, the *AIChE Ammonia Technical Manuals*, have been published since 1959 [3]. More than 1,800 papers have been published since 1959, with more than 440 incident-related papers identified. As shown in Figure 2, the rolling average of incident-related papers since 1959 is more than six per decade. The ten-chapter process safety book will cover 89 of these ammonia incidents. The current titles of the ammonia process safety book's chapters are listed in Table 2.

The most significant incident for the ammonia industry was a toxic release with fatalities that occurred in 1992 in Dakar, Senegal [4]. Although Dakar had over 120 fatalities and over 1,100 injuries, other significant process safety-related incidents across all industries have played an important role in establishing the Process Safety Management Systems of today.

A significant incident in the global chemical industry includes a tragic toxic gas release in Bhopal, India, at the end of 1984 [5], [6]. Another significant incident, an explosion with fatalities, occurred early in 2005 at a refinery in Texas City, Texas. [7], [5]. These significant incidents are shown on the timeline in Figure 3. The average number of incidents discussed in the next four years after Bhopal and Texas City was ten papers per year.

Highlights from an ammonia incident

This section presents the highlights from an ammonia incident discussed in the ammonia process safety book: An explosion in the primary reformer fire box.

For consistency, each incident is presented in the ammonia process safety book using the following format:

1. Title of ammonia process safety incident
   a. Incident summary
   b. What happened: The sequence of events
   c. Why it happened
   d. Investigation action items
   e. Lessons learned using an RBPS system analysis

Each "Lessons learned" section, Section e, contains the RBPS Elements, which, had they existed at the time, could have helped prevent the incident from occurring.

Ammonia Incident: Explosion in the primary reformer firebox, 2009, [8]

Summary

A fire box in an ammonia plant's primary reformer exploded, injuring two operators. Had the explosion caused the rupture of an ammonia pipe on the nearby pipe rack, the consequences could have been worse for people on-site and off-site. For eleven months after the explosion, the damaged, inoperable plant was demolished, and the new one was constructed. Included in the reconstruction was the installation of a new automated Burner Management System (BMS).

What happened: The sequence of events

The initiating event occurred when a severe thunderstorm knocked out the external power supply to the ammonia plant, causing the plant to trip. Since the burner was operated manually, some of the burner valves in the firebox remained open after the trip, introducing a significant amount of fuel into the unlit burners. When an operator attempted to manually re-light the furnace, the firebox exploded.

As shown in Figure 4, the east wall of the reformer was destroyed. Debris was thrown towards the control room, about 25 m (~80 ft) from the furnace. The North rack shown in Figure 5, supporting ammonia and natural gas piping, was deformed, but no releases occurred from the damaged pipelines.

Why did it happen

The investigation team's root cause analysis identified that the plant relied on procedures for starting up and shutting down its furnace. They were relying on administrative controls to manage their furnace's process safety risks. In this incident, the start-up procedure checklist was not followed—the barrier, an administrative control, had failed. The fuel gas entered through the unlit burners, collected in the fire box, ignited, and exploded, causing significant damage to the fire box.

Investigation action items

To prevent a recurrence, the following actions were taken:

1. A new Burner Management System (BMS), an engineering control, was installed and implemented for the start-up lighting procedure of the primary reformer.
2. The reformer furnace hazards identification (HAZID) and hazard and operability (HAZOP) studies were updated to document the updated controls for the furnace start-up.
3. The Start-up Operating Procedure (SOP) was updated with the BMS instructions, and the operators and maintenance personnel were trained on their updated SOPs for the new BMS.

Lessons learned using an RBPS system analysis

Had the following RBPS Elements been implemented, they could have helped prevent the explosion or reduce the damage to the furnace.

1. ***Operational Discipline (Conduct of Operations)***: Had better OD existed, the start-up procedure's checklist would have been followed, detecting the flammable atmosphere inside the firebox before attempting to light the furnacet.
2. ***Process Knowledge Management (PKM):*** An effective review of, and understanding of updated engineering standards, part of a PKM program, may have determined that better burner designs existed (i.e., the BMS).
3. ***Hazards Identification and Risk Analysis (HIRA):*** An effective, updated barrier analysis (through the HIRA program) may have determined that an administrative control (the start-up procedure) was not as robust as an engineering control (the BMS). The "Hierarchy of Controls" methodology applies [9].

Mapping the RBPS systems to each incident

The ammonia incident book concludes with a matrix that maps each of the book's ammonia incidents to a few major systemic RBPS root causes (Table 3). The check marks represent the management systems that did not exist or that were inadequately implemented at the time of the incident.

By presenting this RBPS matrix, the book will provide an invaluable resource for people working in the ammonia industry for relevant incidents that have happened in the past. The management systems in place today exist for a reason.

> *Those who cannot remember the past are condemned to repeat it [10, p. 284]*

Closing Discussion

A goal of the compilation of ammonia incidents is to help keep the memories of ammonia-related incidents alive for people new to the ammonia industry. The incidents occurred during each of the plant's stages, as was shown in the process flow diagram in Figure 1 and listed by chapter in Table 2. People new to the ammonia industry may wonder why a part of the process has been designed as it is. This book may answer the "why" by explaining the "what went wrong," the engineering or administrative solutions selected at that time, and the CCPS RBPS management systems, had they existed and been implemented, that could have helped prevent the incident from occurring.

The authors hope readers understand why the twenty CCPS RBPS Elements exist today. Incidents can occur when management systems are inadequate or missing. Implementing them will help plant personnel understand and reduce their process safety risks.

References

[1] CCPS, Center for Chemical Process Safety Guidelines for "Risk Based Process Safety," Hoboken, NJ USA: John Wiley and Sons, 2007.

[2] CCPS, "CCPS Process Safety Glossary," Center for Chemical Process Safety, 2025. [Online]. Available: www.aiche.org/ccps.

[3] AIChE, "AIChE Ammonia Technical Manuals," The American Institute of Chemical Engineers, The Ammonia Committee, aiche.org, 1959-2024.

[4] S. Dharmavaram and V. Pattabathula, "Learning from the Worst Ammonia Accident," *Chemical Engineering Progress (CEP),* pp. 47-52, July 2023.

[5] CCPS, Incidents That Define Process Safety, Center for Chemical Process Safety, Hoboken, NJ USA: John Wiley and Sons, 2008.

[6] K. Bloch, Rethinking Bhopal: A Definitive Guide to Investigating, Preventing, and Learning from Industrial Disasters, www.sciencedirect.com: IChemE, Elsevier, 2016.

[7] CSB, "Refinery Explosion and Fire, Report No. 2005-04-I-TX," US Chemical Safety and Hazard Investigation Board, Washington, DC USA, 2007.

[8] A. Flamme, S. Van Den Broeck and W. Versteele, "Primary Reformer - Firebox explosion new BMS and HAZOP Actions Overview," in *AIChE Ammonia Technical Manual*, San Francisco, CA , September, 2010.

[9] NIOSH, *Hierarchy of Controls,* cdc.gov/niosh: US National Institute for Occupational Safety and Health, 2025.

[10] G. Santayana, The Life of Reason, Vol. 1, Reason in Common Sense, New York: Charles Scribner's Sons, 1905.

[11] R. Majeed and S. Ali, "Ammonia Converter Reaction Loss followed by Fire Incident," in *AIChE Ammonia Technical Manual*, Denver, CO , 2016.

Number	RBPS Element	RBPS Element Intent
1	Process Safety Culture	The Process Safety Culture Element validates management's commitment to process safety, combining group values and behaviors that determine how to manage it.
2	Compliance with Standards	The Standards Element is the system used to manage the standards, codes, and regulations for the design and maintenance of the process equipment.
3	Process Safety Competency	The Process Safety Competency Element is the system used to ensure that the workers understand and can apply the RBPS Systems essential to their role.
4	Workforce Involvement	The Workforce Involvement Element is a system that ensures the active participation of company and contractor workers in the design and management of the RBPS Systems.
5	Stakeholder Outreach	The Stakeholder Outreach Element establishes the company's commitment to share with the surrounding communities how the facility safely manages its process safety risks.
6	Process Knowledge Management	The Process Safety Knowledge Element is the collection of documents that specify how the process equipment is designed, fabricated, installed, and maintained.
7	Hazard Identification and Risk Analysis	A Hazard Identification and Risk Analysis (HIRA) is used to identify hazards and evaluate whether their risks are tolerable.
8	Operating Procedures	The Operating Procedures Element is the system that ensures the development, review, and use of the facility's operating procedures.
9	Safe Work Practices	The Safe Work Practices Element helps control hazards associated with maintenance and other non-routine work.
10	Asset Integrity and Reliability	The Asset Integrity and Reliability Element ensures that the process equipment is properly designed, fabricated, installed, maintained, and is fit-for-use until it is retired.
11	Contractor Management	The Contractor Management Element is the system that controls the selection, acquisition, use, and monitoring of safe contracted services.
12	Training and Performance Assurance	The Training Element is the system that demonstrates worker proficiency, documenting their performance from the beginning of and throughout their career.
13	Management of Change	The Management of Change Element ensures that changes to a process do not introduce new hazards or increase the risk of the existing hazards.
14	Operational Readiness	The Operational Readiness Element ensures the safe startup of processes and equipment from their commissioning, through their maintenance, and to the end of their useful life.
15	Conduct of Operations	The Conduct of Operations Element reflects the company's process safety culture, providing evidence that the workers deliberately structure and perform their tasks.
16	Emergency Management	The Emergency Management Element is the written Emergency Response Plan (ERP) and defines how everyone inside and outside the fence line responds to an emergency.
17	Incident Investigation	The Incident Investigation Element is the procedure describing the formal steps when investigating incidents, analyzing for trends, and managing its recommendations.
18	Measurement and Metrics	The Measurement and Metrics Element is the process for establishing and managing RBPS performance indicators.
19	Auditing	The Auditing Element is used to evaluate the effectiveness of the RBPS management systems.
20	Management Review and Continuous Improvement	The Management Review and Continuous Improvement Element is the routine management review, tracking, and evaluation of the RBPS System's performance.

Table 1. The intent of each RBPS element

	Chapter
1	Primary and Secondary Reformers
2	Boiler Feed Water Heat Recovery
3	Carbon Monoxide Converter
4	Carbon Dioxide Removal
5	Methanator
6	Ammonia Synthesis and Refrigeration
7	Ammonia Storage
8	Fired Heaters
9	Pumps and Compressors
10	Catalysts

Table 2: List of book chapters

Incident Number	Risk Based Process Safety Element (see CCPS RBPS Figure 12)	Year of incident	Pillar 1: Commit to Process Safety	Process Safety Culture	Compliance with Standards	Process Safety Competency		Emergency Management	Pillar 4: Learn from Experience	Incident Investigation	Measurement and Metrics	Auditing	Management Review and Contin. Imp
	Element Number			1	2	3	...	16		17	18	19	20
1-1	Incident A	2010			√	√							
1-2	Incident B	2009								√			
1-3	Incident C	1998		√				√					
1-4	Incident D	1971								√			√
	etc.												
10-1	Incident AA	1983		√							√		

Table 3: Example part of the matrix showing the significant RBPS elements for each ammonia incident

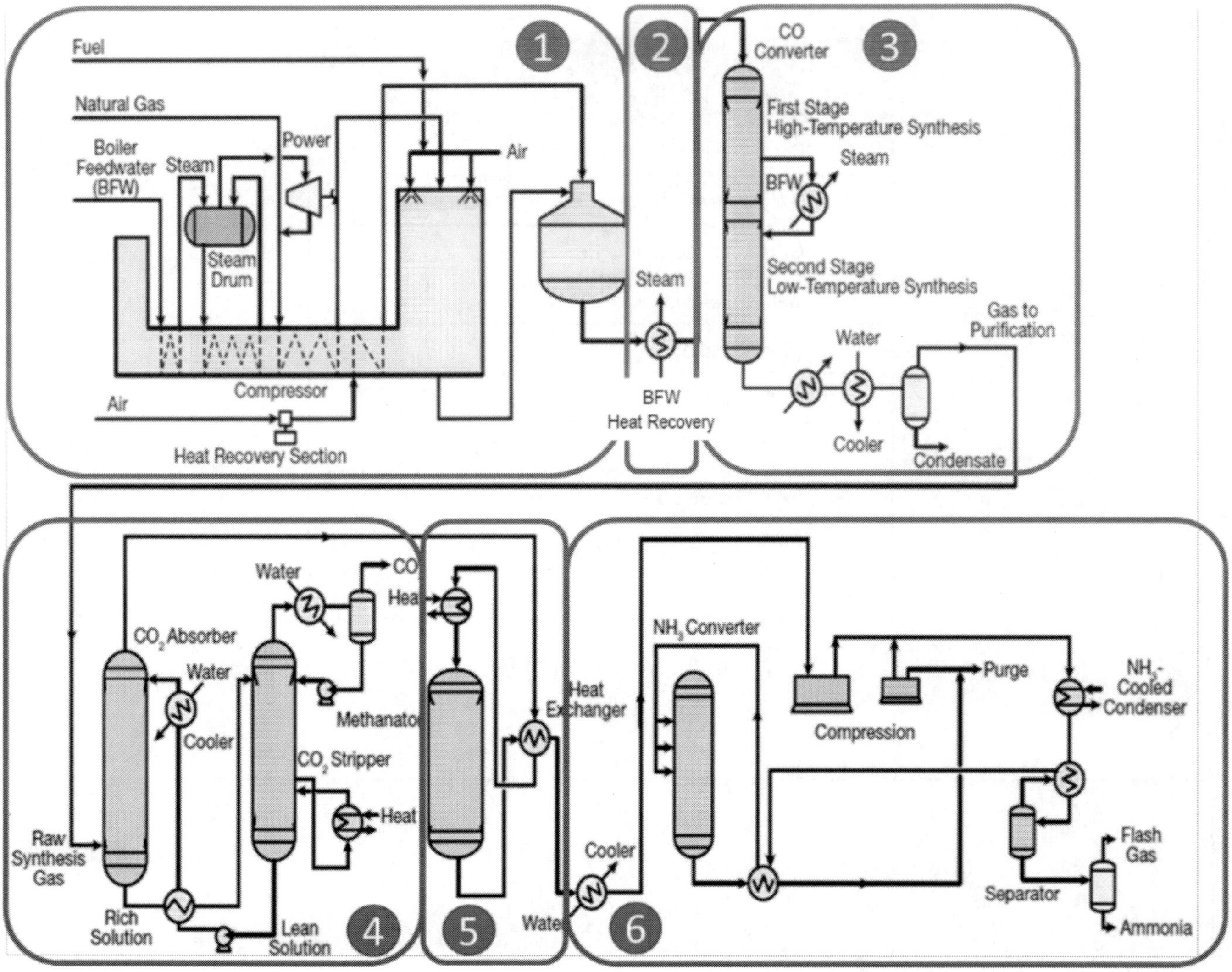

Figure 1: Simplified ammonia process flow diagram

Note for Figure 1**:**
The numbers 1 through 6 represent the chapters in the ammonia incident book (see **Error! Reference source not found.**).

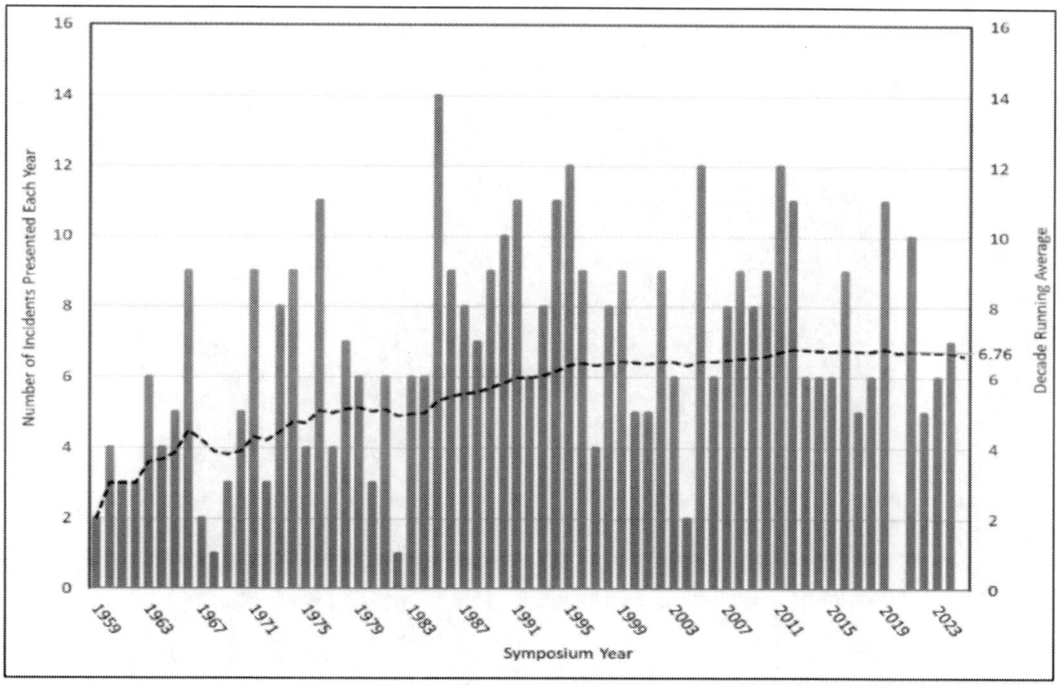

Figure 2: Running average for ammonia incident-related papers

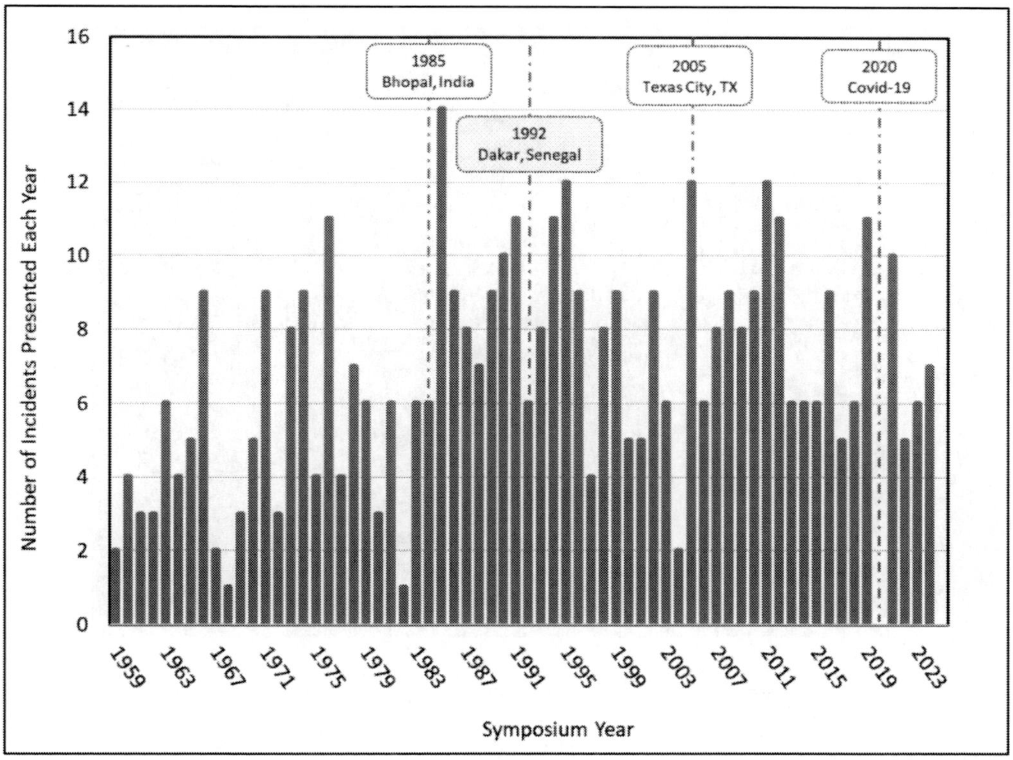

Figure 3: Timeline of significant industrial incidents

Figure 4: Radiant section of the reformer after the explosion

Figure 5: View of the north rack after the explosion

Nitridation Resistance of High-temperature Resistant Cast Alloys – The Effect of Alumina Scales on High-temperature Ammonia Nitridation

Energy vectors, like ammonia as a "hydrogen carrier", appears to be the ideal solution to reduce greenhouse gas emissions in the energy sector. The carrier vector ammonia has to be cracked into its elements N and H in an endothermic process named Ammonia Cracking, which shares similarities with the steam reforming process. Therefore, the application of centrifugally cast tubes of austenitic high temperature resistant Cr-Ni-Fe alloys is highly probable for ammonia cracking plants due to their high corrosion and excellent creep resistance. This study focuses on the suitability of centrifugally cast austenitic Cr-Ni-Fe-Al alloys forming protective Al_2O_3 scales under nitridation in ammonia. The influence of alumina scale properties is also addressed. Metallurgical investigations focus on revealing the degradation mechanisms in gas atmospheres of high nitrogen activity.

Benjamin Aymans[1], Pedro Imizcoz[1], Pablo Cardin[1], Dietlinde Jakobi[1]
[1] Schmidt + Clemens GmbH + Co. KG, Kaiserau 2, D-51789 Lindlar, Germany

Matteo Villa[2,3], Marcel A. J. Somers[2]
[2] Technical University of Denmark, Department of Civil and Mechanical Engineering, Produktionstorvet b. 425, DK-2800 Kgs. Lyngby, Denmark
[3] Department of Mechanics, Mathematics and Management, Polytechnic University of Bari, Via Orabona 4, IT-70125 Bari, Italy

Introduction

Today´s energy sector relies vastly on fossil fuels, leading to large greenhouse gas emissions and largely contributes to global warming. A solution for the transition to low-emission energy supply is hydrogen, widely expected to be the successor of fossil fuels such as LNG and coal. [1] The transition to H_2 requires major changes in infrastructure, its synthesis and on-site supply. While the low-emission synthesis process is widely established and proven by means of water electrolysis, transportation, storage and supply of hydrogen will be a major obstacle.

Due to those hurdles, transport and storage vectors for H_2, such as ammonia (NH_3), are of interest. As ammonia is already widely used in chemical industry, especially in the fertilizer industry, the infrastructure for storage and transportation via vessels and pipelines is well established. Liquefaction can be performed under rather mild conditions ($T_{liq.}$ = -33 °C [-27.4 °F]) and the ammonia molecule does not contain carbon, unlike methanol or LNG. [2] In the literature, reported energy efficiencies for possible liquefied transport vectors are given in Table 1[3]. These showed that transportation of H_2 as a chemically synthesized vector will be largely more efficient than using purely liquefied H_2.

Table 1: Energy efficiency for the offshore supply of possible energy vectors based on their production process. [5]

Origin	NH_3	CH_3OH	H_2	LNG
LNG	62.31 %	65.67 %	47.78 %	87.38 %
Renewable	97.37 %	98.02 %	89.10 %	96.23 %

To regenerate H_2 for energy production ammonia has to be cracked. Ammonia cracking, the reverse reaction of the Haber-Bosch process, is endothermic and can be done catalytically or autothermal (non-catalytic). Ni catalysts are of interest for large-scale plants, as those are already used in conventional Steam Methane Reformers (SMR). Thus, the plant design, technical requirements and regulations are expected to be similar for both processes, leading to the use of similar alloys when manufacturing the reaction tubes. [4] The industrial standard in Steam Methane Reforming is centrifugally cast tubes from austenitic high-temperature Cr-Ni-Fe alloys. These combine a high corrosion resistance with excellent creep resistance. In particular Al-alloyed Cr-Ni-Fe alloys are by far superior to comparable grades of wrought and extruded tubes. [5]

Corrosion mitigation occurs by in-situ formation of oxide scales. Two different alloying concepts are established regarding oxide protection against corrosion. Cr-Ni-Fe alloys rely on formation of a Cr-scale. In the last decades, a new alloy concept of Cr-Ni-Fe-Al alloys, pioneered by Schmidt+Clemens, has been established. This relies on formation of an alumina scale, which dramatically improves the corrosion resistance as compared to Cr-Ni-Fe alloys. [6, 7]

The major advantages of alumina are its thermodynamic and the thermal stability compared to chromia. Al_2O_3 has a higher thermodynamic stability, is therefore formed at lower oxygen partial pressures of the surrounding gas phase (cf. $p(O_2)_{form.}$ (Al_2O_3, 1000 °C [1832 °F]) = 10^{-34} bar [$1{,}45\cdot10^{-33}$ PSI] to $p(O_2)_{form.}$ (Cr_2O_3, 1000 °C [1832 °F]) = 10^{-22} bar [$1{,}45\cdot10^{-21}$ PSI]). Furthermore, chromia scales are prone to volatilization under presence of water traces and temperatures beyond 950 °C [1742 °F], which reduces the protectiveness of such scales in typical applications compared to alumina. [8, 9]

While the tubes typically exhibit carburization or metal dusting during Steam Cracking or Steam Methane Reforming, ammonia cracking leads to a different corrosion mechanism, named nitridation. Nitridation results in the formation of nitrides of alloying elements such as Cr, Fe and Al, comparable to carbide formation (for Cr) by carburization. Nitridation is highly dependent on the N-activity of the gas phase and the temperature.

NH_3 shows a roughly 10^6 times higher nitriding potential than elemental N_2 at similar temperatures. [10] Therefore, uncracked ammonia, likely present at inlet components and upper tube parts can induce major corrosion problems and might compromise the integrity and lifetime of an ammonia cracker.

Temperature changes the mechanism of corrosion upon thermodynamic stabilization or destabilization of different nitrides. Fe-nitride formation for example can be highly critical at lower temperatures. Fe-nitrides can form even at rather low temperatures and will decompose at higher ones generating nitrogen gas. [11] This dissociation can lead to significant pore formation and embrittlement of components, which is considered highly critical for their mechanical stability.

Even though Al also tends to form stable nitrides (AlN), it is reported that a dense oxide scale of alumina can inhibit nitridation. Previous investigations featuring Cr-Ni-Fe alloys suggest that a chromia scale also enhances nitridation resistance. [12,13] Nevertheless, its stability with respect to $p(O_2)$ in the reaction gas and temperature likely decreases its long-term stability compared to an alumina scale. This is reported analogously for carburization in applications such as Steam Cracking. [5]

The present study investigates the behavior of four commercially available centrifugally cast Cr-Ni-Fe-Al alloys under realistic ammonia cracking conditions. This article addresses the comparison of these commercial materials considering oxidized and non-oxidized states. From the comparison of those alloys, compositional trends are established, which influence the resistance of the material against nitridation. Furthermore, the effect of different oxidation procedures was studied and optimized to improve alloy performance in service.

Experimental

Table 2 presents the nominal compositions of the alloys investigated in this study, which were designed to inhibit high-temperature corrosion while having excellent creep resistance. [5]

Rectangular samples from these materials (20 x 10 x 1.5 mm^3) were machined. The machined inner surface of the samples was retained on the sample, to evaluate the influence of a surface condition, typically used in industrial large-scale cracking furnaces.

Table 2: Nominal compositions of the alloys tested in this study.

No.	C	Cr	Ni	Fe	Al	add.
s001	0.45	30	45	bal.	3-5	MAE
s002	0.45	27	60	bal.	2-3	MAE, RE
s003	0.45	27	60	bal.	3-5	MAE, RE
s004	0.45	33	35	bal.	3-5	MAE

MAE: Addition of micro alloying elements which could be for example Ti, Zr, Nb, Ta of rare earth element such as Y.

The samples were preoxidized under controlled conditions to test different oxide layer conditions. A summary of those conditions is shown in Table 3.

Table 3: Experimental conditions of the preoxidation treatment.

	Heating	Cooling-	T [°C]	t [h]	Flow rate (Ar)	T_{steam} [°C]
Exp. 1	50 K/h	25 K/h	1050	12	4 Nl/min	60.4
Exp. 2	50 K/h	100 K/h	1050	12	4 Nl/min	60.4

The preoxidation was performed at 1050 °C [1922 °F] in order to form the thermodynamically most stable polymorph of Al_2O_3 (corundum) on all samples. Phase transformations of the metastable polymorphs into corundum typically requires roughly 1000 °C [1832 °F]. [14] Furthermore, the thickness of the scale is crucial. Oxidation at 1050 °C [1922°F] produces an oxides scale of sufficient thickness for investigation with the applied analysis techniques (see below).

Samples were exposed isothermally for 336 hours at 900 °C [1652 °F] under a continuous gas stream of ammonia (containing < 0.2 ppm H_2O, < 0.1 ppm O_2). After exposure the samples were cooled in the NH_3 until a temperature of 500 °C [932 °F] was reached, further cooling was done in N_2 (g). Thus, additional nitridation can occur during the cooling phase.

With respect to the gas temperature, auto-thermal decomposition of ammonia is anticipated. The equilibrium reaction of the decomposition of ammonia (Equation 1) is thermally activated and shifts towards the formation of molecular N_2 and H_2 at higher temperatures.

$$2\,NH_3\,(g) \rightleftharpoons N_2\,(g) + 3\,H_2\,(g);$$

$$\Delta H^0 = 92\ kJmol^{-1}$$

Equation 1

Literature [10] reports that the static equilibrium at 900 °C is shifted strongly towards the molecular products, so that the used ammonia gas stream rather consists of ammonia, nitrogen and hydrogen. The chosen exposure temperature of 900 °C [1652 °F] can further lead to partial dissociation of molecular N_2 leading to additional nitridation by atomic nitrogen. The respective citation further reports the dissociation mechanism of N_2 at minimum temperatures of 750 to 800 °C [1382 to 1427 °F]. [10]

Nitridation damage progression on the samples was evaluated by various techniques. Mass changes were determined, by measuring the sample masses before and after the exposure. Samples were investigated with backscatter electron imaging (BSE) and energy-dispersive X-ray spectroscopy on cross-sections using a field-emission scanning electron microscope (SEM) (Jeol JSM-7800 F). X-ray diffraction (XRD)

(Bruker D8 diffractometer, λ = 1.5480 Å) was used to identify oxides and nitrides formed at the surface during the experiment. Surface morphology of the sample was investigated by light optical microscopy in top-view as well.

Results

Figure 1 provides an overview of the appearance of the tested surfaces, comparing the metallic and oxidized samples as well as their condition after the exposure to the nitriding atmosphere.
The bare-metallic surfaces show typical machining marks on the inner diameter of a tube. After the exposure experiment, the metallic surface appeared dark grey (s001 to s004) corresponding to nitride formation.

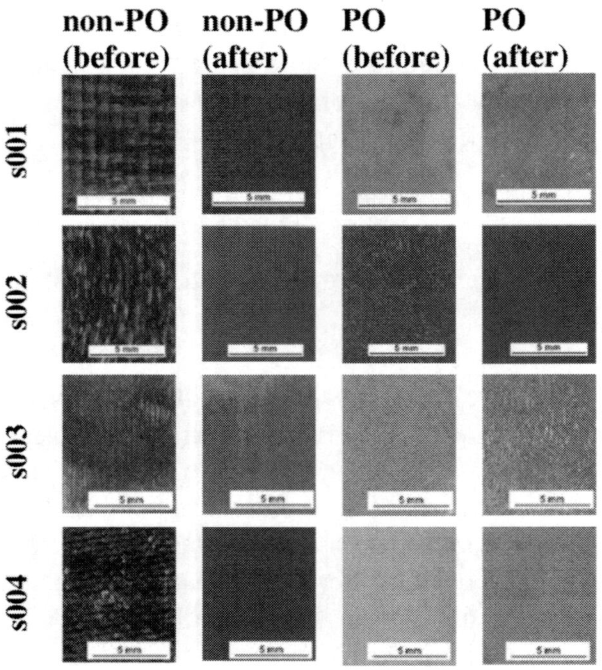

Figure 1: Images of the sample surfaces for the different alloys in their bare-metallic and oxidized state, before and after nitridation.

The preoxidation of the surface led to formation of a light grey oxide scale for samples s001, s003 and s004, having a higher Al-content than for sample s002. After preoxidation, the surface of alloy s002 shows formation of dark-grey and light-grey regions. After the exposure the surfaces of the high Al-containing alloys (s001, s003 and s004) appear slightly darkened, possibly corresponding to slight nitridation, while the surface of alloy s002 (low Al-content) is dark grey, corresponding to severe nitridation. Figure 2 presents the measured weight change for the specimens during the experiment.

The bare-metallic samples show a larger weight gain (average: 8.8 mg) during the experiment compared to the preoxidized ones (average: 1.1 mg). This corresponds to the appearance of the samples observed by light microscopy. Alloy s004 shows a larger weight gain than the remaining alloys (s001 – s003). The most significant difference for this alloy is the reduced content of Cr and Ni (cf. Table 2). If the materials are preoxidized, the weight gain during nitridation is strongly reduced. Alloys s001, s003 and s004 show comparable weight gains of roughly 0.65 mg.

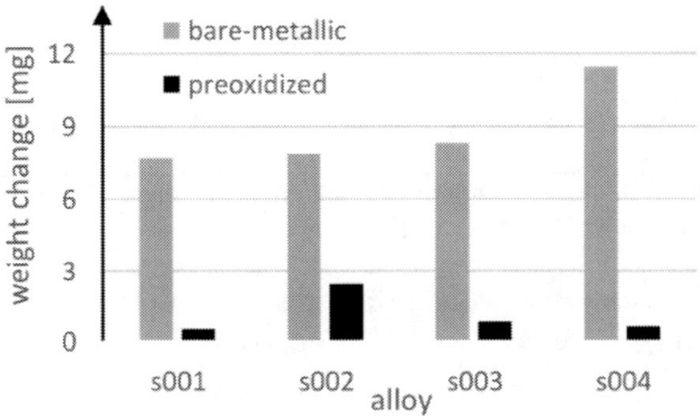

Figure 2: Weight change of the samples after experiment for the bare-metallic (grey) and preoxidized (black) condition.

The weight change of alloy s002 after nitridation of the preoxidized sample is larger (2.4 mg). Such differences might be related to the lower Al-content of s002. Note that s003 has a similar composition, despite some small differences on Fe-content (cf. Table 2).

Alloy s001

Alloy s001 showed the lowest weight change for the metallic and the preoxidized state. Figure 3 shows the result from X-ray diffraction on the samples before and after the experiment. In metallic condition, the diffraction pattern corresponds to the face-centered cubic (FCC) matrix

(A). The pattern of the preoxidized sample (B) confirms the presence of Al_2O_3, in its thermodynamically most stable modification corundum, as well as $Cr_{23}C_6$ carbides, which are typical precipitates in cast Ni-Cralloys. After nitridation, the material forms CrN, Cr_2N, AlN and some oxides (Al_2O_3 (corundum), Cr-Al-spinel) (C). The observation is in accordance with the previously described dark greyish appearance of the surface and the weight change (7.7 mg). In contrast, the preoxidized sample does not show significant changes on the diffraction pattern after nitridation (compare B/D). The Al_2O_3 is retained, and no additional phases are identified with XRD. This is in accordance with the small weight change during exposure (cf. Figure 2).

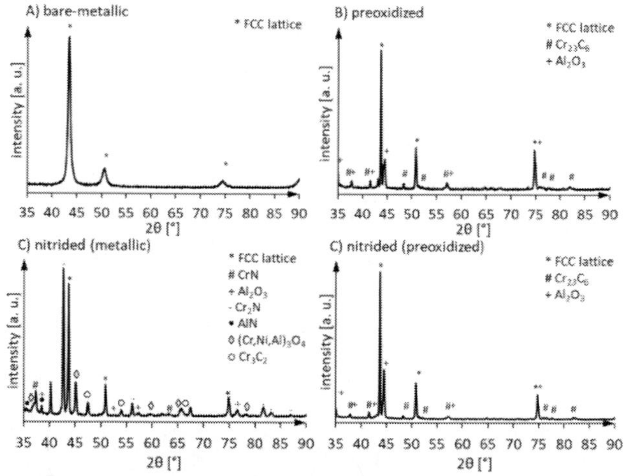

Figure 3: Diffraction patterns (Bragg-Brentano, λ = 1.54080 Å) of the samples of alloy s001: A) metallic, B) preoxidized, C) nitrided (metallic), D): nitrided (preoxidized).

Figure 4 presents the comparison of the cross-section SEM-BSE images for the nitrided metallic sample (A) and the preoxidized (B) after the experiment.

The bare-metallic sample shows formation of an Al_2O_3 oxide scale on top (dark grey), partially interrupted by particles of higher density, appearing lighter in the BSE image. Those were identified as CrN in accordance with the result from XRD. Below the surface, different types of precipitates were observed. Dark appearing needles/plates are AlN, while greyish precipitates correspond to Cr-carbides (cf. A). The preoxidized sample does not show severe precipitation of nitrides in the subsurface zone. Cr-carbides, typically present due to the centrifugal casting process, are observed (greyish precipitates), which is in accordance with the XRD analysis. On the surface, an Al_2O_3 scale of roughly one μm thickness is observed. In contrast to the non-preoxidized sample (A), the material does not show any evidence of nitridation. The chemical composition of the scale, precipitates and matrix in the investigated subsurface region were confirmed by elemental analysis (Table 4).

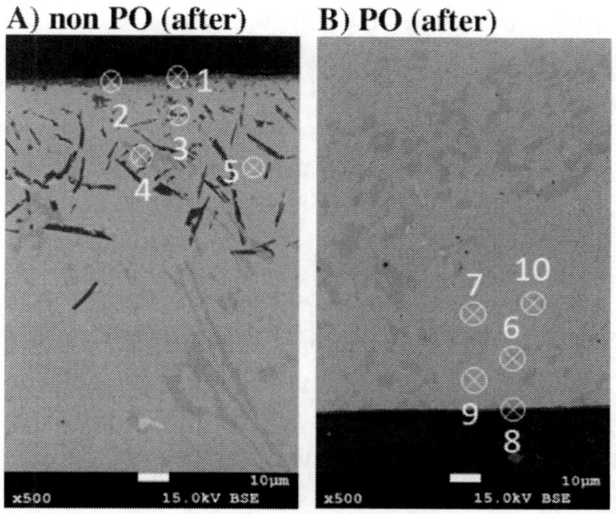

Figure 4: Cross-section BSE images for the samples of alloy s001 after nitridation: A) as received metallic, B): preoxidized.

Table 4: Elemental analysis results for the cross-section SEM samples of alloy s001 (1 – 5: metallic after nitridation, 6 – 10: preoxidized after experiment).

#	N	O	Al	Si	Ti	Cr	Mn	Fe	Ni	Nb
1	9	13	11	*	-	65	*	1	1	*
2	-	62	34	-	-	*	3	*	*	-
3	44	*	40	*	*	5	*	4	6	-
4	-	9	*	*	-	82	-	4	4	*
5	-	3	*	*	*	73	1	9	13	*
6	-	5	6	-	-	67	-	10	12	-
7	-	-	3	-	-	61	-	15	20	-
8	-	62	35	-	-	2	-	-	-	-
9	-	-	6	*	-	26	-	21	46	*
10	-	-	7	*	-	25	-	21	46	*

*: Traces (< 1 at.-%) observed during elemental analysis by EDS. Due to the rather low accuracy of the technique the calculated

quantification is not sufficiently reliable and the calculated values therefore omitted.

Alloy s002

Figure 5 shows the XRD pattern measured for the specimens before and after exposure. As expected, the XRD pattern of the sample before exposure only shows presence of the FCC matrix (A). By XRD, its mass gain (7.9 mg) as well as the formation of a dark grey scale on the surface during the experiment were proven to be CrN, Cr_2N and Al_2O_3 (B). Additionally, signals resulting from the metallic FCC lattice are observed. The XRD pattern of the preoxidized sample (C) of s002 confirms the presence of Al_2O_3. Additional peaks correspond to chromia Cr_2O_3, indicating that the Al-content of roughly 2.3 wt.-% is not sufficient to form a continuous alumina scale. After exposure the observed Cr_2O_3 is absent, reflections of different chromium nitrides (CrN and Cr_2N) are observed. The FCC reflections as well as those belonging to Al_2O_3 remain present on both samples (D).

The detrimental changes observed in the XRD pattern of the bare-metallic sample (A, C) as well as the smaller change for the preoxidized ones (B, D), are in line with their respective weight changes (7.9 vs. 2 4 mg).

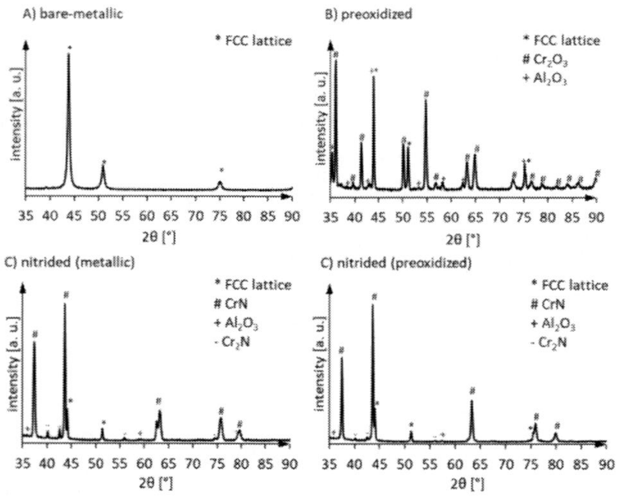

Figure 5: Diffraction patterns (Bragg-Brentano, λ = 1.54080 Å) of the samples of alloy s002: A) metallic, B) preoxidized, C) nitrided (bare-metallic), D): nitrided (preoxidized).

Figure 6 presents the SEM-BSE cross section images for the metallic (A) and the preoxidized sample (B) after exposure. The metallic sample (A) forms a continuous top-layer of CrN (dark grey scale, point 1 + 2) as well as Al-nitrides internally (point 3). Moreover, light-grey appearing participates are identified as niobium-rich carbides, normal in such castings. Other precipitates are observed in the material, which have been identified as Cr_2N (point 4). The matrix material shows significant depletion in Al and Cr (point 5), in accordance with the formation of the respective nitrides. The cross-section of the preoxidized sample shows a continuous top-layer of two phases. The upper part of the scale consists of CrN (point 10), while the lower part consists of Al_2O_3 (point 9). The observation is in line with the XRD-pattern and confirmed by elemental analysis. Needles of AlN (black, point 6)) are observed in the subsurface region, the subsurface zone is depleted in Al (point 7). The depletion can be explained by oxidation during the preoxidation process, recognizing the very low partial pressure of oxygen necessary to form this oxide. Further dark grey precipitates were identified as Cr_2N (point 8), also confirmed by XRD. The elemental analysis results confirming SEM investigations are given in Table 5.

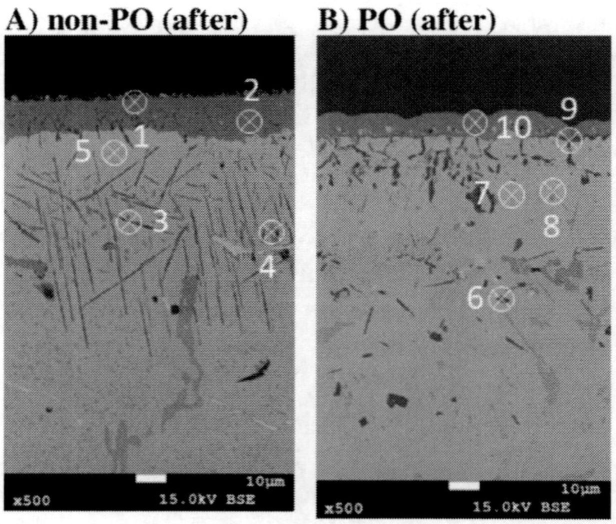

Figure 6: Cross-section BSE images for the samples of alloy s002 after the experiment: A) metallic, B): preoxidized.

Table 5: Elemental analysis results for the cross-section SEM samples of alloy s002 (1 – 5: metallic after nitridation, 6 – 10: preoxidized after nitridation).

#	N	O	Al	Si	Ti	Cr	Mn	Fe	Ni	Nb
1	40	5	2	-	-	48	-	*	3	*
2	40	4	2	-	-	54	-	-	*	*
3	11	4	12	-	-	14	-	8	51	-
4	25	4	*	-	-	68	-	1	2	-
5	4	4	-	*	-	6	1	13	71	-
6	37	-	56	-	-	2	-	1	4	-
7	-	-	-	1	-	20	-	12	68	-
8	20	-	-	-	-	80	-	-	-	-
9	-	53	42	-	-	1	-	1	3	-
10	23	6	1	-	*	68	6	2	10	-

*: Traces (< 1 at.-%) observed during elemental analysis by EDS.

Alloy s003

Figure 7 shows the XRD patterns measured for the specimens before and after exposure. Before exposure the pattern of the non-preoxidized sample, shows reflections corresponding to the FCC lattice of the material and $Cr_{23}C_6$ typically precipitating during casting of the material (A). Upon oxidation, additional reflections of Al_2O_3 are observed (B). After exposing the non-preoxidized sample to ammonia, reflections corresponding to CrN, AlN and Cr_2N are found, as well as the reflections of the FCC lattice (C).

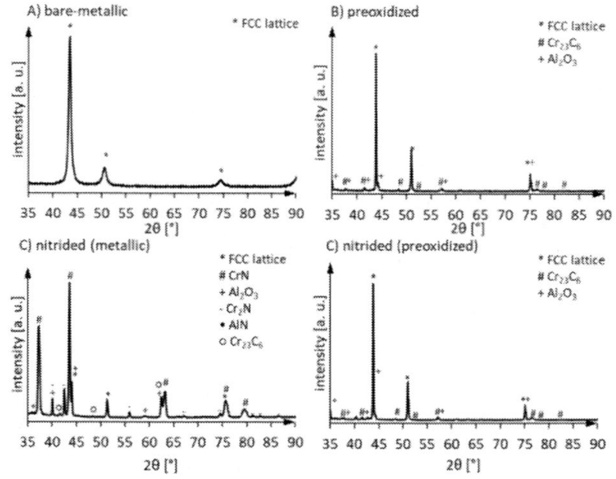

Figure 7: Diffraction pattern (Bragg-Brentano, λ = 1.54080 Å) of the samples of alloy s003: A) bare-metallic, B) preoxidized, C) nitrided (bare-metallic), D): nitrided (preoxidized).

When the preoxidized material is exposed to ammonia gas, weak reflections of CrN appear (D). Nevertheless, changes in the pattern are minor for the non-preoxidized samples (cf. A-C, B-D). This is in line with the significantly lower weight change reported for the preoxidized sample (cf. Figure 2).

The cross-section SEM images of the samples after exposure are shown in Figure 8 / Table 6. The non-preoxidized sample (A) exhibits formation of a CrN scale (point 1), while internal AlN needles are formed by ingress of nitrogen into the base material (point 3). Besides this, Cr-rich carbides from casting are converted into Cr_2N (point 4). The base material is locally depleted from Cr and Al, leading to an enrichment in Fe and Ni (point 2). Below the nitrided region, the typical carbide network is not affected by nitrogen ingress (point 5). SEM findings were supported by XRD as well as mass change (8.5 mg). The preoxidized sample (B) shows a thin layer of Al_2O_3 on top (point 6), which was confirmed to be of corundum structure (XRD). Small areas showed scale discontinuities, consisting from Cr_2N, which could be a result of surface located Cr-carbides being present after machining (point 7).

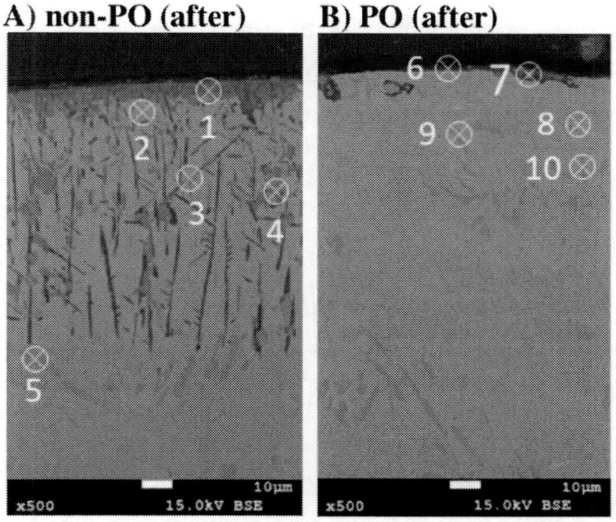

Figure 8: Cross-section BSE images for the samples of alloy s003 after the experiment: A) metallic, B): preoxidized.

Table 6: Elemental analysis results for the cross-section SEM samples of alloy s003 (1 – 5: metallic after nitridation, 6 – 10: preoxidized after nitridation).

#	N	O	Al	Si	Ti	Cr	Mn	Fe	Ni	Nb
1	39	-	-	-	-	61	-	-	-	-
2	-	-	-	-	-	16	-	13	71	-
3	32	-	38	-	-	7	-	4	19	-
4	25	-	-	-	-	75	-	-	-	-
5	-	-	-	-	-	80	-	3	5	-
6	-	65	35	-	-	*	-	-	*	-
7	25	5	1	-	-	67	-	1	1	-
8	-	3	6	-	-	24	-	11	57	-
9	-	-	-	-	-	92	-	2	6	-
10	-	-	-	-	1	90	-	2	7	-

*: Traces (< 1 at.-%) observed during elemental analysis by EDS.

Those are observed in XRD as well. The base material does not show depletion of Cr and Al, resembling the expected composition of the alloy (point 8). The precipitates in the subsurface base material were identified as Cr-carbides, which are expected casting products (point 9/10). Results indicate that dense Al_2O_3 layer leads to increased resistance against ammonia-induced nitridation.

Alloy s004

Figure 9 shows the XRD patterns measured for the specimens before and after exposure. The diffraction pattern of the non-preoxidized sample shows reflections belonging to the FCC lattice of the matrix (A).

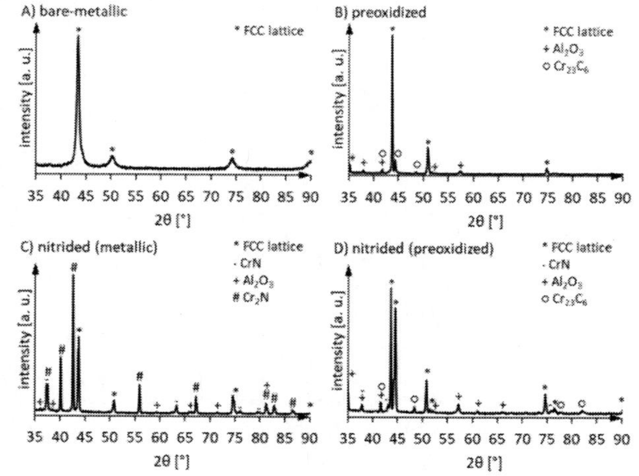

Figure 9: Diffraction pattern (Bragg-Brentano, λ = 1.54080 Å) of the samples of alloy s004: A) metallic, B) preoxidized, C) nitrided (metallic), D): nitrided (preoxidized).

After exposure to the nitriding atmosphere, formation of CrN and Cr_2N according to expectations was reported (C). Results conclude that the oxygen partial pressure of the ammonia containing gas phase is sufficient to stabilize the alumina.

The preoxidized sample of alloy s004 shows reflections of Al_2O_3 (corundum), FCC matrix and precipitated carbides ($Cr_{23}C_6$) in the cast alloy (B). After the experiment, the diffraction pattern of the sample shows weak reflections corresponding to CrN (D). The results from XRD confirm the vast differences of the weight change and the optical appearance of the samples.

Table 7: Elemental analysis results for the cross-section SEM samples of alloy s004 (1 – 5: metallic after nitridation, 6 – 10: preoxidized after nitridation).

#	N	O	Al	Si	Ti	Cr	Mn	Fe	Ni	Nb
1	21	-	*	-	-	76	-	2	1	-
2	-	-	*	1	-	14	-	43	42	-
3	18	-	2	*	-	68	-	7	5	-
4	-	-	-	1	-	28	-	37	35	-
5	19	-	51	-	-	11	-	10	9	-
6	-	65	33	-	-	1	-	1	1	-
7	-	-	5	*	-	25	-	36	33	-
8	-	-	-	-	-	89	-	7	4	-
9	-	-	5	1	-	26	-	36	33	-
10	-	-	23	*	-	24	-	18	36	-

*: Traces (< 1 at.-%) observed during elemental analysis by EDS.

Cross-section SEM was performed on the samples. Figure 10 shows cross section BSE images of the metallic sample (A) and the preoxidized sample (B). Without preoxidation, the alloy forms a continuous scale of CrN, which is in accordance with the result from XRD. Needle-shaped AlN crystallites have precipitated in the subsurface region of the sample. Those could not been observed from XRD, which might be a result of the depth of their precipitation or the unfavorable orientation of the crystallites with respect to the incident X-ray beam. Deeper in the subsurface region, the $Cr_{23}C_6$ carbide network

was gradually transformed into Cr_2N. Elemental analysis was performed to verify the approximate composition of the respective phases.
Preoxidation of the sample leads to formation of a dense Al_2O_3 scale of corundum (B). This scale is observed on top of the sample in the cross-section BSE image. The subsurface zone does not exhibit uniform nitridation, with a larger precipitates of Cr-carbides as well as smaller NiAl, according to elemental analysis findings. No depletion of Cr and Al was observed, which is in accordance with the protective behavior of the oxide. The result of SEM are consistent with the data from X-ray diffraction and the differences in weight gain (see Table 7).

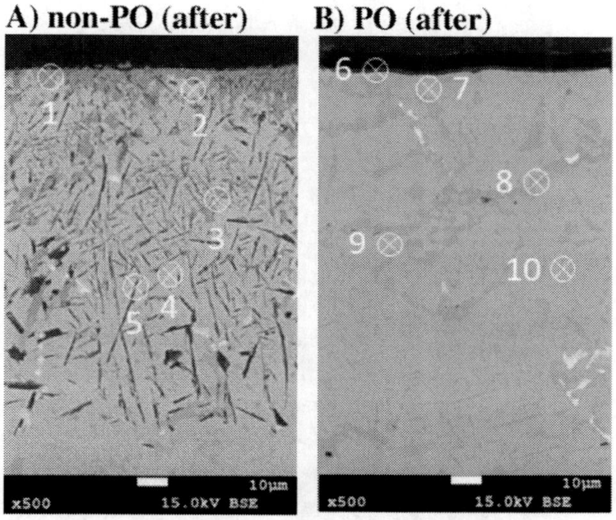

Figure 10: Cross-section BSE images for the samples of alloy s004 after the experiment: A) metallic, B): preoxidized.

Effect of preoxidation process (alloy s001)

Samples of alloy s001 were exposed to different cooling rates during the preoxidation procedures. Figure 11 presents the oxidized surfaces of the samples after cooling, with cooling rates of 100 K/h (A) and 25 K/h (B).

The investigation of the samples after exposure revealed, that the spalled areas of the surface were severely nitrided, while the intact oxide, did not show any signs of degradation. Figure 13 shows the comparison of top-views and cross-sections on both samples after nitridation. The specimen cooled at 100 K/h forms CrN when alumina is spalled. From those surface-covering slabs of CrN, nitrogen diffuses into the material, leading to internal formation of AlN precipitates.

Those precipitates lead to depletion of Al in the base alloy, which is expected to hinder later formation of an alumina scale (A). The specimen cooled at 25 K/h does not show signs of spallation. Furthermore, neither formation of CrN slabs on the surface nor inner nitridation damage, in terms of internal nitride formation or depletion of Al and Cr are observed (B). Overall, the experiment varying the cooling rate reveals, that optimization of the preoxidation process is crucial for the resistivity of the alloy against nitridation.

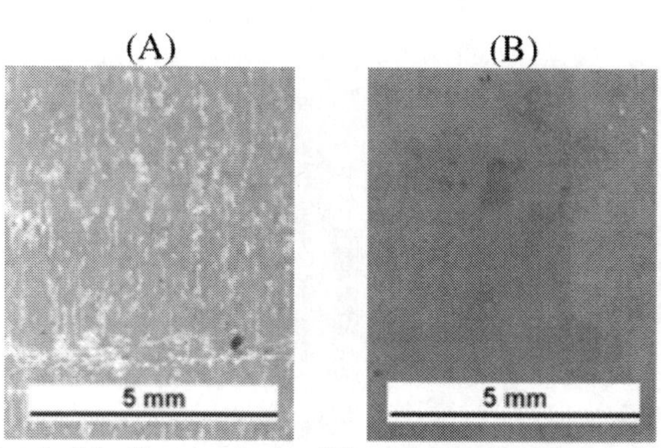

Figure 11: Comparison of the oxidized surface of two samples from alloy s001: A) cooling rate: 100 K/h, B): cooling rate: 25 K/h.

Discussion

The nitridation experiments have proved that significant protection of alloys against nitridation by ammonia-containing gas phases can be achieved by oxide scale protection in form of alumina. Figure 12 presents the correlation between the weight change and the metallographic investigation for both specimens of alloy s001. The weight change difference is clearly correlated to the ingress of nitrogen by internal AlN formation.

Referring to the figures in the result section, this observation similarly is confirmed for the other alloys (cf. Fig. 4, 6, 8 and 10).

The combination of the analytic methods of SEM and XRD with the weight change additionally revealed that certain compositional variations alter the corrosion mechanisms of the alloys, especially the Ni-, Cr- and the Al-content.

The metallic samples of alloy s001 to s003 without preoxidation do not show significant differences in their weight change. A similar result is obtained by plotting the depth of internal nitridation, which was determined by statistical evaluation of the precipitation depth of AlN (Figure 14).

Alloy s002 and s003 show the smallest depth of internal nitridation (roughly 88 µm). On alloy s001 a slight increase is observed (roughly 91 µm). A clear change is observed for alloy s004 which shows a significantly enlarged depth of internal nitridation, consistent with larger weight change. The difference is shown in Figure 14.

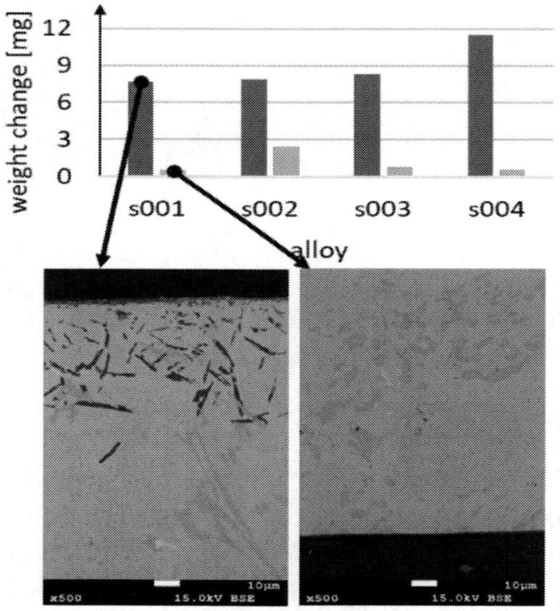

Figure 12: Correlation of the cross-section SEM images of alloy s001 to the weight change observed during the exposure.

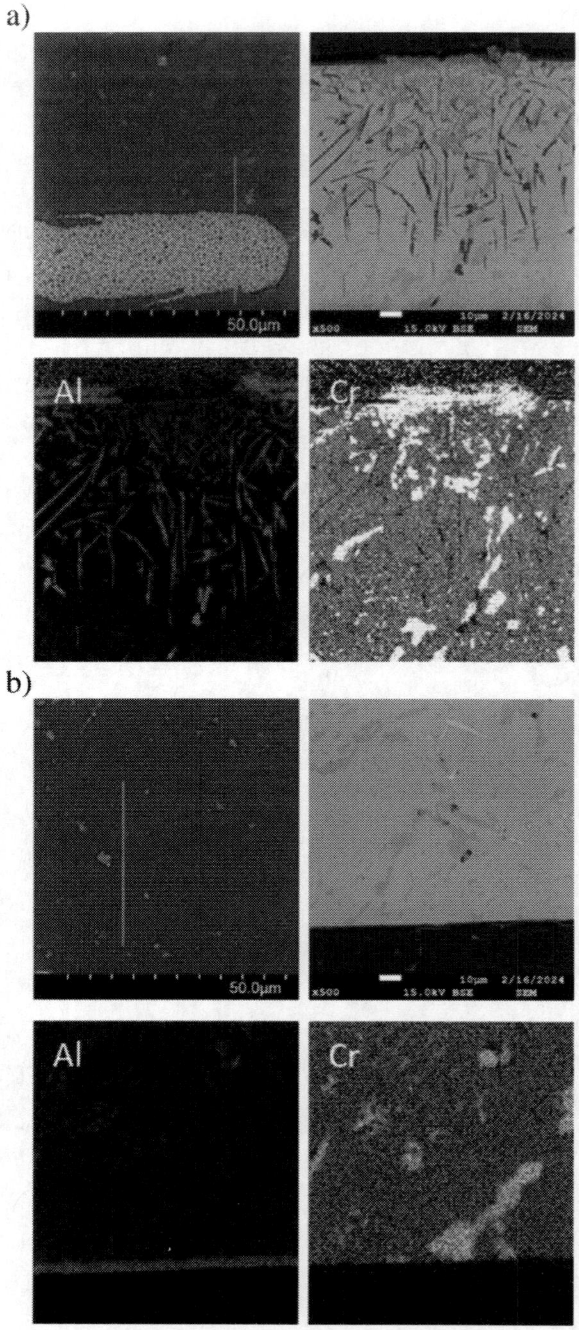

Figure 13: SEM investigation of the sample bodies of alloy s001, preoxidized under different cooling rates (a: 100 K/h, b: 25 K/h), after the nitridation experiment.

It is expected that the increased tendency for internal nitridation results from the Cr- and Ni-content of the alloy. Similar tendencies have been mentioned in literature as well. [15]

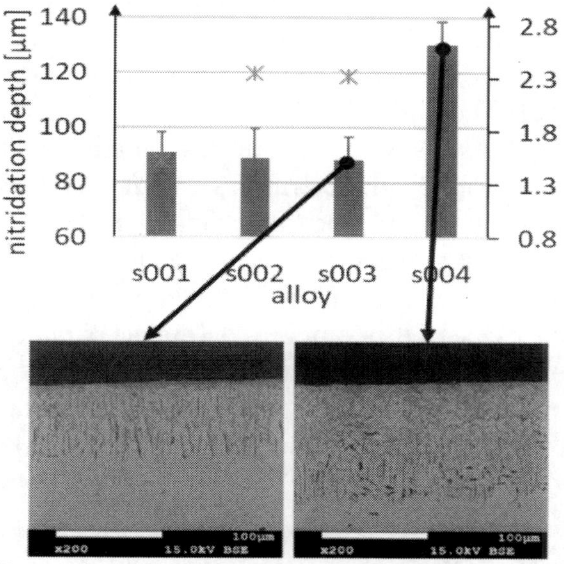

Figure 14: Correlation of the nitridation depth [µm] and the Ni/Cr-ratio to the cross section SEM images for alloy s003 and s004.

Even though the dataset of this study is rather small, it indicates, that a threshold in the Ni/Cr-ratio increases nitridation resistance of an alloy even without any preoxidation. According to our work it is expected, that Cr-contents up to 30 wt.-% with simultaneous Ni-contents > 45 wt.-% will lead to higher resistance of alloys against nitriding atmospheres.

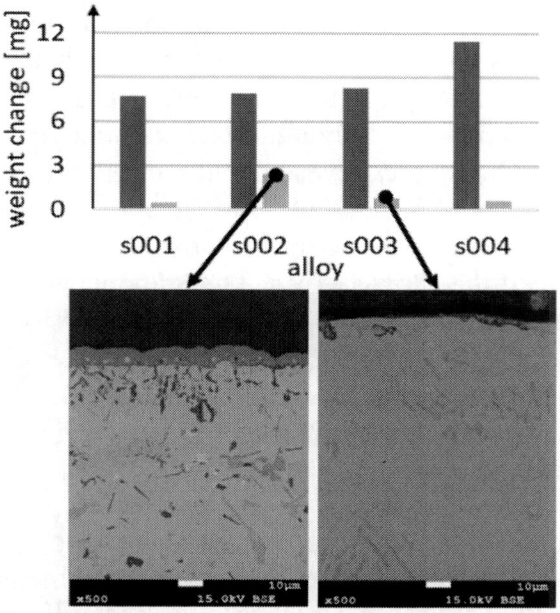

Figure 15: Comparison of the preoxidized samples of alloy s002 and s003 with respect to their weight change during the experiment.

The protective behavior of Al_2O_3 scales was presented in this study upon oxidation of commercially available Schmidt + Clemens Al-containing alloys. A difference in weight change was observed for alloys s002 and s003, which showed a variation of the Al-content. The remaining composition is nearly similar; therefore, the observed effect is expected to be a result of the reduced Al-content. Figure 15 represents the comparison of the weight changes to the cross-section SEM results for the preoxidized samples of alloy s002 and s003. Preoxidation of alloy s002 with rather low Al-content, leads to formation of an oxide scale consisting from Al_2O_3 as well as mixed Al-Cr-oxides. The Al-content is not sufficient to produce a dense alumina scale. The XRD-pattern of the oxidized specimen of alloy s003 only shows reflections belonging to Al_2O_3. The scale does not incorporate Cr or mixed-metal oxides. Upon nitridation the alloys behaved differently. The oxide scale on alloy s002 is partially transformed into CrN, being present as a top layer. Below that, a residual alumina layer is present. Internally, nitridation in the form of AlN precipitates and even inner oxidation were detected. Alloy s003 in contrast, shows a dense alumina scale after exposure to ammonia. No indications of nitrides were found during the investigation of the sample.

Therefore, it can be concluded that a dense Al_2O_3 scale is necessary to protect the alloy against internal nitridation. It further proves that a minimum amount of > 3 wt.-% of Al is necessary to form a dense Al_2O_3 barrier scale that protects against ammonia-containing gas phases. The result is in accordance with literature[16], which typically reports the formation of a dense and protective scale from alumina on high-temperature resistant cast alloys when the Al-content exceeds roughly 2.5 wt.-%.

Apart from formation of the dense oxide scale, a second criterium for protective behavior of the scale is its thermodynamic stability. As described in the introduction, Al_2O_3 can be stabilized under oxygen partial pressures as low as 10^{-34} bar (1000 °C). As long as the partial pressure of oxygen is larger than this threshold, the oxide remains stable during the process. The experiments in this study showed that alumina is retained while being exposed to the ammonia gas stream. Additionally, all non-preoxidized specimens formed Al_2O_3 during the exposure phase, even if no oxygen was deliberately added to the gas stream. Results demonstrate that Al_2O_3 remains stable under NH_3. It is likely, that the gas phase composition of the lab experiment is comparable to the real process conditions. Consequently, it is expected that the oxide scale formed by preoxidation can reduce nitridation in large-scale ammonia cracking units and even more critical, is likely to be retained unless mechanically stripped by erosion or by other types of mechanical damage.

Beside compositional variation, the scale morphology of the alumina itself is detrimental for the resistance of the alloy against nitridation. The experiments in this study revealed, that specifically the procedure of the preoxidation process is critical in obtaining a fully protective oxide. Samples being exposed to improper preoxidation parameters were severely nitrided, particularly in areas where the oxide scale had spalled. The effect was measured as an increasing weight gain and confirmed by scanning electron microscopy (SEM).

In theory, several factors can be varied during the preoxidation process to optimize the scale's properties. It is well reported in literature that oxide scales show increased spallation tendency with increasing thickness. [17] This can be attributed to geometrically induced stresses during growth of alumina. Therefore, the targeted scale should be as thin as possible in order to reduce spallation. The thickness of the scale can be controlled by the temperature and duration of the preoxidation process.

A second variable in order to modify the properties of the scale is the cooling rate. The cooling rate influences the spallation behavior by reduction of time, which is present to relax thermally induced stresses during cooling. Those stresses originate from the differences in thermal expansion coefficients between formed oxides and base alloy and lead to spallation when a critical temperature drop is reached within a certain time interval. The ability of the substrate to accommodate the stresses by plastic deformation within this time interval, determines whether spallation occurs of not. [17] Our study concludes that rather low cooling rates should be achieved in order obtain a continuous and undamaged oxide scale on the surface of the component. If those parameters are chosen correctly, it is likely that preoxidized components would exhibit good resistance against nitridation during the ammonia cracking process.

Conclusion

The presented study has investigated the fundamentals of the protective behavior of an alumina scale on cast Cr-Ni-Fe-Al-alloys against nitridation. Various results regarding the influence of the alloy composition, the thermodynamic stability of the alumina scale and the morphology of alumina were presented.

Overall, alumina is proven to increase the resistance of the alloys investigated against nitridation by ammonia. Furthermore, it was shown that the oxygen partial pressure in the gas phase, which is comparable to the one present in an industrial ammonia cracker, is sufficient to stabilize the protective alumina scale thermodynamically. Our studies revealed that a minimum content of about 3 wt.-% of Al in the alloy is necessary to form a dense and completely protective scale. If the alloy is exposed to ammonia without any protective oxide layer, because of erosion or mechanical damage for example, a higher alloy Ni-content as well as Cr-levels below 30 wt.-% would reduce the nitridation of the material significantly.

The influence of improper conditions during preoxidation was evaluated. Processing parameters, affecting oxide scale thickness, were identified to be the major factors leading to material damage, in particularly scale spallation. Generating less-protective barriers against nitridation, leading to internal nitridation damage (CrN and AlN formation). Those effects are expected to hinder later rehealing of the alumina scale and should therefore be minimized by optimized preoxidation processes.

Our work only featured testing at 900 °C [1652 °F]. Nevertheless, the atmosphere in the cracker is expected to have different temperature regimes ranging from 550 to 950 °C [1022 to 1742 °F]. In this temperature interval, the gas phase equilibrium of ammonia, especially under use of a cracking catalyst, changes significantly. Therefore, our results can only be considered for specific conditions in an industrial cracker. As a result, further testing under different temperatures, pressures, ammonia purities and under presence of the catalyst materials is likely to be necessary in order to establish guidelines for the optimal material selection in large-scale ammonia cracking units.

References

[1] M. Yu, K. Wang, H. Vredenburg: Insights into low-carbon hydrogen production methods: Green, blue and aqua hydrogen, *Int. J. Hydrogen Energy*, **2021**, *46*, 21261 – 21273.

[2] Y. Kojiima, M. Yamaguchi: Ammonia as a hydrogen energy carrier, *Int. J. Hydrogen Energy*, **2022**, *47*, 22832 – 22839.

[3] Q. Song et al.: A comparative study on energy efficiency of the maritime supply chains for liquefied hydrogen, ammonia, methanol and natural gas, *Carbon Capture, Sci. Technol.*, **2022**, *4*, 100056.

[4] B. Mutz, R. Baumgarten: Unlocking high-pressure ammonia cracking, *Nitrogen and Syngas international conference and exhibition*, **2024**.

[5] D. Jakobi, P. Karduck, A. von Richthofen: The high-temperature corrosion resistance of spun-cast materials for steam cracking furnaces – A comparative study for alumina- and chromia-forming alloys, *NACE corrosion conference and exhibition*, **2013**, 2287.

[7] D. Jakobi, J. Weigandt: Benefits from educated selection of advanced materials for radiant coils – A 10 year field performance review of alumina forming alloys, *26th Ethylene Producers Conference*, New Orleans, LA, USA, **2014**.

[8] A. Yamauchi, K- Kurokawa, H. Takahashi: Evaporation of Cr_2O_3 in atmospheres containing H_2O; *Oxid. Met.*, **2003**, *59*, 517 – 527.

[9] H.J. Grabke, D. Jakobi: High-temperature corrosion of cracking tubes; Mater. Corros., 2002, 53, 494 – 499.

[10] E.J. Mittemeijer, M.A.S. Somers: Thermodynamics, kinetics and process control of nitriding, *Surf. Eng.*, **1997**, *13*, 483 – 497.

[11] H. Jacobs, D. Rechenbach, U. Zachwieja: Untersuchung zur Struktur und zum Zerfall von Eisennitriden; *HTM J. Heat Treat. Mater.*, **1995**, *50*, 205 – 213.

[12] B. Aymans, P. Imizcoz, D. Jakobi, M. Villa, M.A.J. Somers: Evaluation of the corrosion resistance of commercially available alloys in ammonia containing atmospheres – effectiveness of oxide scales, *under review*, **2025**.

[13] B. Li, L. M. Pike: High-temperature nitridation resistance of several modern high temperature Ni- and Co-based alloys, *AMPP Annual conference*, **2013**, 19011.

[14] Y. Xie, et al.: The effect of novel synthetic methods and parameters control on morphology of nano-alumina particles, *Nanoscale Res. Lett.*, **2016**, *11*, 259.

[15] U. Krupp, H.-J. Christ: Internal nitridation of Nickel-base alloys. Part II. Behavior of Quarternary Ni-Cr-Al-Ti Alloys and computer-based description; *Oxid. Met.*, **1999**, *52*, 299 – 320.

[16] J. Engkvist, et al.: High-temperature oxidation of FeCrAl-alloys – influence of Al-concentration on oxide layer characteristics, *Mater. Corros.*, **2009**, *60 (11)*, 876-881.

[17] M. Schütze: Mechanical aspects of high-temperature oxidation, *Corros. Sci.*, **1993**, *35*, 955 – 963.

Life Extension of SCC Affected Urea Reactor through Inspection, Repair & Replacement

Fauji Fertilizer Company Limited (FFC) operates four Urea plants, with Plant-III (SAIPEM, 1980) being the oldest. Urea Reactor, R-101 developed chloride-induced stress corrosion cracking (Cl-SCC) due to residual chlorides from flushing fluids over 42 years, leading to multiple leaks in the corrosion resistant lining. Extensive SCC rendered repairs difficult and uneconomical, prompting a reactor replacement in 2026. To sustain operations, critical lining segment were addressed through repair / replacement, to keep the reactor operational until replacement. The reactor now operates without any leakage. This paper outlines damage mechanisms establishment through advance inspection techniques, repair strategies, and mitigation measures ensuring continued safe operation until replacement.

Syed Hassan Raza
Fauji Fertilizer Company Limited (FFCL), Pakistan

Hafiz Abdul Rahman Awan
Fauji Fertilizer Company Limited (FFCL), Pakistan

Introduction to FFC

Fauji Fertilizer Company, Limited is the largest Urea Manufacturing Company in Pakistan, with a significant role in the country's agricultural sector. Established in 1978, the company operates four ammonia urea complexes in Pakistan and managing O&M of a fifth plant as well. FFC holds 47% share of the urea market in Pakistan.

Urea Process

The Urea plant is based on SAIPEM's Total Recycle technology. Urea synthesis process begins in the Urea Reactor, where liquid ammonia at 160 kg/cm^{2}g (2,276 Psi) pressure, 30°C (86 °F) temperature; recycled carbamate solution and CO_2 gas at 159.5 kg/cm^2 (2,269 Psi) pressure & 104°C (219 °F) temperature are introduced at the bottom of the urea reactor for urea formation reaction. Within the reactor, Ammonia and CO_2 reacts to form ammonium carbamate, which subsequently dehydrates to produce Urea. Given the highly corrosive environment inside the reactor, where the corrosion rate (CR) for carbon steel might reach nearly 900 mm/year (35,433 mpy), the use of specialized corrosion-resistant materials (liner, which is in contact with the process fluid) is imperative. In order to keep the corrosion resistant liner passivated and reduce its corrosion rate to a minimum, passivation air is added to the urea reactor at 4 kg/cm^2g (57 Psi). Corrosion rate of the liner in the urea reactor varies significantly through its length, with highest being in the vapor zone (top of the reactor) 0.15 to 0.18 mm/yr., mild in the middle zone (0.08 to 0.10 mm/yr. and minimal in the bottom portion (0.02 to 0.04 mm/yr.

The reactor effluent, comprising unconverted ammonium carbamate and urea solution, is directed to the high-pressure urea stripper, where ammonium carbamate is efficiently decomposed and recycled back to the reactor. The process stream then advances through downstream equipment, where further concentration and refinement ultimately yield the final urea product.

Figure 1. Internal view of the Urea Reactor

Construction of Urea Reactor (Leakage path & Weep Holes Network

Urea Reactor is one of the most critical equipment of any Urea Plant. The reactor installed at FFC Plant-III is 42m tall, 2.35m in diameter, has a coiled layer pressure bearing carbon steel shell (79mm thick), originally lined with 5mm thick SS-316L Urea grade material. The material is specially designed for urea process having significantly less corrosion rate compared to regular SS-316L. The reactor under discussion was built by M/s Mitsubishi Heavy Industries, Japan in 1977 and was taken into service with plant commissioning in 1980.

With the depletion of corrosion-resistant liner, phase-wise relining of the reactor with superior lining material Cr.Ni.Mo 25.22.2 of 5.0mm thickness was initiated in turnaround (TA)-1992 and completed in TA-2017, except for the bottom dome. The reactor is constructed of 31 shell courses and two hemispherical heads (top head and bottom head). In case of any pressure boundary breach through the corrosion-resistant liner, the leakage manifests outside the reactor through a system of weep holes drilled into the carbon steel shell. Except for top and bottom domes, each pair of shell courses (starting from the top) is provided with four weep holes (180° apart at two elevations) for leakage communication and flushing. (weep hole diameter: 8mm). Weep holes are monitored for any leakage on weekly basis using Draeger pump and ammonia sensitive Draeger tube.

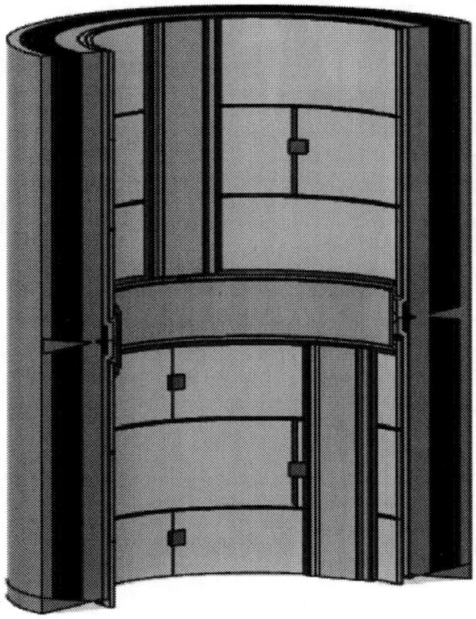

Figure 2. Schematic view of the reactor construction

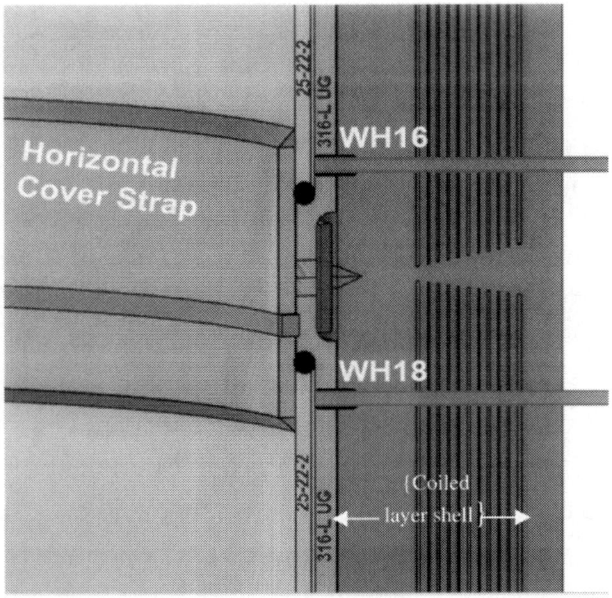

Figure 3. Schematic view of the carbon steel shell, liner and weep holes configuration

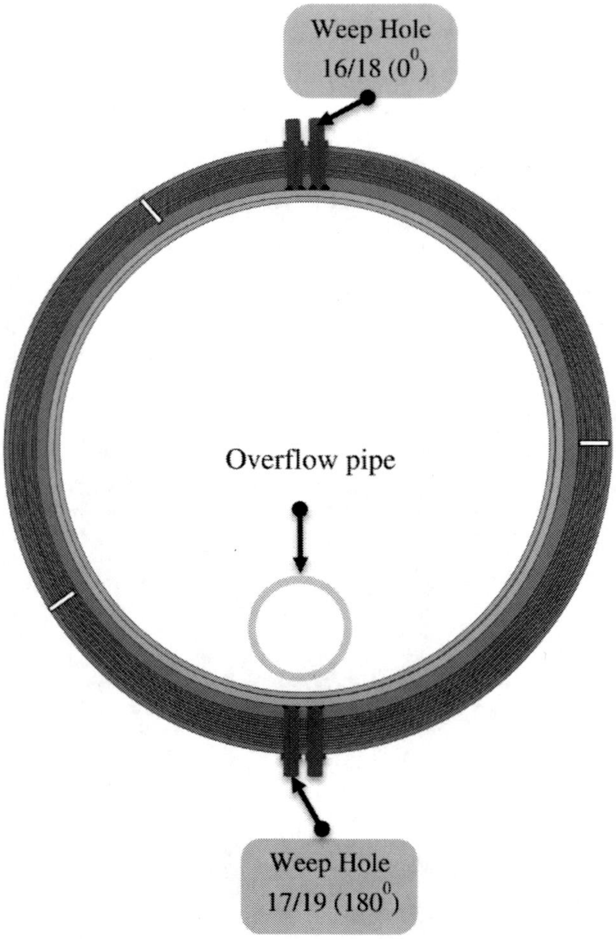

Figure 4. Orientation of weep holes

History of Leakages

Throughout this paper, the word leakage will be used for leakage through the corrosion-resistant liner. The reactor operated trouble-free for 11 years since commissioning in 1980. The first leakage through the corrosion-resistant liner appeared in 1991. For addressing all the leakages, the plant is shutdown, the reactor is opened, the leakage is identified and rectified using process licensor techniques. To date, 24 leakages and shutdown events are on account of this reactor.

Leakage through Horizontal Cover Strap's Base Metal in Shell Course No. 4:

A leakage in the reactor appeared on 23rd September 2022 through weephole no. 14, which corresponds to shell course nos. 3 & 4. Leakage management of the affected shell courses, in terms of continuous flushing with low pressure steam (LS) or condensate, monitoring and trending was carried out until the leakage was confirmed and reached a predefined detectable rate by the end of October 2022. For flushing, LS is connected to one weep hole, with two plugged and one left open to flush a specific shell course. The process is repeated by rotating the open and plugged weep holes until all paths are flushed. Another leakage through weep hole no. 18 (corresponding shell course no. 5 & 6) started on 3rd October 2022, which was also kept under monitoring.

The plant was shut down on November 06, 2022 to address the leakages. Two leak points were identified in an Ammonia leak test using ammonia-sensitive paper and confirmed through ammonia detection spray (Figure. 5).

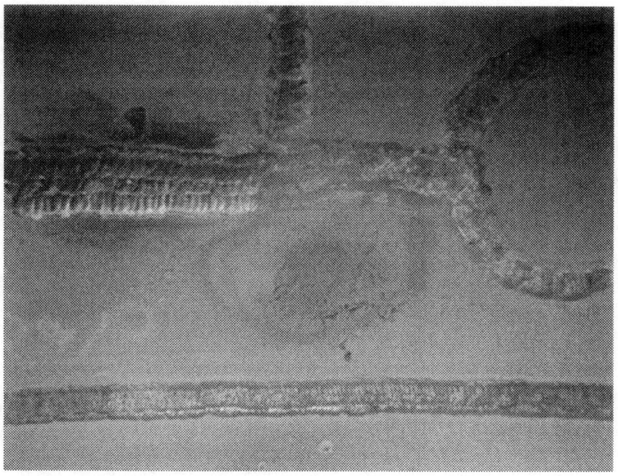

Figure 5. Ammonia test result showing the leakage location on horizontal cover strap in Shell course no. 4

Figure 6. DPT result after grinding the leakage site on shell course no .4

The first leakage was through a fine crack in the base metal of the horizontal cover strap (East Side); approximately 2" away from the weld between horizontal & vertical cover straps of shell course no. 4 (Figure. 6). Lining thickness at and around leakage site was measured and found to be 4.1~4.2mm. Progressive grinding of the leakage location with DPT (dye penetrant testing) was carried out. After grinding the leakage site by 3.0mm in depth, a cluster of cracks indicative of SCC (stress corrosion cracking) were observed. Further grinding was stopped and a 430mm long circumferential piece of cover strap plate was removed. Performing DPT of the removed cover strap plate from the back/ non-process side, the exposed 25.22.2 re-lining plates under the cover strap, and the 316L UG original lining plates, revealed fine branched cracks in all the locations described above (Figure. 7).

The removed 430mm long cover strap plate was replaced with a new cover strap (Material: 25.22.2; Thickness: 5.0 mm) with slightly larger cover straps in order to change the weld locations from the previous ones. Fine cracking in the base metal, adjacent to the weld seams, was observed by DPT. The cracking was addressed using a larger cover strap to avoid welding in the vicinity of repair welds. The wider cover strap welds were checked using DPT showing satisfactory results.

Figure 7. Branched cracking in SS-316L UG and Cr.Ni.Mo 25.22.2 lining material

Leakage through Vertical Cover Strap's Base Metal in Shell Course No. 5:

The second leakage was observed through a fine crack in the base metal of the vertical cover strap in shell course no. 5 (adjacent to weephole no. 18). The leakage location was 1" away from the weld between the horizontal cover strap of shell course no. 4 (bottom side) and vertical cover strap of shell course no. 5 (Figure. 8). Progressive grinding followed by DPT of the defect site was carried out and the crack in the lining base metal was similar to the one observed in the shell course no. 4, detailed above (Figure. 9).

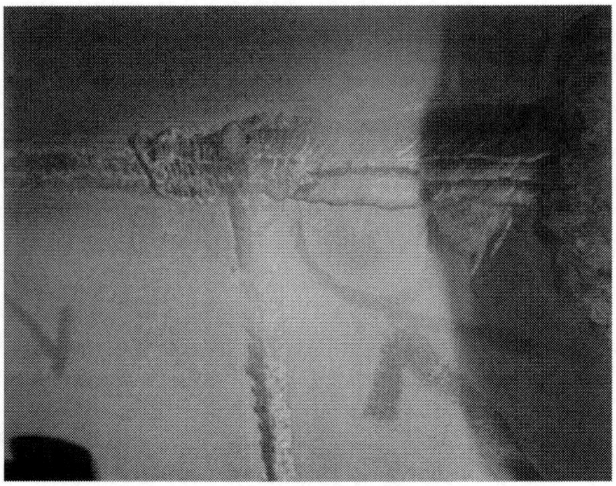

Figure 8. Leakage location in shell course 5

Figure 9. View of the crack in SC-5

Exploratory Work

In order to look for the extent of cracking behind the cover straps and the lining plates, horizontal cover strap plates in front of other weep holes (WH No. 16 and 17) were removed. White deposits were found accumulated in the cracks. DPT of the area under the cover strap plate revealed a spider web of fine branched cracks (Figure 10). A similar observation was made on the cover strap replaced against 1st and 2nd leak points in shell course no. 4 & 5.

The cracking was more pronounced behind the interruption strips in the form of circles, (Figure 11) thus highlighting the fact that the cracking attack had followed the path of the flushing medium and got severely close to possible salt concentration locations.

The back side of the removed cover strap (non-process side / exposed to flushing fluids) was also checked with DPT for the presence of cracks. Fine branched cracking was also observed on removed cover plates (Figure. 12), suggesting a widespread cracking on the relining plates exposed to the flushing media.

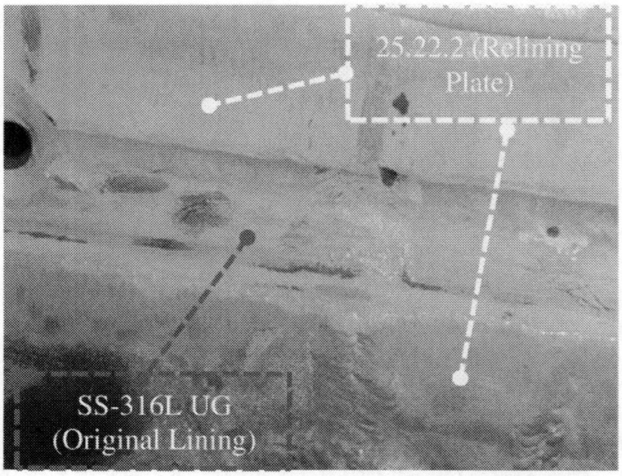

Figure 10. Fine branched cracks in welds, SS-316L UG and Cr. Ni. Mo 25.22.2 lining plates underneath the cover strap

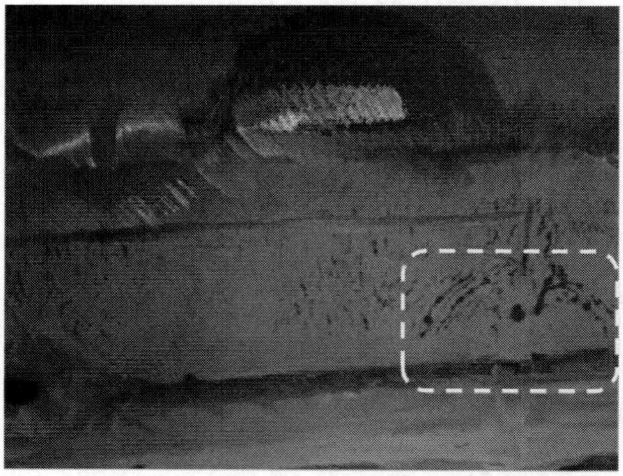

Figure 11. The circular pattern is the location where interruption strip was removed.

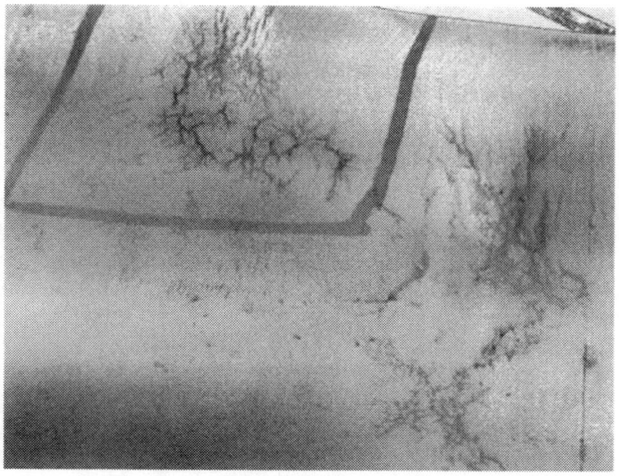

Figure 12. Fine branched cracks on rear side (non-process side) of cover strap plate

Damage Mechanism Identification

The fine branched cracking observed behind the stainless-steel lining was of prime concern and was observed for the first time in the Urea reactor. The morphology of cracks found at the leakage locations, behind the cover strap plates, and on the original lining material (SS 316L UG) raised serious concerns on the reliability and weldability of the reactor lining. The following important conclusions were drawn based on data obtained from the leakages and the exploratory work described above;

1. The indications behind the cover straps were suspected to be SCC due to residual salts containing chlorides (as seen accumulated in the cracks) present in the flushing fluids, which were steam and condensate.
2. These cracks grew with stress (especially in cyclic loadings, both mechanical and thermal) and are more pronounced in the 3rd layer of the lining, which in this case was the cover straps.
3. The extent and location of SCC behind the lining plates could be widespread, as shell courses three to six had remained in operation with flushing for a prolonged duration during their operational history.
4. Further exploratory and repair work was stopped to assess the inspection findings and viability of initial repairs.

NDTs on the Removed Cover Strap Plates

The circumferential cover strap plates removed from WH No. 16 & 17 were subjected to NDTs (nondestructive tests) to extract data from the sample plates. Results of the NDTs are given below.

a. Visual Inspection of the removed cover strap plates illustrated surface breaking cracks (*Figure 13*) and white deposit on the sections (*Figure 14*).
b. The chemical composition of all the plates matched to Cr. Ni. Mo. 25.22.2.
c. The measured thickness of all the plates ranged from 4.1mm to 4.2mm against nominal thickness of 5.0mm.
d. Ferrite content of the welds was less than 0.1% (Max. permissible value for SMAW is 0.6%; Ref. SAIPEM Spec.).
e. DPT of the removed cover strap plates was also performed. Fine tight branched cracks were observed on the plate's surface exposed to flushing fluids (steam and condensate) Figure. 12.

Figure 13. Cracks behind removed cover strap (non-process side) seen in visual inspection

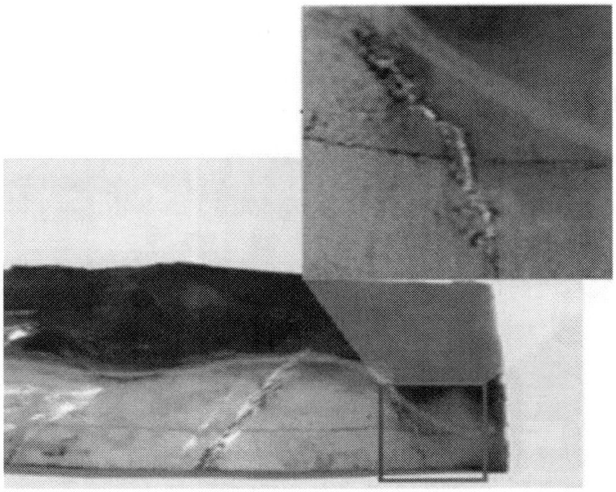

Figure 14. Deposits are magnified

f. Replica Metallography

Replica Metallography of the removed plates was performed. Examination of extracted replicas from surface and cross section revealed a number of cracks that originated from liner backwall i.e. OD and exhibited branching that traversed in transgranular manner. The microstructure comprised of equiaxed austenite grains exhibiting twins. No significant evidence of carbide precipitates was noted. The observed structure was typical of the material in solution annealed condition.

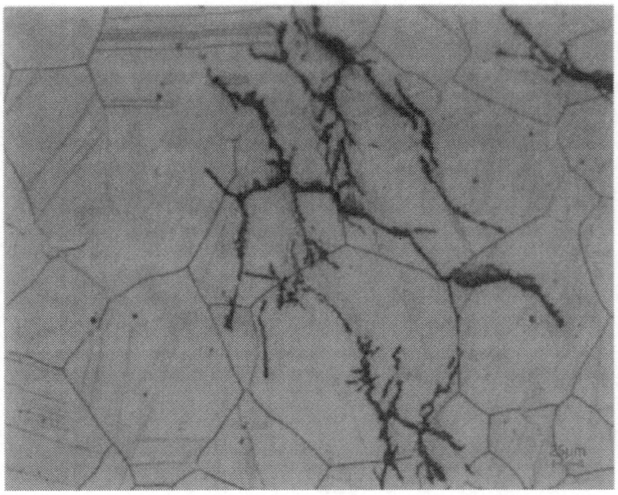

Figure 15. Branched cracking noted on surface. Cracks were transgranular; Etched: 10% Oxalic Acid

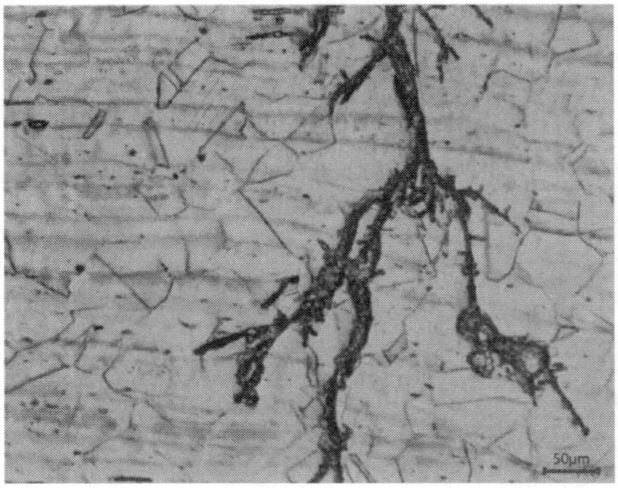

Figure 16. Branched, trans-granular cracking noted on plate cross-section

g. Energy Dispersive Spectroscopy (EDS) Analysis:

Energy Dispersive Spectroscopy (EDS) analysis was performed on the intact surface location remote from the deposit, and areas with deposits provided elemental compositions (Table 1). The presence of Ni, Cr, and Fe noted in deposit-free area is typical of constituents of urea-grade stainless steel. The presence of Na, K, Mg and Cl indicated the presence of water-soluble salts.

Table 1: EDS Analysis Results (Wt%)										
Location	O	Cr	Fe	Ni	Na	Si	Cl	K	Ca	Total
Deposit Free	6.58	30.22	45.04	18.15	-	-	-	-	-	100
Deposit	55.72	12.18	20.39	7.62	1.61	0.43	0.78	0.84	0.42	100
Note: Carbon peaks removed and data normalized over the compositional range for better / accurate assessment of results. Details normalized elemental composition after removal of carbon.										

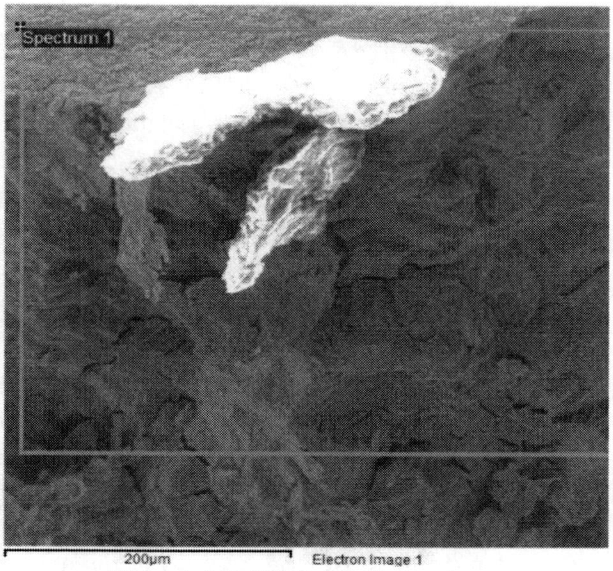

Figure 17. EDS Analysis scan

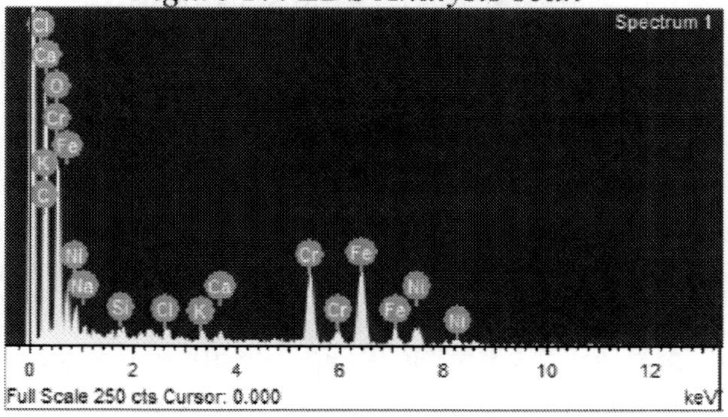

Figure 18. Fracture face, including deposits

Energy Dispersive Spectroscopy (EDS) analysis results at the deposit location indicated a chloride concentration over 5000 ppm, which appears to be significantly high for austenitic stainless steels to operate with. Although chlorides in the steam and condensate (flushing mediums) are kept to a bare minimum value, the concentration mechanism collected them in the narrow channels as found in the weep hole passages over the period of 42 years.

h. Scanning Electron Microscopy (SEM) Analysis:

Examination of the fracture was carried out close to OD. Examination of the surface topography exhibited a step like feature on a portion of the fracture, corresponding to the original crack face. These are typically characteristic of a cleavage type of fracture. Further examination within the grains exhibited feather markings that are an array of very fine cleavage steps on large facets. The observed features were typical of the cracks, propagating in a transgranular manner.

From metallography, SEM and EDS analysis results, it is evident that the liner material failed as a result of chloride-assisted stress corrosion cracking (Cl-SCC). The brittle nature of fracture and propagation of crack in branched transgranular manner are typical characteristics of stress corrosion cracking.

Like all stress corrosion cracking, Cl-SCC occurs only if the right combination of tensile stresses, susceptible material and environment is present. Likelihood of Cl-SCC increases with an increase in acidity, oxygen concentration, temperature and chloride concentration. Chloride tends to concentrate at localized areas where crevices are present or under deposits or at areas where wet-dry conditions are prevalent. In fact, one of the very common causes of stress corrosion cracking is the evaporation of water within the crevices. Whilst the chloride content within the bulk water maybe within the desired ppm concentration, the local concentration after evaporation can be significantly high to cause Cl-SCC.

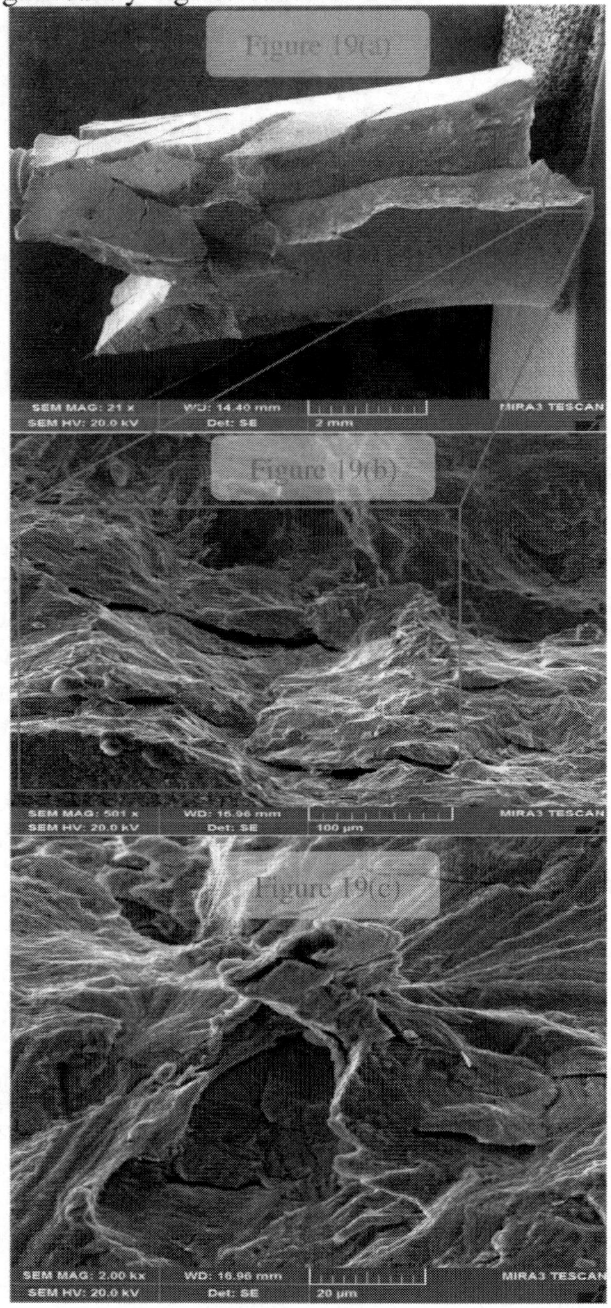

Figure 19. SEM photograph of fracture face
(a) Low mag image of forced open crack
(b) Fracture face exhibiting a significant amount of secondary cracking
(c) Cleavage fracture morphology indicative of a transgranular type of fracture

Though there is no definitive threshold for the right concentration of chlorides taking into consideration all the permutation of temperature, pH, and oxygen concentration, the obtained concentration in excess of 5000 ppm present a significantly higher risk for Cl-SCC in austenitic stainless steel.

Life Extension Plan - TA-2023 (October 2023)

Multiple leakages from the reactor, confirmation of wide spread stress corrosion cracking in the liner plates, probability of further leakages from the SCC affected liner, difficulty in the weld repair of the liner and time required to replace all the SCC affected regions of the reactor (which was economically unjustifiable) necessitated the replacement of Urea Reactor. Based on these facts, the already decided inspection and repair plan of the urea reactor for the turnaround 2023 was altogether changed. The new plan was focused on replacing the severely damaged cover straps and addressing the reliability concerns only within the approved turnaround duration while ensuring the reliable operation of the reactor until a replacement reactor could be constructed. Because the scope of damage was extensive, selection of severely affected cover straps for replacement was a challenging task within the available turnaround window.

To determine the most affected locations, the reactor's leakage history since commissioning was analyzed, focusing on the extent and duration of operation with leaks. Reactor operation with leakage requires extensive flushing of the annular space between the carbon steel shell and the stainless-steel liner with steam/condensate to keep the weep holes from clogging and reduce the corrosion damage from the melt urea. Moreover, the lining thickness data (remaining thickness), the condition of weld beads, the age of lining plates and weld rehabilitation history were also critical factors for identifying the inspection-repair plan. After an extensive and rigorous analysis, the most susceptible locations of stress corrosion cracking were identified, and the replacement plan was chalked out.

During the turnaround, five horizontal and four vertical cover straps, based on susceptibility of SCC damage, were replaced with new wider ones of 25Cr. 22Ni. 2Mo. material (5.0mm thickness) in different regions of the reactor. The wider cover straps helped change the weld location for the new cover straps from the previous ones, thus avoiding the exposure of welding heat to the SCC-affected liner plates at previous weld lines and heat affected zone. Weld rehabilitation of a few horizontal and vertical cover straps was also completed in the turnaround.

Depositing a sound weld and maintaining segregation between two shell courses on a SCC-affected base metal was also a challenge. Weld integrity was ensured through strict quality control (control on weld technique, bead profile, low welding current, inter-pass temperature) and extensive NDTs (DPT after each weld pass, the ferrite content check) of the weld, while the option of maintaining the segregation was dropped to limit the welding scope.

A thorough risk assessment of the pressure-bearing carbon steel shell, from a third-party specialist, was conducted to ensure the safe operation of the urea reactor. The risk assessment was based on extensive radiographic testing of the coiled-layer shell to look for any thickness loss due to carbamate corrosion or environmental cracking. No significant metal loss or cracking was noted on the CS shell, declaring it fit to maintain the pressure boundary.

The urea reactor was successfully put back into service and has been operating without any leakage to date. The leakage-free operation of the reactor is a testament to the thorough and precise homework in selecting the repair and replacement sites along with the careful execution of the corrosion-resistant lining. A new Urea reactor is on order, and the existing one will be replaced with the new one during the next turnaround planned for 2026.

Conclusion

1. Prolonged operation of the Urea Reactor with leakage shall be avoided.
2. Preventive flushing of the weep hole system with steam or condensate should be avoided, as these fluids may contain residual chlorides that concentrate over time and contribute to chloride-induced stress corrosion cracking. Instead, a vacuum-based online leakage monitoring system for the urea reactors shall be explored.
3. Vintage urea reactors with a history of frequent steam/condensate flushing through the weep hole system should have their corrosion-resistant liners inspected against Cl-SCC to prevent unexpected leakages in service.
4. For new urea reactors, increasing the thickness of the corrosion-resistant liner in the vapor zone, where corrosion rates are highest, can extend the service life before relining becomes necessary.

Repair Strategies for Tube-Sheets in HP Urea Heat Exchangers

The corrosiveness of the process fluid in the high-pressure synthesis section of a urea plant is well known and possesses potential risks. One of the risks is corrosion of the tube-to-tube-sheet welds in the shell and tube heat exchangers such as the HP Strippers, HP Carbamate Condensers (HPCC) or HP Scrubbers. The corrosion could result in a leak, and when not noticed or not attended, the ammonium carbamate could corrode the carbon steel tube sheet severely in a relatively short period. Several cases are reported where a small leakage in a tube-to-tube-sheet weld (a small pinhole or wormhole) created a huge cavity underneath the stainless-steel weld overlay, compromising the structural integrity of the vessel.

The repair options to restore the integrity and reliability of said carbon-steel tube-sheet are hampered. Through the years, Stamicarbon applied creative solutions to repair tube sheets. This paper describes possible failure scenarios and repair options, as well as two case histories.

Alex Scheerder
Stamicarbon BV, The Netherlands

Introduction

Stamicarbon is a global leader in licensing urea plants. Also, full life cycle support is offered to our customers. Amongst others, Stamicarbon provides mechanical services, inspections, repairs and troubleshooting. For over 60 years, vast experiences and know-how have been gained with respect to the repair of critical high-pressure urea equipment. It is well known that in the high-pressure synthesis section of a urea plant, the process conditions are highly corrosive. To combat these conditions several mitigation strategies are implemented to minimize the risk of failures and damages. Amongst others these mitigation actions entail:

- ✓ Proper design of the equipment
- ✓ Use of special developed stainless steels
- ✓ Scrupulous quality control during manufacturing
- ✓ Defining a proper integrity operating window
- ✓ Safeguarding by leak detection monitoring systems
- ✓ Inservice inspections

Despite mitigation strategies, corrosion damage cannot be completely excluded, and repairs might be needed to restore the integrity and reliability of the equipment. This paper describes possible failure scenarios and repair options for a tube-sheet in a shell and tube urea heat exchanger, as well as two case studies.

Stamicarbon urea process

Both case histories described in this paper relate to older so-called CO_2 stripping urea plants (Stamicarbon design) having a urea reactor (a) and three high-pressure heat exchangers, i.e., HP Stripper (b), vertical falling film HPCC (c), and

HP Scrubber (d), as schematically presented in Figure 1.

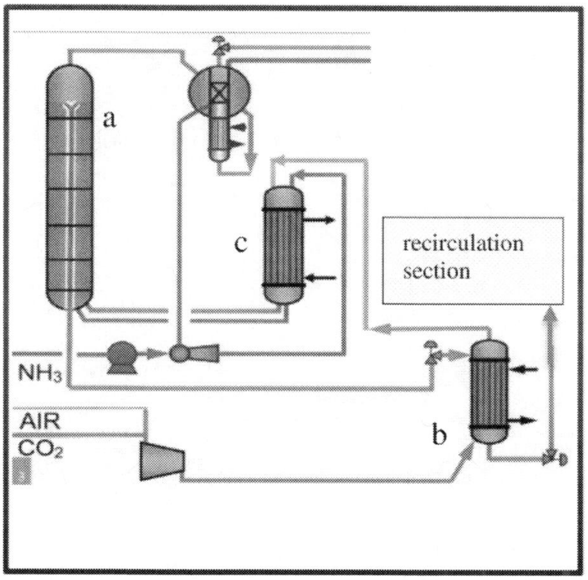

Figure 1: CO_2 stripping synthesis section with urea reactor (a) and three heat exchangers; HP Stripper (b), HPCC (c), HP Scrubber (d)

The ammonia from battery limit (blue line) is fed to the top of the HPCC together with the off-gas (mainly carbon-dioxide, green line) coming from the top of the HP Stripper. In the HPCC the first condensation of ammonia and carbon-dioxide takes place forming ammonium-carbamate and urea. At the bottom of the HPCC the nonconverted ammonia (blue line) together with a mixture of ammonium-carbamate and urea (red line) is fed to the bottom of the urea reactor. In the urea reactor further conversion of ammonium-carbamate into urea takes place. A mixture of ammonium-carbamate and urea (red line) leaving the bottom of the urea reactor is fed to the top of the HP Stripper. In the HP Stripper the non-converted ammonium-carbamate is stripped off from the urea solution by use of carbon-dioxide (containing passivation air, so-called CO_2 stripping) entering in the bottom of the HP Stripper (green line), increasing the concentration of urea in the process stream. The stream leaving the HP Stripper at the bottom (red line) is further processed in the low-pressure recirculation section.

For the three heat exchangers, the corrosive process medium is on the tube-side. The carbon-steel tube-sheet is protected by a corrosion-resistant weld overlay. The stainless-steel heat exchanger tubes are protruding through the tube-sheet and connected by a fillet weld as schematically presented in figure 2.

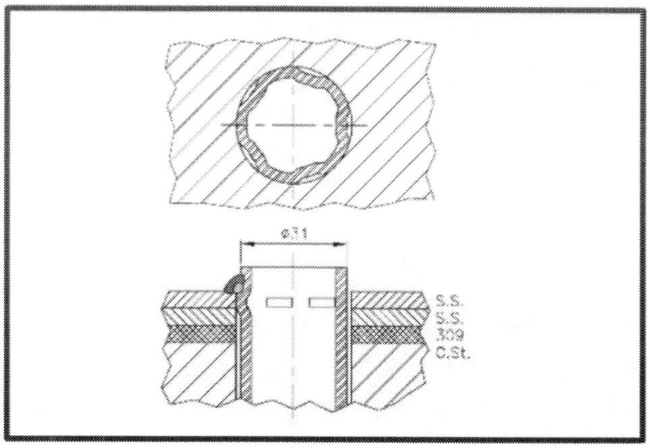

Figure 2: Typical tube-to-tube sheet weld connection. The tube is slightly expanded

The tube-to-tube sheet weld is critical since it is not only sealing off the process media from the carbon steel tube-sheet but is also a strength weld. Typical pressure differences between the process at the tube side (2100 psi, 145 bar) and the shell side (typical 58 psi, 4 bar) are 2030 psi (140 bar).

To ensure proper welding, the tube ends are slightly expanded in the welds overlay area, however the crevice cannot be closed off completely to allow leak detection. An open connection between the crevice in the tube-sheet and the shell side is essential to monitor any leaks in the fillet welds by measuring the conductivity of the steam condensate or cooling water at the shell side. Even though from a corrosion point of view, a crevice should be avoided, Stamicarbon requires this design for leak detection purposes. To mitigate leakages of the tube-to-tube sheet welds, the high quality of this critical weld is secured by observing the following measures during the manufacturing:

- ✓ Welding by orbital welding machines
- ✓ Welding Procedure Qualifications

- ✓ Close control of welding parameters
- ✓ Volumetric UT examination of all TTS-welds
- ✓ Dye Penetrant Testing (DPT) of the TTS-welds
- ✓ Visual inspection for burn-through defects
- ✓ Ammonia leak test after the pressure test

Possible scenarios

Despite optimal design, selection of proper stainless-steel grades such as 316L UG (X2CrNiMo18-14-3), 25-22-2 (X2CrNiMo25-22-2) or super duplex steel grade (Stamicarbon proprietary E-type grade UNS S32906, known as Safurex®), and optimal welding and inspections, corrosion of the tube-to-tube-sheet (TTS) welds and subsequent leakages resulting in corrosion of the carbon-steel tube-sheet cannot be completely avoided. Typical weld defects which can lead to a leakage due to corrosion include:

- o Porosities
- o Worm holes
- o Fusion defects
- o Burn-through defects

Besides corrosion of the tube-to-tube-sheet welds, other possible leak paths to the carbon-steel tube-sheet include:

- ✓ Wall loss of the tube-end by corrosion
- ✓ Cross-cut end attack of the tube-end
- ✓ Local defects in the tube
- ✓ Chloride Stress corrosion cracks

In all these cases, a leak path from the high-pressure tube-side to the low-pressure shell side is possible. The corrosive process fluid (ammonium-carbamate) will enter the crevice between the tube and tube-sheet hole. An example of chloride stress corrosion cracking is schematically presented in figure 3. In case of chloride contamination of the boiler feed water, these chlorides can accumulate in the crevice between the heat exchanger tube and the tube-sheet. Even at low levels of chlorides in the boiler feed water (few ppm) can result in high accumulated concentrations of chlorides in the crevice (up to 1000 ppm). Despite this risk, Stamicarbon does not allow the heat exchanger tube to be rolled in to close the gap, since the gap allows for leak detection in case of a carbamate leak. The possibility of having an early warning for a carbamate leakage outweighs the risk for stress corrosion cracking.

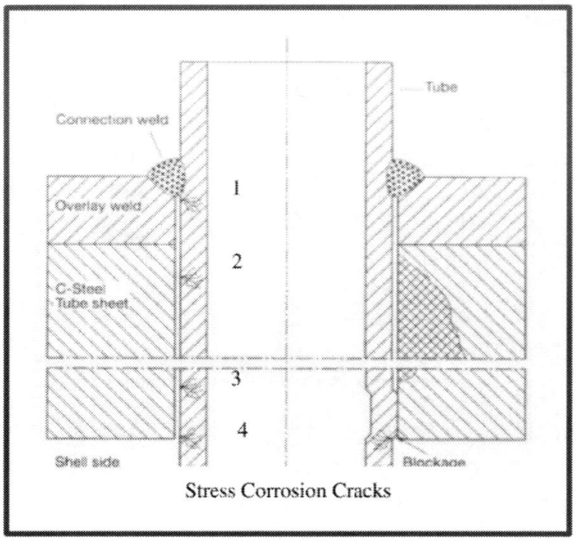

Figure 3: Possible chloride SCC in the top tube-sheet of a HPCC at several locations

As can be seen from Figure 3, the chloride stress corrosion cracks start from the outer tube surface (facing the crevice) and travel towards the inner surface resulting in an ammonium-carbamate leak from tube-side towards the crevice. These cracks can start at several positions inside the crevice, however, the most common are positions 1 and 3. The chloride stress corrosion cracks can be detected and evaluated by an eddy current inspection.

Figure 4 presents two other corrosion damages which also can lead to a leakage of ammonium-carbamate towards the carbon-steel tube sheet. In the upper picture, a severe crosscut end attack is visible in the heat exchanger protruding tube ends in a HP Scrubber. The brownish color of the tube ends is typical for the 316L UG tube material exposed to ammonium-carbamate. 316L UG tube material is prone for crosscut attack, since at

the tube cut end surface area, non-metallic inclusions (silicon oxides) emerging to the surface may create corrosion which results in a rough surface. Subsequently, the corrosion will progress into the tube-to-tube sheet weld, resulting in a carbamate leak.
The lower picture shows the tube-to-tube sheet welds in a HPCC constructed in the X2CrNiMo-25-22-2 steel grade. One of the welds shows a small pin hole (welding defect). Due to oxygen depletion of the ammonium-çarbamate penetrating the pin hole, active corrosion creates a leakage through the tube-to-sheet weld.

Figure 4: Two corrosion examples resulting in leakages.

In principle, the high-pressure ammonium-carbamate will decompose into non-corrosive ammonia and carbon dioxide when leaking to the low-pressure shell side, but this is not always guaranteed. Therefore, if a leak is detected, immediate action is required to avoid severe corrosion of the carbon steel tube sheet.

Repair strategies

Due to an unattended or unnoticed leak, severe corrosion of the carbon-steel tube-sheet may occur in a relatively short period, resulting in large cavities underneath the stainless-steel weld overlay. This cannot be readily repaired by welding since, due to the presence of the stainless-steel weld overlay as well as heat exchanger tubes, a post-weld heat treatment (PWHT) of the repaired carbon-steel tube-sheet is not possible. The temperatures required for this heat treatment will impair the corrosion-resistant properties of the stainless-steel weld overlay and heat exchanger tubes.

For more than 60 years, Stamicarbon has dealt with severe corrosion damage in tube-sheets and some unorthodox repair strategies have been developed and implemented with success.
The principles of these repair strategies are schematically described in Figure 5 till 8. These repair options require no welding on the carbon steel tube-sheet.

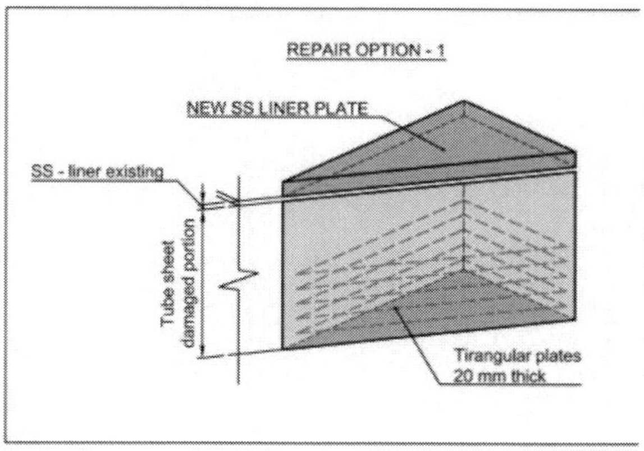

Figure 5: Repair option 1: filling cavities using plates

The first repair option, the cavity is filled with steel plates and covered with a stainless-steel plate which is connected by welding to the existing weld overlay.

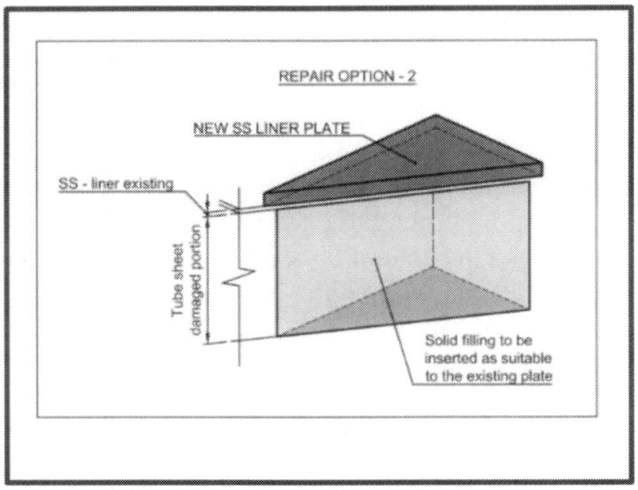

Figure 6: Repair option 2: filling the cavity with a solid steel block

Repair option 2 entails the closure of the cavity by a single steel block machined to fit. Again, the steel block is covered with a stainless-steel plate.

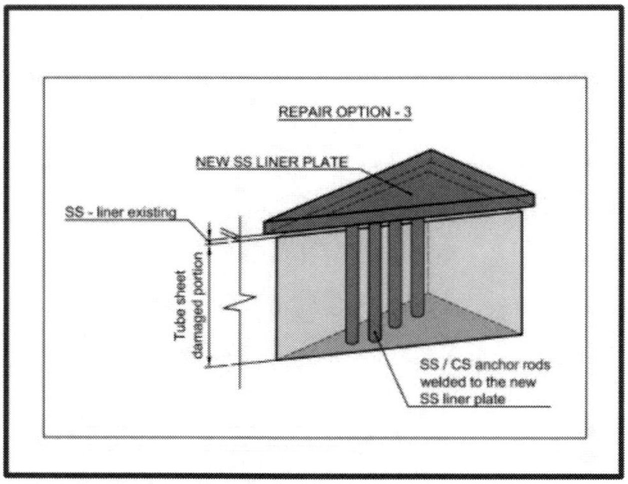

Figure 7: Repair option 3: Cavity is covered by a stainless-steel plate supported by anchor rods

Repair option 3: The cavity is only covered by a stainless-steel plate supported by steel rods anchored in the tube holes.

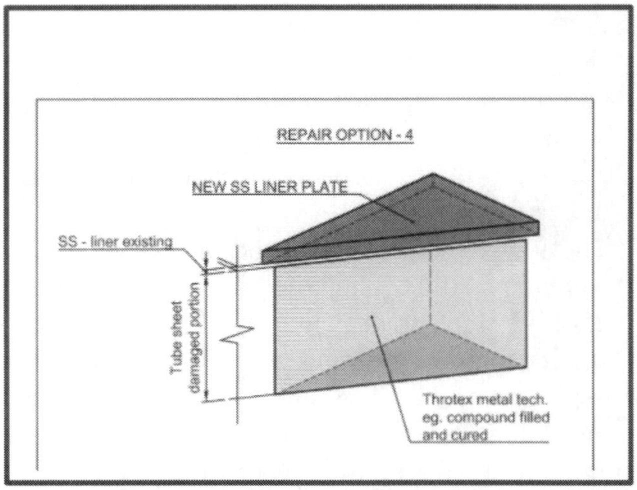

Figure 8: Repair option 4: filling the cavity with metal compound

Finally, the 4th repair option entails the closure of the cavity with a metal compound and again covered by a stainless-steel plate.

In the next section two cases will be presented.

Case History 1: HP Scrubber

The first case history is a HP Scrubber in a Stamicarbon CO_2 stripping urea plant in the Middle East. A few months before the planned turnaround in September 2013, a sudden increase in conductivity in the cooling water at the shell side was noticed, indicating a leak from the process side (tube side).
The HP Scrubber was manufactured in 2002, and the all-wetted internal parts were designed in the X2CrNiMo25-22-2 austenitic stainless-steel grade. The HP Scrubber entails a spherical explosion dome on top of the vertical heat exchanger (see Figure 9).

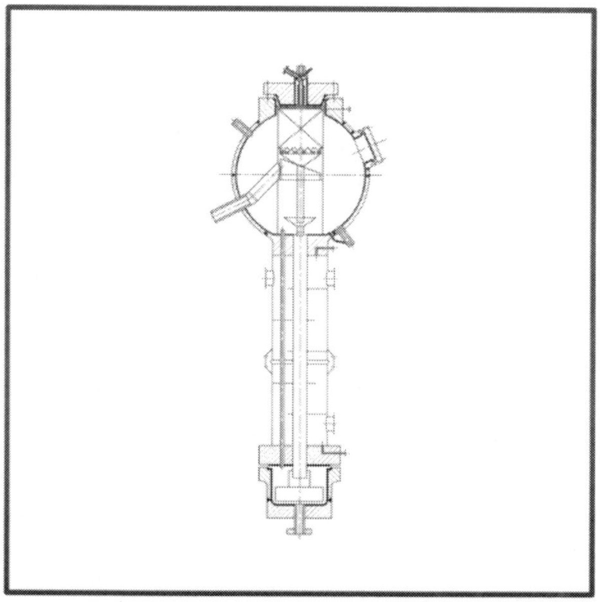

Figure 9: Conceptual design of HP Scrubber

At the tube side, condensation of gaseous carbon-dioxide in the liquid ammonium-carbamate will take place. The condensation heat is removed by circulating cooling water (CCW) at the shell side. The cooling water is continuously monitored for conductivity and an increase could indicate a leakage.

During the planned turnaround, a full inspection was carried out, including an eddy current measurement of the heat exchanger tubes for wall losses and localized stress corrosion cracks. Initially, no abnormalities were found. However, during the subsequent air soap test two leaking tubes were identified, see Figure 10.

Figure 10: Air soap test indicating one of the leaking tubes in top tube-sheet area

The leakages were visible in the top tube sheet. Both tubes were removed for further metallographic examinations (see figure 11).

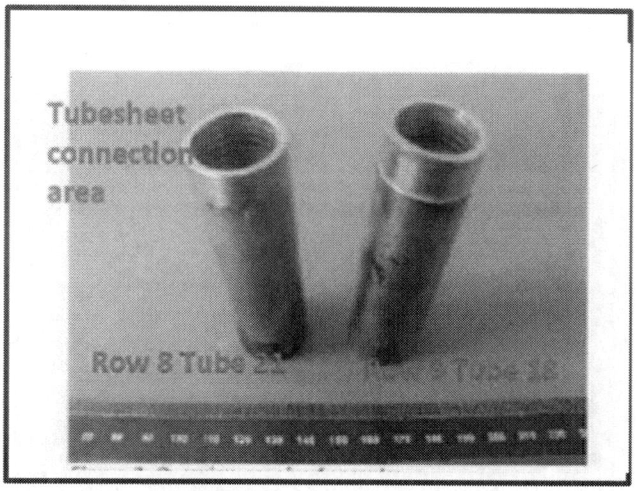

Figure 11: two leaking tubes removed.

Subsequent metallographic examination confirmed that the root cause of the leakage was related to chloride stress corrosion cracking. Cracks started from the outer surface (OD) of the heat exchanger tube, propagating to the inner bore side (ID) resulting in a leak, see figure 12. It is important to determine the source of the chlorides to the CCW system to avoid re-occurrence.

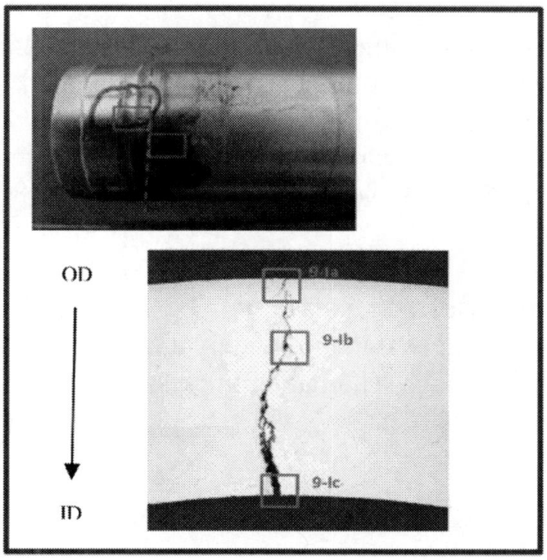

Figure 12: Chloride stress corrosion cracks through wall leading to leakage

After removal of the tube, a cavity underneath the stainless-steel weld overlay was observed around both leaking tubes with a maximum radius of around 4.72-inch (120 mm) mm and a depth of around 2.44-inch (62 mm), see figure 13.

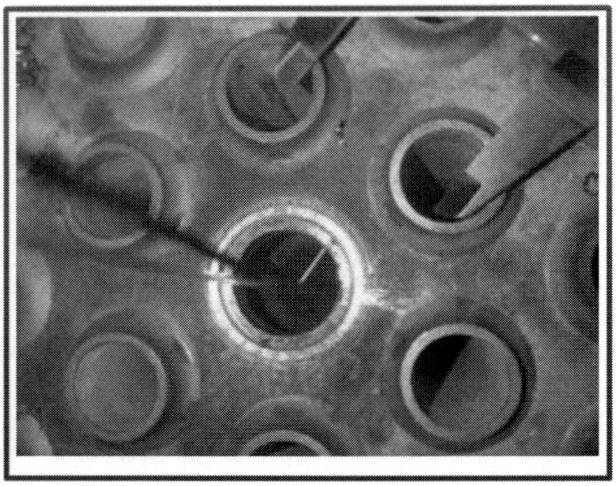

Figure 13: One of the two cavities developed underneath the weld overlay

The size of the cavity was such that a repair by plugging the tube holes was not sufficient to restore the integrity of the tube-sheet. The tube-sheet is made of HII (1.0425) DIN 17155 steel grade and with a design pressure of 2,378 psi (164 bar) and design temperature of 428 °F (220 °C). The cavity had to be filled. The assessment was done according to AD 2000 Merkblatt-B5: 20009-03. Based on this assessment, the thickness of the cover plate over the cavity should be at least 0.866 -inch (22 mm), whereas the actual thickness (weld overlay) is only 0.315-inch (8 mm).

For this case, the repair option 2 using a steel plug was selected to completely close the cavity. The principle is schematically presented in figure 14.

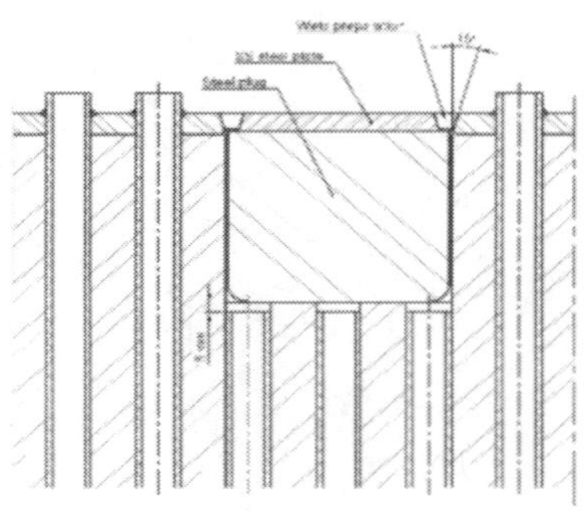

Figure 14: Principle of repair

More in detail following repair procedure was followed, see pictures 15, 16 and 17:

- ✓ Remove six tubes surrounding the leaking tube
- ✓ Smoothen the cavity by grinding
- ✓ Prepare a carbon steel plug with the exact shape and size of the cavity and covered with a stainless-steel plate (urea grade: 25-22-2)
- ✓ Weld the stainless-steel plate to the existing weld overlay
- ✓ Plug the tubes at the bottom tube-sheet

After the repair, the HP Scrubber was pressure tested and subsequently the repair was checked for any leakage using an ammonia leak test. After the repair and subsequent tests, the HP Scrubber was put into operation successfully without any problems.

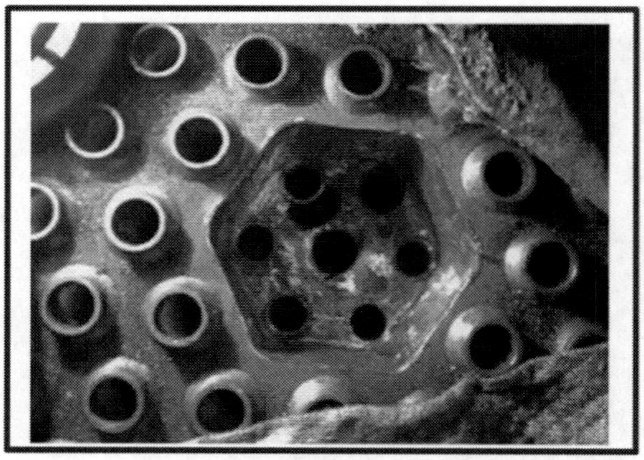

Figure 15: Excavating, smoothening the cavity

Figure 16: Steel plug with SS-steel coversheet

Despite trouble free and safe operations, the HP Scrubber was replaced within two years, since the shell was contaminated with chlorides which are trapped in the crevices of the tube-sheet. Further leakages therefore cannot be excluded.

Figure 17: Plug installed in the cavity

History 2: HPCC

The second case is severe corrosion of the carbon-steel tube-sheet of a falling film high pressure carbamate condenser (HPCC) in a Stamicarbon CO_2 stripping urea plant with a name plate capacity of 1750 MTPD and commissioned in 2000. The HPCC has 2929 tubes made from 316L UG steel grade (X2CrNiMo18-14-3).

The HPCC is a vertical falling film condenser where liquid containing ammonia and ammonium-carbamate together with gaseous carbon-dioxide and ammonia vapor enter from the top. The established liquid film running down the tubes reacts with the co-current gas vapor mixture, condensing the carbon-dioxide and ammonia and forming liquid ammonium-carbamate. The reaction heat is removed by producing low pressure steam at the shell side. The conceptual design of the HPCC is presented in Figure 18.

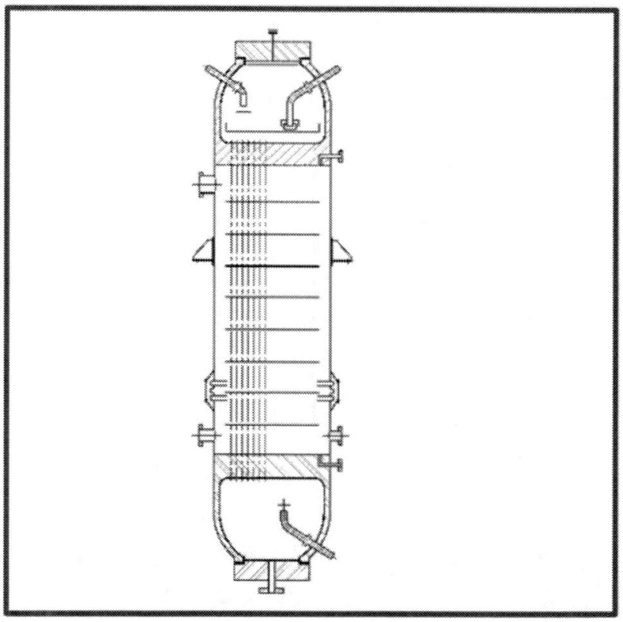

Figure 18: Conceptual design of a HPCC

In 2007, seven years after commissioning, severe corrosion was observed at some tube ends in the bottom tube-sheet (see figure 19).

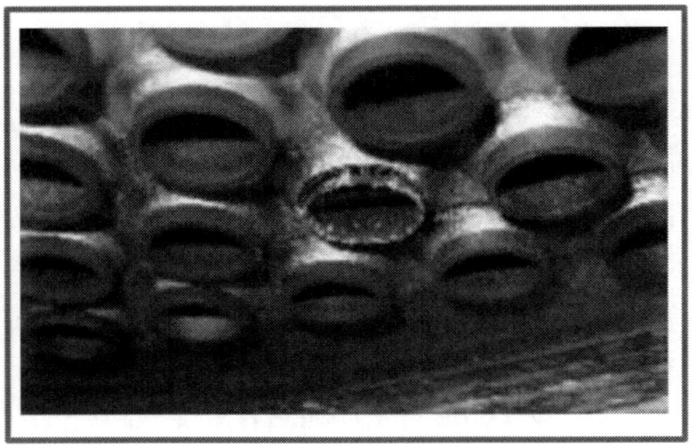

Figure 19: Corrosion at tube ends in the bottom tube-sheet observed in 2007(zone 2)

These tubes were plugged preventatively to avoid a leakage of ammonium-carbamate to the carbon-steel tube-sheet.

In 2020, a steep increase in conductivity was observed in the blow-down of the steam condensate, indicating a large leak in the vessel. The plant was immediately shut down and inspection revealed not only leakages from the tube plugs installed in 2007, but also other tubes clustered together showing leakages at severely corroded burn through defects when pressuring the shell side with water. A burn-through defect is an area inside the tube at the tube-to-tube-sheet weld where the corrosion resistance is compromised by the welding heat. The leakages are found in two areas (zone 1 and zone 2) in the bottom tube-sheet area, see figure 20.

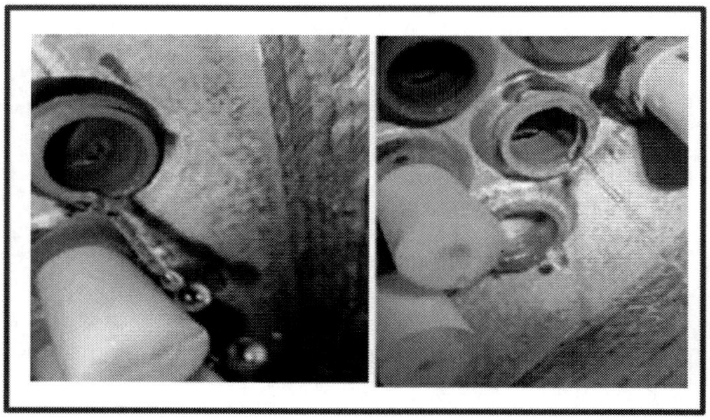

Figure 20: Water seeping out at the severely corroded burn-through defect bottom tube-sheet (zone 2)

Both clusters of leaking tubes are located close to the knuckle radius of the tube-sheet opposite each other.

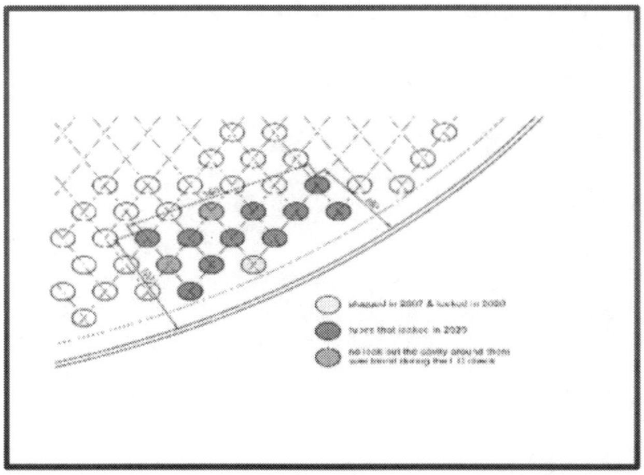

Figure 21: Cluster (zone 2) of leaking tubes close to knuckle radius of the bottom tube-sheet.

In figure 21 the yellow marked tube was already plugged in 2007. The red marked tubes were found leaking in 2020 and both green marked tubes were not leaking, but the cavity underneath the weld overlay extended to these tubes.

After removal of the tubes from both clusters a big cavity was found underneath the weld overlay as presented in figure 22.

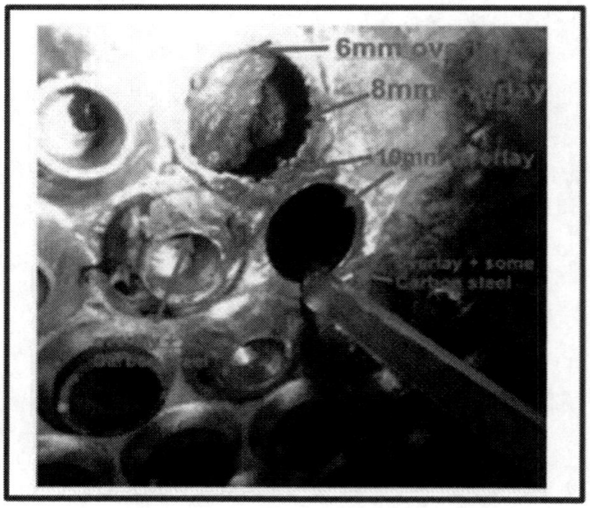

Figure 22: cavity underneath weld overlay; cluster near knuckle radius (zone 2)

Since the knuckle radius of the tube-sheet is a highly stressed area, the integrity of the tube-sheet could not be safeguarded by just plugging the tube holes and it was decided to reinforce the tube-sheet by installing a steel plug in both cavities in both clusters.

Since the repair had to be carried out at the bottom tube-sheet, it was decided to install a carbon-steel plate supported by stud rods (repair option 3), see figure 23.

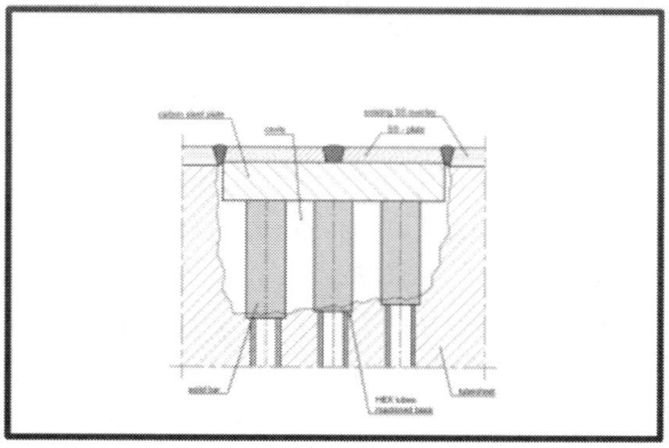

Figure 23: Concept of repair by using a carbon-steel plate supported by stud rods

The thickness of the carbon steel plate is calculated to withstand the synthesis pressure without the studs. The studs are additional supports anchored in the tube holes of the tube sheet.

The repair procedure entailed the following steps (see also figures 24, 25, 26 and 27):

- ✓ Removing the unsupported weld overlay over the cavity
- ✓ Grinding smoothly the cavity
- ✓ Install the carbon steel stud rods
- ✓ Install the carbon-steel plate
- ✓ Cover the carbon-steel plate with a stainless-steel plate (urea grade: 25-22-2)
- ✓ Weld the stainless-steel plate to the existing weld overlay
- ✓ Perform a hydrostatic test and an ammonia leak test

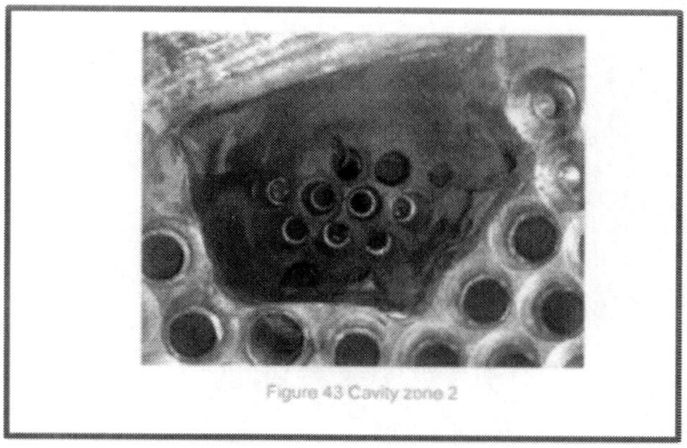

Figure 24: Removal of weld overlay and smoothing the cavity next to the knuckle radius

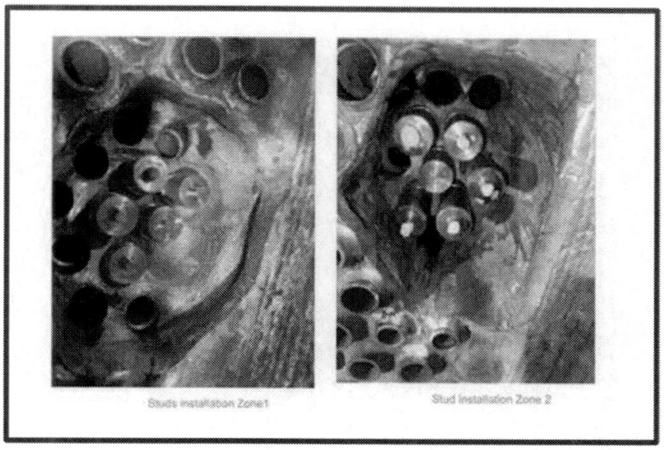

Figure 25: Installation of stud rods

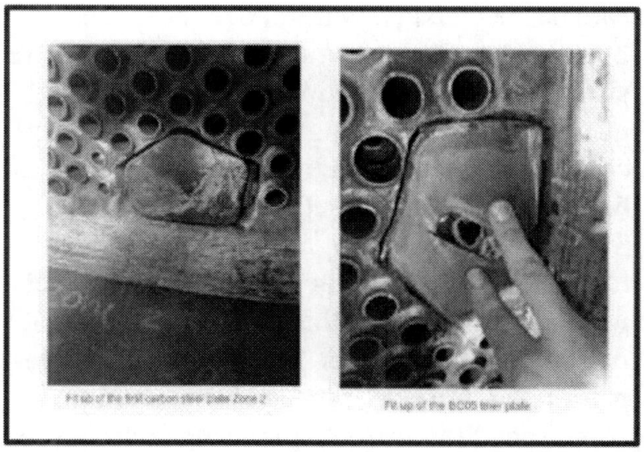

Figure 26: installation of CS and SS plates

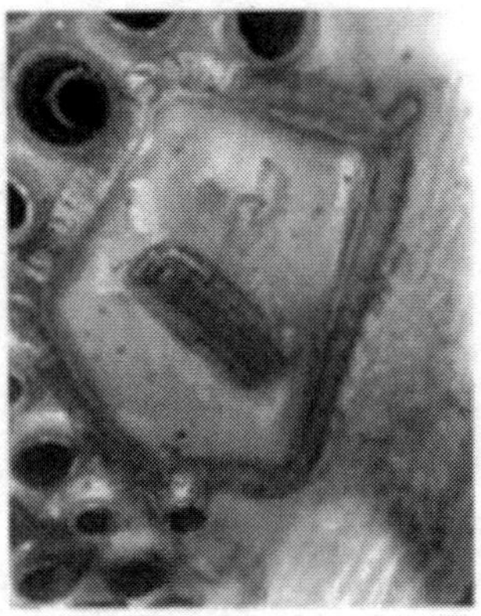

Figure 27: welding the SS-steel plate to the existing weld overlay

After this repair the HPCC was successfully put into operation without problems for a couple of years before the vessel was replaced.

Concluding remarks

Stamicarbon, as licenser of Urea technology, provides full life-cycle services to its customers for more than 60 years. Following failures, challenging repairs have had to be done on carbon steel tubes-sheets of high-pressure urea synthesis equipment, with creative solutions having to be developed. Failures still occur even though care has been taken to minimize these mishaps by proper design, selection of high corrosion resistant materials, selecting and observing adequate welding technologies and careful quality control during manufacturing of this critical high-pressure equipment.
Since repair welding of a damaged carbon-steel tube-sheet is not feasible (no possibility to perform a local-PWHT), this necessitates alternative solutions for repair. In this paper, some repair methods are described. This is possible by combining Stamicarbon knowledge with respect to ammonium-carbamate corrosion, urea grade materials, equipment design, welding technologies and extensive knowledge on the process conditions in the high synthesis section. In this way, critical equipment could be safely put back into operation, at least for a couple of years until replacement.

References

[1] From "potentially" unsafe to unsafe by inadequate tube plugging, by Joost Roes, Stamicarbon. Proceedings AIChE Ammonia Safety Conference 2015, Boston, USA.

[2] Operating Urea Plants with Small Leaks Pose Potential Safety Risks by Alex Scheerder Stamicarbon. Proceedings AIChE Ammonia Safety Conference 2014, Vancouver Canada.

[3] Repair of a Pool Condenser Tube-sheet Severely Damaged by Corrosion by Alex Scheerder, Robert |Rozek, Stamicarbon and Richard Jandl and Robert Bunzl, SBN. Proceedings AIChE Ammonia Safety Conference 2017, New York, USA.

Failure Analysis and New Design of Secondary Waste Heat Boiler (102-C) of Ammonia Plant at IFFCO Kalol

Waste heat boilers (WHBs) are crucial for recovering heat from the hot process gases to generate high-pressure steam at ammonia plants. Secondary waste heat boiler at front end of ammonia plant plays a significant role in recovering heat from the hot exhaust gases of the secondary reformer after primary waste heat boilers. This WHB is Important for two purposes i.e., to generate high pressure steam and to control the inlet process gas temperature of the high temperature shift (HTS) converter at IFFCO Kalol Unit. Failure in WHB caused by various factors, including design deficiencies, defects introduced by manufacturing or fabrication, service-related deterioration and/or upset in plant operation. In this article, efforts have been made to share the operating experience, failures analysis and benefits achieved with the replacement by a new improved design of the secondary waste heat boiler at Ammonia plant of IFFCO Kalol Unit.

Bhupen Mehta
IFFCO Kalol

Introduction

Kalol Unit, the oldest unit of Indian Farmers Fertilizer Co-operative (IFFCO) Limited, commissioned in the year 1974 to produce 910 metric tons per day (MTPD) ammonia and 1200 MTPD urea. The ammonia plant process was based on MW Kellogg, USA, technology and Urea Plant process technology was Stamicarbon CO_2 Stripping Process. In August 1997, the capacity of the ammonia and the urea plant has been enhanced to 1100 MTPD and 1650 MTPD respectively under a Capacity Enhancement Project (CEP).. Progressive improvement in productivity of ammonia plant with passage of time in a decade is shown at **Figure 1**.

This provides the maxium average of daily ammonia production since commissioning of plant during the decade and as mentioned ammonia plant capacity was enhancd to 1100 MTPD in 1997, but this capacity level has been achieved only after 2006-07 on a consistent basis. Also, two Energy Saving Projects (ESP) have been implemented during 2004-05 and 2014-17 to improve the productivity of Ammonia Plant. Presently, ammonia plant is in operation at 1140 MTPD.

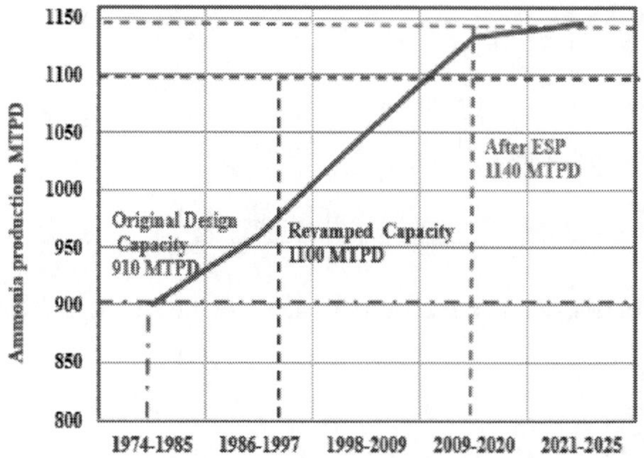

Figure 1: Progressive ammonia production

Heat Recovery and Secondary Waste Heat Boiler (S-WHB)

102-C is a vertical shell and tube boiler and has a bypass line with temperature control valve to maintain the inlet process gas temperature of high temperature shift (HTS) converter. High pressure steam is generated in the boiler utilizing the heat recovered from process gas. This gas enters the inlet channel through two separate inlets from 101-CA and 101-CB waste heat boilers. The bypass flow controlled by temperature control valve TRCV-10 by diverting part of the process gas. The remaining gas flows through the tubes of the boiler and gets cooled and goes out through manual butterfly valve to mix with the bypass flow and then to HT shift converter as shown - **Figure-2**.

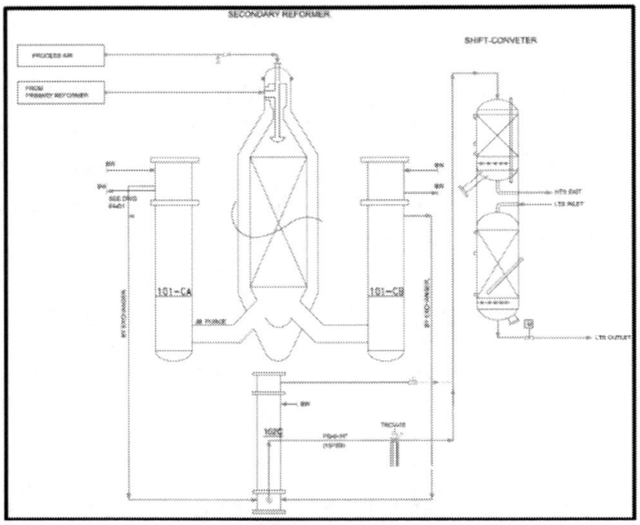

Figure-2: Waste heat boiler at Ammonia Plant

The water flow through the boilers is by natural circulation. The water from the steam drum flows down through downcomers, gets heated up at the heating surface of the boilers and steam bubbles are formed. These steam bubbles and water together are displaced through the risers to steam drum by heavier water from the steam drum in the downcomer. Temperature indicators located on the risers and on the bottom and top surfaces of the steam drum indicate the metal skin temperature and enable to maintain the temperature difference within the designed limit during heating and cooling of the system. Steam drum and associated waste heat boilers of ammonia plant are shown in **Figure-3**.

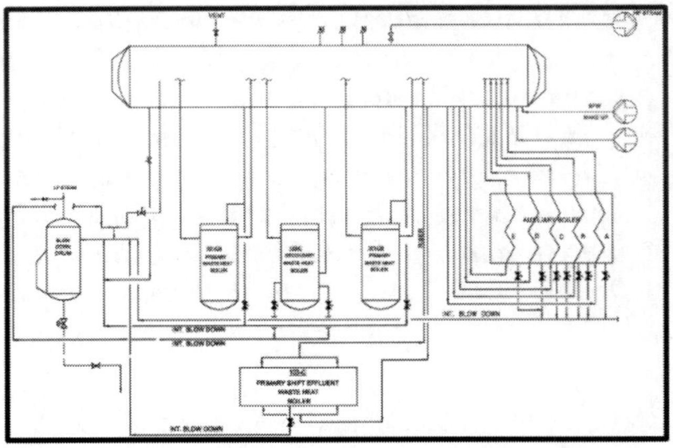

Figure-3: Steam drum and waste heat boilers of Ammonia Plant

Operating Experience of Secondary Waste Heat Boiler

The secondary waste heat boiler (S-WHB) is proprietary design of MW Kellogg and was in service since April 1974 with 910 MTPD ammonia production till 2008 and has been replaced in 2008 with the original design without any modifications even though plant capacity has been increased to ammonia production rate of 1100 MTPD. The WHB has a bypass line with a temperature control valve, having a 38- bar steam injection for gland sealing. As shown at Figure-2, there are two inlet process gas lines at the bottom channel of the WHB and one bypass tapping for the TRCV , which mixes with the exit process gas to control the inlet process gas temperature of HTS.

The first tube leakage in this WHB was observed after ten years of service. The chronology of 102-C leakage is as listed below:

Year 2017: Remote Field Eddy Current Testing (RFET) carried out and found satisfactory

Year 2019: Leak observed during running plant, one tube found leaking in hydrotest and leaky tube plugged. (1st failure).

Year 2021: Leak observed during running plant, one tube found leaking in hydro test and leaky tube plugged. (2nd failure).

Year 2022: Leak observed during running plant, 3 tubes found leaking in hydro test. Also, RFET was carried out, 49 tubes were having thickness reduction of 20 to 40%. Total 54 tubes were plugged in year 2022. (3rd failure)

Year 2023: Leak observed from 1 tube, so this tube was plugged and 12 more tubes in nearby area are also plugged as a preventive measure. Total 13 tubes were plugged in year 2023. (4th failure)

There was a particular pattern of tube failure as shown at **Figure-4**. Failure zone was at the opposite side of the bypass nozzle. Total 71 nos. tubes were plugged (out of 750) in the WHB as of April 2023. As the heat duty of this WHB increased due to capacity enhancement and with above plugged tubes, operational bottlenecks have been faced in maintaining the inlet process gas temperature of HTS at start of run (SOR) condition as it required throttling of bypass temperature control valve (TRCV) to divert the process gas through this WHB. This mode of operation also increased the front-end pressure drop and increased the discharge pressure of process air compressor.

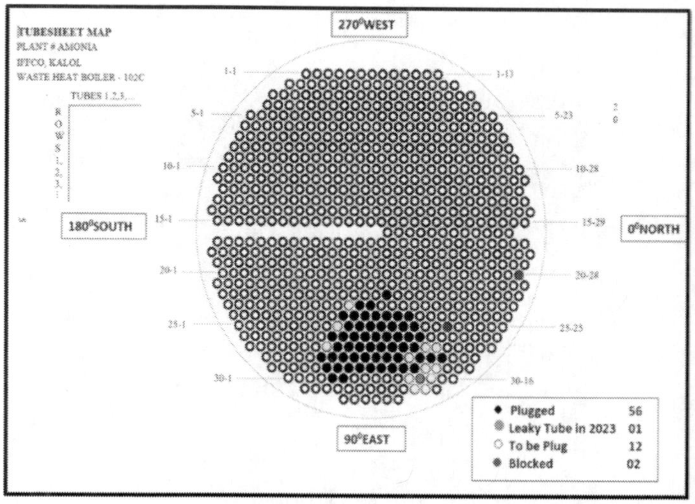

Figure-4: Tube failure locations

Tube failure photographs and defect zone are shown in **Figure 5** and **Figure-6** respectively. Defect zone was about 2485 mm from top tube sheet and exactly at the 3rd baffle from the top tube sheet as shown in Figure-5.

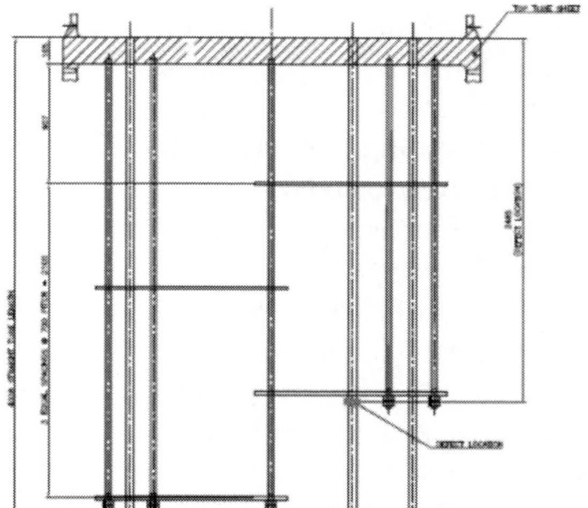

Figure-5: Defect location of 102-C

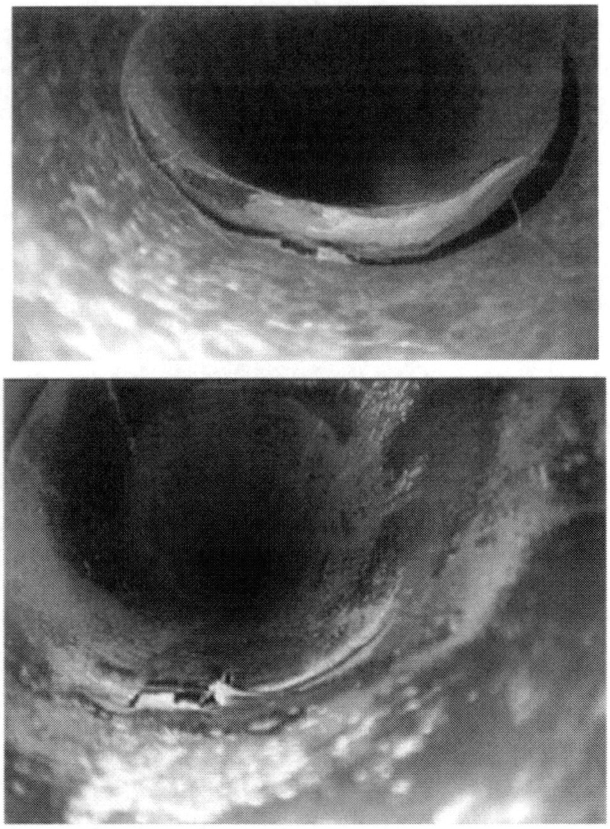

Figure-6: Tube failure photographs

Major observation regarding the failure pattern were as under:

- All defected tubes were observed having baffle fretting like indication at approx. 2.5 meter from top tube sheet.
- 5 nos. of tubes were observed plugged before RFET inspection.
- 21 nos. of tubes were observed blocked during RFET inspection.
- 16 nos. of tubes were observed having 31-40% wall thickness loss compared to calibration curve fit. All 16 nos tubes plugged.
- 33 nos. of tubes were observed having 21-30% wall thickness loss compared to calibration curve fit. 71 nos of tubes plugged.
- Rest of all inspected tubes were observed having <20% wall thickness loss compared to calibration curve fit.

Cause of Failure and improved design of 102-C

Work order placed on M/s KBR India to conduct the study for root cause analysis of tubes failure and engineering services for new improved design in Sept 2022 with following scope of work:

- Investigation on cause of failure of tubes/ loss of metal
- Thermal performance evaluation of existing WHB as per actual plant operating data
- Evaluation of BFW quality
- Recommendations for new design of WHB based on KBR experience in past.

KBR submitted the performance of front-end waste heat boilers based on actual operating data of ammonia plant to the best possible fit and clean case scenario as shown at **Table-1**.

Parameters	Primary WHB (101-C A/B)		Secondary WHB(102-C)	
	Fouled	Clean	Fouled	Clean
Process temperature in/out (°C)	970 / 459	970 / 411	459 / 338	411 / 322
Process side fouling (hr-m²-C/kcal)	0.00043	0.00000	0.00026	0.00000
BFW side fouling (hr-m²-C/kcal)	0.00010	0.00000	0.00010	0.00000
Fraction of Process flow in exchanger	100.0%	100.0%	72.9%	45.0%
Fraction of Process flow in bypass	NA	NA	27.1%	55.0%
Process side pressure drop (kg/cm²)	0.171	0.164	0.174	0.069
Duty (Gcal/hr)	40.1	43.8	6.5	2.9
Steam generated (kg/hr)	131400	143200	21400	9600
Vaporization	8.6%	9.4%	6.4%	3.0%
BFW Friction pressure drop (kg/cm²)	0.497	0.507	0.390	0.337

Table-1: Performance of Waste Heat boilers based on actual plant data

Major observations of the study report were:

- Heat fluxes were higher in the range of 500-1500 mm from bottom tube sheet.
- Tube outer diameter (OD) under-deposit solids in low velocity areas on the shell side.
- All leaked tubes found are at same location, at second baffle from shell inlet. Fretting damage was the suspected reason for tube failure.
- Since close to 10% of the total tubes plugged hence replacement of the 102-C with a new one with improved design proposed.

- No piping modification on tube side inlet and outlet nozzle envisaged by the WHB replacement.
- WHB replacement shall have same shell diameter and tube length for the required 1140 MTPD ammonia production.
- Flow at the shell inlet nozzle should sweep the tube sheet and avoid stagnant areas. Sweep baffle at bottom space to provide high water velocity.
- Design improvements include improvements to internals to provide better blowdown abilities and increased water velocities in the lower portion of the WHB. Incorporate improvements to the internals to minimize deposits from collecting in the lower portion of the WHB.
- Improved metallurgy and weld overlay on the tube sheets will also make any field repairs easier.
- For the existing 102-C, welding the tubes to the tube sheet in the field is difficult due to tubes and tube-sheet of low alloy. For new WHB Inconel overlay of the tube sheet is proposed.
- Forged hubs provided on both sides of the tube sheet.
- Three blow-down nozzles (2") provided at bottom.

In brief, new WHB has replaced with fixed tube sheet configuration with no size change to retain same tube side nozzle locations with following proposed improvements:

- Upgraded materials.
- Weld overlay on the tube sheet face to allow easier field repairs when required.
- Modified tube layout and blowdown arrangement to reduce the potential of boiler water settlement and corrosion of the tubes in the lower tube sheet face area.
- Improved shell side baffling to minimize the potential of solids settling.

The comparison of old and new improved design of WHB (102-C) given at Table-2.

S.No	Descriptions	Unit	Original Installed 102-C in 1974		Existing 102-C installed in 2008		New 102-C installed in 2025	
A	Process Parameters							
			Shell Side	Tube Side	Shell Side	Tube Side	Shell Side	Tube Side
1	Design Pressure	kg/cm2g	117.76	34.09	118.00	34.10	118	34.1
2	Operating Pressure	kg/cm2g	105.46	30.65	105.46	30.65	108.1	30.65
3	Operating Temperature	degC	314 (IN) / 314 (OUT)	482 (IN) / 371(OUT)	314 (IN) / 314 (OUT)	482 (IN) / 371(OUT)	315 (IN) / 315 (OUT)	459(IN) /
4	Corrosion Allowance	mm	3.18	3.18	3.00	NIL	3.00	3.00
5	Surace Area	m2	237 m2		237 m2		243 m2	
B	Constructional Parameters							
1	Tube Size		25.4 MM X 4.19 MM Thk		25.4 MM X 4.19 MM Thk		25.4 MM X 4.19 MM Thk	
2	Tube Material		SA 213 T11		SA 213 T11		2-1/4 Cr - 1 Mo	
3	Tube Length		4028.00		4028 MM		4028	
4	No. of Tubes	No	750.00		750		774	
5	Top Tube Sheet Material &Thickness		SA 182 Gr F1, 165 MM		SA 336 Gr F11 Class-2, 165 MM		1-1/4 Cr- 1/2 Mo with 10mm Inconel	
6	Bottom Tube Sheet Material &Thickness		SA 182 Gr F11, 165 MM		SA 336 Gr F11 Class-2, 232 MM		2-1/4 Cr - 1 Mo with 10mm Inconel Overlay.	
7	Bottom (hot end) Channel Material & Thickness		SA 387 Gr C (Annealed), 54 MM		SA 387 Gr 11 Class-1, 60 MM		1-1/4 Cr- 1/2 Mo with 10mm Inconel Overlay.	
8	Top (Cold End) Channel material & Thickness		SA 204 Gr B, 25.4 MM		SA 387 Gr 11 Class-2, 30 MM		1-1/4 Cr- 1/2 Mo with 10mm Inconel Overlay.	
9	Shell ID & Thk	mm	1193.8 X 66.68		1194 X 70		1194x30	
10	Shell Material		SA 516 Gr 70 (Normalised)		SA 516 Gr 70		1-1/4 Cr- 1/2 Mo	
11	No. of Baffles	No	4.00		4.00		5 + 2 sweep baffles	
12	Baffle Material & Thickness		SA 515 Gr 70		SA 516 Gr 70 X 16 Thk		1-1/4 Cr- 1/2 Mo	
13	Baffle Spacing	mm			1st 907 from top and then @ 700		562	
14	Bypass Nozzle Reducer Size		20" X 14"		20" X 14"		20" X 16"	
15	Shell Nozzles Size (In/Out)		12" / 12"		12" / 12"		12" / 12"	
16	Tube Size Nozzles(In/Out)		2 X 28" / 18"		2 X 28" / 18"		2 X 28" / 18"	

Table-2: Comparison of Secondary Waste Heat Boiler (102-C)

As new WHB have more numbers of tubes and have bigger size bypass line and temperature control valve, stress analysis of nozzles and piping system associated with WHB has also been carried out for process side and utility (BFW/steam) side. Additional supports have been provided as per analysis. Some additional supports for piping have been provided.

Conclusion

A single failure of the waste heat boiler can easily result in monetary loss equal to total cost of the boiler. Hence, reliable and cost-effective design of WHB is a crucial step to improve the productivity of Ammonia Plant. Higher surface area with bigger bypass line size provides the additional operational flexibility. The numbers of baffles increase from 4 to 7 (including 2 nos of sweep baffles). Synergy of sharing the operating experience of WHB with OEM helped to improve the design of new WHB at Ammonia plant of IFFCO Kalol Unit. Installation of the latest design temperature control valve on bypass line helps to eliminate the requirement of 38-bar steam for gland sealing, which saved about 0.5 MT per day of 38-bar steam, and a greater number of tubes helps to reduce the front-end pressure drop of ammonia plant by about 0.04 to 0.05 kg/cm2 at same plant load. The new WHB and associated system is working satisfactorily at 1140 MTPD ammonia plant load from 7th May 2025.

Enhanced Safety and Reliability in Fertiberia Palos Ammonia Plant with the Replacement of Obsolete Boilers with Casale-Arvos Double-Tube Design

Fertiberia Palos ammonia plant has successfully restarted after a significant intervention with the objective of improving safety and reliability of the process gas cooling section. As part of the upgrade, outdated 101-CA/B and 102-C boilers downstream of the secondary reformer were replaced with three new double-tube boilers designed and manufactured by Arvos. This paper outlines the installation of the Casale-Arvos boilers and the plant's successful start-up, emphasizing the safety and reliability improvements introduced by the new system. The robust double-tube design, selected to reduce the risk of failures providing an increased safety and reliability of the ammonia unit, enhanced also plant's overall performance. The replacement was completed with minimum plant modifications, reducing investment costs while ensuring safe procedures.

Felipe Ramòn Fernandez Latorre
Fertiberia

Heinrich Ormann
Arvos GmbH | SCHMIDTSCHE SCHACK

Francesco Baratto, Marco Mazzamuto Carlucci
Casale SA

Background

In conventional ammonia plants, particularly those following the classic Kellogg design, process gas cooling has historically been a point of vulnerability. The cooling system typically consists of two primary waste heat boilers (WHBs) of bayonet water tube type and a secondary WHB of vertical fire tube type, located downstream of the secondary reformer. These components are exposed to extreme temperatures exceeding 1000°C (1832°F) and rely on boiler feed water (BFW) for efficient cooling and heat recovery.

Despite their long-standing use, these WHB designs have exhibited persistent reliability issues:

- **Bayonet water tube primary WHBs**: The design suffers from sludge deposition in the bayonet end cap, leading to localized overheating and blockages. Additionally, the unsupported tube structure is prone to flow-induced vibrations, which accelerate tube wear and necessitate frequent maintenance and costly replacements.

- **Vertical fire tube secondary WHBs**: These units also experience sludge deposition, particularly on the lower tube sheet, leading to inefficient heat exchange.

Poor riser nozzle placement contributes to steam blanketing, forming vapor pockets that compromise thermal efficiency and increase stress on tube walls, leading to failures.

These recurring issues resulted in frequent shutdowns, increased maintenance costs, and elevated safety risks due to the potential for sudden equipment failure [1, 2, 3]. The need for a more robust and reliable cooling solution became critical.

The Schmidt'sche® Double Tube System: A Technological Breakthrough

Recognizing the need for a superior alternative, the engineering team at Fertiberia Palos opted for the Schmidt'sche® double tube system, a well-established solution in high-temperature industrial applications such as ethylene production and gasification plants. This innovative system was adapted for ammonia applications to address the unique challenges faced in process gas cooling [4].

The Schmidt'sche® double tube system (see *Figure 1*) combines the best features of fire and water tube designs, offering:

- Enhanced mechanical stability: Coaxial double tubes are securely welded to oval headers, providing flexibility to accommodate the extreme thermal stresses encountered in reforming.

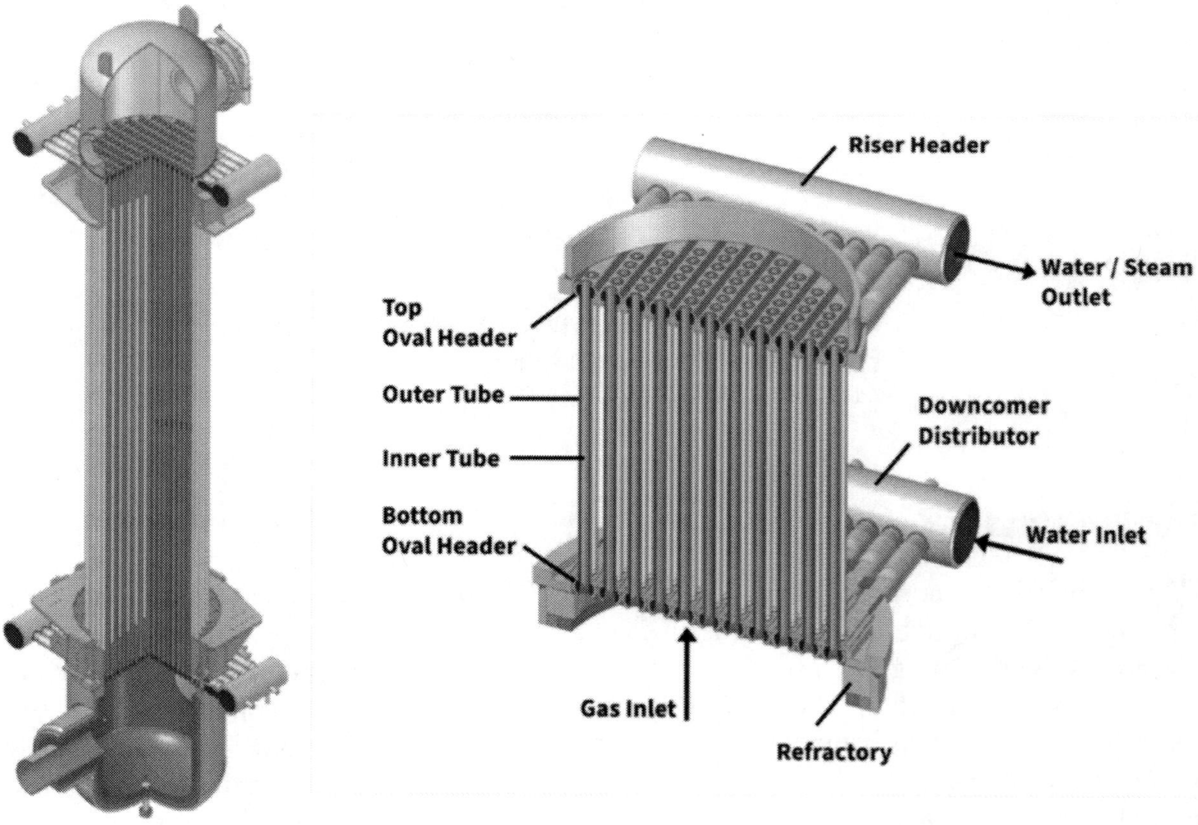

Figure 1. Sketch of double pipe system

- Optimized heat transfer: Water flows between the inner and outer tubes, ensuring efficient phase transformation into steam while maintaining stable temperatures across the system and intensive cooling of the "tube sheet" and tubes (no need for refractory on the tubesheet and ferrules)
- Elimination of common failure modes: high fluid velocity of BFW prevents sludge deposition and steam blanketing.
- Straight flow path on gas and water side: no risk of vibration-induced tube damage, significantly extending equipment lifespan.
- Reduced maintenance demands: Crevice-free, full-penetration welds eliminate corrosion risks, while modular construction allows for easy replacement of individual tube elements without requiring full system overhauls.

The robustness of this system has been demonstrated in industries operating under even harsher conditions than ammonia production. Given that ammonia plants operate at relatively lower pressures and less aggressive environments compared to ethylene cracking and gasification, the Schmidt'sche® double tube system was expected to deliver substantial reliability and safety improvements at Fertiberia Palos.
The Schmidt'sche® double tube system has proven its robustness in severe environments.

- Ethylene Plants
  - Operates cyclically with process gas temperatures of ~900°C and pressures up to 10 bar.
  - Water-steam systems handle pressures of 100-120 bar.
- Gasification Plants
  - Handles gas at up to 900°C and 27-50 bar with water-steam pressures of 50-115 bar.
  - Successfully operates under high fouling, metal dusting, and hydrogen attack conditions.
- Adaptation for Ammonia Plants
  - The conditions in ammonia plants are less severe, making the design highly reliable.
  - The double tube system's experience in harsher environments ensures confidence in ammonia applications.

Implementation at Fertiberia Palos: A Case Study

Fertiberia Palos, a MW Kellogg-designed ammonia plant with a daily production capacity of 1175 metric tons, had faced persistent challenges associated with its legacy vertical WHBs.
The introduction of a proven technology from adjacent industries provided an ideal opportunity to modernize the facility without necessitating a complete redesign of the cooling section.

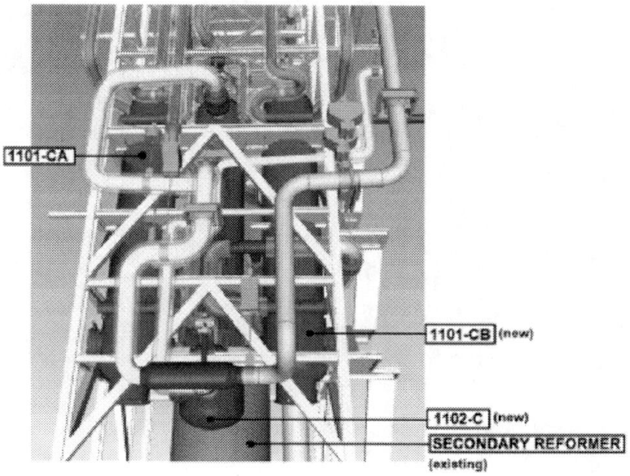

Figure 2. 3D Modeling of boiler section

Casale and SCHMIDTSCHE SCHACK spearheaded the project, with SCHMIDTSCHE SCHACK supplying the new double tube WHBs and reconfiguring risers and downcomers to optimize circulation. Casale managed the detailed engineering phase, which encompassed:

- Structural assessments: A comprehensive laser scan of the plant enabled precise adaptation of the new components within the existing infrastructure.
- Piping and civil modifications: Ensuring compatibility of the new units with pre-existing piping and support structures while minimizing additional investments (see *Figure 2*).

- Instrumentation upgrades: Enhancing monitoring and control capabilities to maximize efficiency and safety of the cooling section.

Fertiberia was responsible for dismantling the old boilers, the installation of the new ones and commissioning/start-up of the plant.

Figure 3. Laser scan of boiler section

One of the biggest challenges encountered during implementation was the constrained space around the secondary reformer—a common limitation in older ammonia plants (see *Figure* ***3***). To mitigate this, the replacement WHBs were designed to fit onto the positions of the obsolete 101-CA/CB and 102-C units. This strategy minimized the need for extensive civil work and preserved the thermal efficiency of the plant by avoiding heat loss through excessive piping modifications.

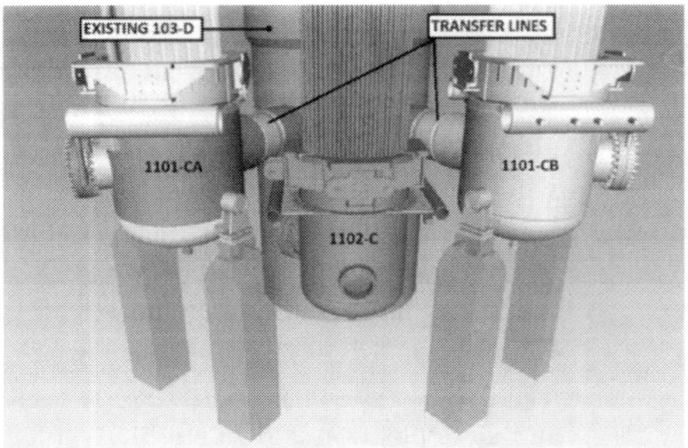

Figure 4. 1101-CA/CB and 1102-C supporting

The new configuration included three double tube WHBs (see *Figure* ***4***):

- 1101-CA and 1101-CB: Installed on newly constructed plinths adjacent to the secondary reformer.
- 1102-C: Supported by the existing steel structure to optimize the utilization of existing space.

A key design improvement was the elimination of external water jackets, which were notorious for being prone to hidden cracks. Instead, the new 1101-CA and 1101-CB units featured lined refractory in the hot channels. This modification simplified failure detection through the use of sensitive paint systems, allowing for early intervention and prevention of extensive damage. To connect the new boilers seamlessly to the existing system, a transition piece was added downstream of the secondary reformer outlet.

The hot gas from the secondary reformer now follows a streamlined flow path. It enters the hot channels of the 1101-CA and CB units, where it is cooled as it rises, before being directed into the top channel of 1102-C for further cooling as it descends. A new bypass line of 1102-C was designed (reusing the existing mixing device) to provide greater flexibility in controlling the downstream high-temperature shift (HTS) inlet temperature, which plays a crucial role in optimizing reactor management throughout the life of the catalyst.

Installation and Start-up: A Seamless Transition

The installation of the new boiler was carried out during an extended turnaround in late spring 2024, which also included other activities, such as the replacement of the refractory lining inside the secondary reformer.

In the absence of other activities, the WHB replacement could be completed in less than four weeks, with around-the-clock work facilitated by effective project planning and monitoring.

The installation sequence involves lifting and positioning 1101-CA/CB vertically. They are initially supported at the base with wooden logs and secured by suspending them from the crane. Once this is done, the reinforcing frame, which is only required for transportation, is dismantled, and the exchanger is lifted, inserted into the plant structure, and placed on its pillar foundation.

Figure 5. Item 1101-CB vertically

Figure 6. Reinforcing frame dismantling

Figure 7. Item 1101-CB lifting

Figure 8. Item 1101-CB lowering

Figure 9. Item 1101-CA placed on foundation

Figure 10. Nozzle alignment

The installation of item 1102-C was carried out as follows: first, the spring supports were hung from the beams of the steel structure, and a temporary steel support was constructed on the ground. Next, item 1102-C was lifted and positioned vertically, placed on the ground on its skirt, and secured while hanging from the crane. Subsequently, its transportation reinforcing frame was dismantled. Following this, item 1102-C was lifted again, inserted into the plant structure, and placed on the temporary steel support with its skirt, while remaining secured by the crane. The spring hangers were then connected to the exchanger lugs, and their lengths were adjusted to achieve the correct elevation. Finally, the lower temporary support was removed.

The criticality of the entire operation is due to the narrow space and the varying height levels of operation. Thanks to proper planning and strict safety procedures, the entire installation was completed without any accidents or injuries.

Figure 11. Spring hangers of 1102-C

Figure 12. Reinforcing frame dismantling

Figure 13. Item 1102-C lifting

Figure 14. Item 1102-C on temporary support

Figure 15. Item 1102-C hanging on spring supports

The spools were prefabricated prior to installation in accordance with the applicable isometric drawings. They were then lifted, fitted, welded, and subjected to NDT testing. Spring hooks and supports were installed progressively as the work advanced. Insulation for both the equipment and the piping was carried out concurrently with the piping installation. Once the welding of the transfer line joints was completed, PWHT and NDT were performed.

Prior to proceeding with the dry-out and boiling-out processes, the entire circuit was filled with water, and the spring hangers were unlocked. The dry-out and boiling-out processes were performed simultaneously to allow for the drying of the secondary reformer at the same time. The boiling-out process lasted a total of 74 hours, after which the chemical solution was drained, and the circuit was rinsed twice with clean water.

Fertiberia requested an inspection of the steam drum, necessitating 6-7 additional flushes to cool it down. Once the steam drum was sufficiently cool, it was opened for inspection. The internal surfaces were found to be clean, with no debris, scales, or other residues. The steam drum was then closed and prepared for startup.

Next, another procedure, recommended by SOLARCA (the chemical cleaning subcontractor), was carried out. This procedure, called Magnetite Accelerated Formation, involved re-boiling the circuit at approximately 40 bar and 250°C for a specific period.

Finally, the dry-out using process gas continued until the operating temperature of 1050°C was reached, at which point the plant was started directly.

After completing boil-out process, Fertiberia prepared the waste heat boilers for operation and started the ammonia plant up. In parallel with the startup activities, Fertiberia performed the steam blowing of high-pressure steam line to syngas compressor turbine. After 3 days, on July 29th 2024, Fertiberia Palos plant resumed ammonia production at 85% capacity. Despite economic constraints preventing operation at full load, the new boilers demonstrated outstanding performance (see *Figure **16***). Key performance indicators observed during initial operations included:

- Stable final temperature of 345°C downstream of 1102-C.
- Optimal bypass valve positioning at 27% open, indicating efficient heat transfer.
- Absence of startup complications, reinforcing the reliability of the new design.

Future-Proofing the Plant

Beyond the immediate improvements in safety and reliability, the new WHB system has been designed with long-term sustainability in mind. One of the standout features of this retrofit is its compatibility with green hydrogen integration. The new boilers are engineered to accommodate a feedstock composition with up to 40% hydrogen content, paving the way for a transition to low-carbon ammonia production.

This proactive approach aligns with Fertiberia's broader sustainability objectives and the industry's shift towards decarbonization. By future-proofing the plant, Fertiberia ensures that its ammonia production remains both economically and environmentally viable in the years to come.

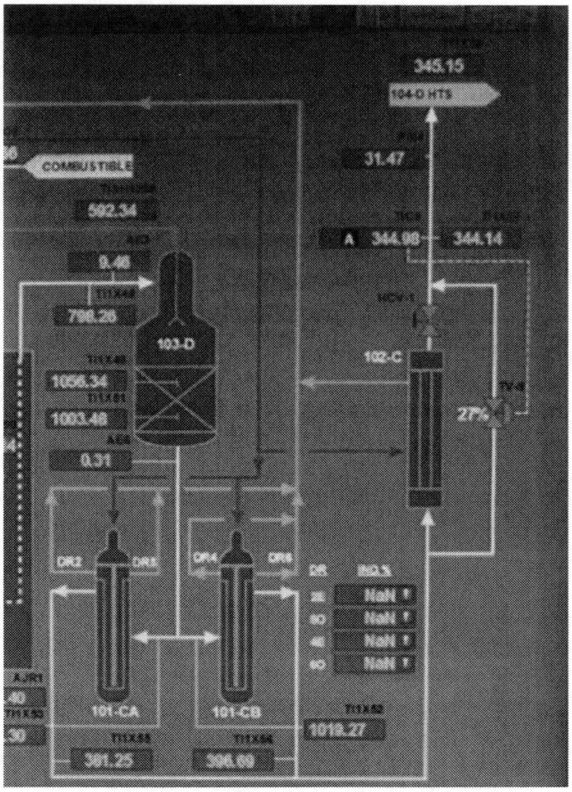

Figure 16. DCS screenshot at 90% load

Conclusions

The replacement of obsolete vertical WHBs with the Schmidt'sche® double tube system represents a transformative step in the modernization of ammonia plant infrastructure. By leveraging proven technology from high-performance industrial sectors, Casale and SCHMIDTSCHE SCHACK have delivered a solution that significantly enhances operational stability while reducing maintenance costs and risks. The successful implementation at Fertiberia Palos serves as a benchmark for other ammonia plants seeking to improve reliability, efficiency, and sustainability in an increasingly demanding industrial landscape.

This project not only addresses immediate operational concerns but also positions Fertiberia Palos at the forefront of technological advancements in ammonia production. By embracing innovative solutions and anticipating future industry trends, the plant has solidified its role as a leader in safe, reliable, and sustainable chemical manufacturing.

Key outcomes and benefits are the following:

- Enhanced Reliability
  - The double tube system eliminates common failure modes associated with traditional WHBs, such as sludge deposition, tube vibration, and welding failures.
- Lower Maintenance and Costs
  - Reduced need for frequent tube replacements and spare part inventory.
  - Modular components simplify maintenance and repairs.
- Environmental and Process Efficiency
  - Optimized thermal performance improves overall plant efficiency.
  - Accommodates green hydrogen integration, supporting decarbonization efforts.

References

[1] I. Ahmad Gondal: Waste Heat Boiler (101-C) Leakages & Possible Causes. 50th Annual Safety in Ammonia Plants and Related Facilities Symposium, Ammonia Technical Manual Vol. 46, 2005

[2] A. Din, S. Hassan and A. Ahmad: A Waste Heat Boiler Replacement – From Bayonet to Fire Tube Design. 53rd Annual Safety in Ammonia Plants and Related Facilities Symposium, Ammonia Technical Manual Vol. 49, 2008

[3] S. Singh, M. Srivastava and M. Masood Ahmad: Failure of Waste Heat Boiler in a Kellogg Designed Ammonia Plant. 57th Annual Safety in Ammonia Plants and Related Facilities Symposium, Ammonia Technical Manual Vol. 53, 2012

[4] J. Weidenfeller, F. Baratto: Reliability Improvement by Replacing the Vertical Process Gas Cooler with Proven Design. 63rd Annual Safety in Ammonia Plants and Related Facilities Symposium, Ammonia Technical Manual Vol. 59, 2018

Nitriding and Crack Propagation in Waste Heat Boiler Lip seals

Ammonia synloop equipment is inherently subject to nitriding, a high-temperature degradation mechanism related to syngas composition, temperature, and material of construction. Conventional use of UNS[1] S34700 (Nb) and S32100 (Ti) stainless steels in synthesis circuits is generally considered an acceptable material choice since nitriding is limited by diffusion, however these materials are not immune. Nitriding presents particular challenges in equipment, such as Waste Heat Boilers (WHB), where the nitride film may be mechanically broken during thermal cycling. Superficial nitride films, commonly of 100-400 □m thickness on stainless steels, are hard, brittle layers which act as crack initiation sites from which propagation occurs into the more ductile underlying steel. A critical function of thin cross-section bellows and lip seals (viz. welded gaskets) in syngas circuits is to mechanically flex, accommodating thermal expansion and contraction cycles. These components are particularly subject to fatigue. The strong but brittle nitride film, once cracked, forms a notch geometry relative to the underlying, weaker, alloy matrix into which cracks propagate during normal fatigue-type cycling. In an ammonia plant synloop, a jet fire started from the stainless steel UNS S34700 lip-seal joint in a WHB which was found nitrided and subject to extensive cracking. Inspection of stainless steel lip-seals (UNS S34700) in several plants indicated that nitriding was a common damage mechanism. Other plants in the MENA[2] region were subject to similar fires from identical lip-seals at precisely the same location.

Abdul Aziz Al-Shehhi, Saeed Al Mehairbi, & Ganesan Balasubramanian,
Fertil Ammonia plant, Al Ruwais, Abu Dhabi, UAE

Rob Gommans,
Materials & Corrosion Know-How, Geleen, Netherlands

Nicholas J.E. Dowling
Fertiglobe, Al Sila tower, Abu Dhabi, UAE

Introduction

In March, 2023, a fire broke out in the syngas circuit of an ammonia plant at the location of the inlet nozzle lip-seal of a waste heat boiler (WHB). While the extent of the fire damage was limited, the unusual nature and location of the fire triggered an extensive root cause analysis (RCA) considering the number of such joints in ammonia-hydrogen syngas service and the cumulative risk of a leak to atmosphere. Communication with other, similar generation, ammonia plants in the MENA region indicated that this was not an isolated incident. Two other plants reported near-identical cases. Lip-seals (otherwise known as *welded gaskets*) exist to provide reliable high-temperature joints between two materials of widely different thermal

[1] UNS: Unified numbering system for systematic identification of metal alloys
[2] MENA: Middle East, North Africa

expansion coefficients. Piping materials of different classes, when welded in a dissimilar joint, are subject to a number of well-known cracking mechanisms. In this specific case, the WHB pressure shell was manufactured in SA336 F22 (2¼Cr–1Mo) while the inlet piping flange was in SA965 F347 (18Cr-8Ni-Nb or UNS 34700). The only real alternative for this material choice is to install a dissimilar weld or pup piece of a metallurgy with intermediate expansion properties and fully weldable. The options are further described, including the common dilution issues, in the AWS handbook Volume 4, Section WHC4.06.

Primary function of Lip seal joints

Conventional, loose-type, metallic gaskets in the chemical and petrochemical industries are treated by ASME B16.20 or ISO 7483. In some high temperature services however, common spiral-wound graphite, kammprofil gaskets or ring-type joints (RTJ) are insufficiently reliable in flange pairs with widely mismatched thermal expansion properties. In such cases, German standard DIN 1591-1 and ASME PCC-2 (article 205) provide a lip seal (*welded ring gasket*) option where both sides of the flange are seal-welded together through a metallic ring: similar to a bellow with single convolution. The ring takes geometric advantage of the small inner diameter which permits use of a thin wall of great flexibility although with limited displacement. In syngas service, tolerance of any leak (toxic, explosive) is considered unacceptable; here lip-seal gaskets find a useful engineering niche by compensating for minor lateral and longitudinal movements, which are normal in dissimilar metal flanges operating at high temperature. The material of ring construction is usually a ductile stainless steel or nickel-based alloy, both of which are less subject to strain hardening (and thereby resistant to cracking under fatigue).

Chronology of the incident

Prior to the incident, the WHB had been in service for 6 years. The WHB incurred only 4 full shutdowns (thermal cycles) since commissioning, however an unknown number of minor process cycles did occur over the same period. The actual fire lasted only 31 minutes from the first indication (by external skin thermocouples on the converter No. 2) until being extinguished by the emergency services. As a result, minor damage was seen to walkways, instrument cables and similar smaller parts were replaced. No injuries occurred. The plant was restarted some 16 days later and resumed production without further upset. One important difference was the *external insulation* and weather shielding installed during the last shutdown before the incident. This insulation covered the entire joint but was not reinstated afterwards. Figures 1 and 2 illustrate the insulation coverage before and after the fire.

Figure 1. WHB inlet nozzle prior to incident

During the previous replacement of the WHB the whole lip seal flange, including stud bolts, were entirely encased in external insulation which proved to be a significant change as explained later. Detail of the lip-seal in Figure 3 shows the

Figure 2. WHB inlet nozzle after the incident

physical relation of the main flange components including the A193 B16 bolts. On the WHB side the lip-seal component is welded to an Alloy 600 pad, avoiding the necessity of an inconvenient PWHT (Post-Weld Heat Treatment) while the pipe-side lip-seal half is welded to the stainless steel tp 347 forged flange. The two lip-seal halves were subsequently *field-welded* together in the Gas Tungsten Arc Welding (GTAW) process (99.99% Argon gas, TIG wire ϕ=2.3 mm) for a final thickness of 5 mm of the pressure retaining ring.

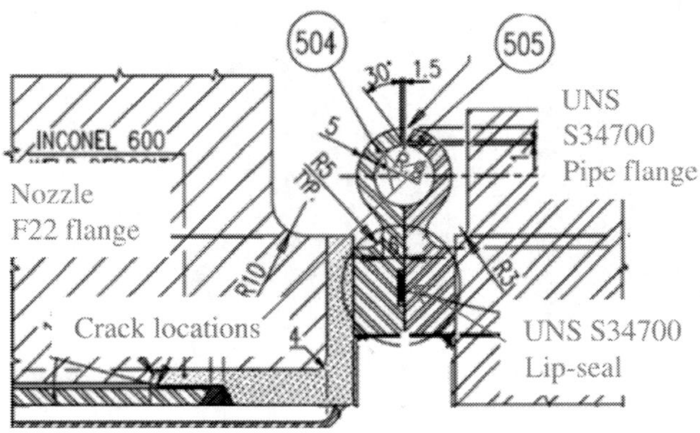

Figure 3. Detailed geometry of lip-seal

Metallurgical findings

Examination of the removed UNS S34700 lip-seal after the fire showed a large population ratcheted cracks (Figure 4) originating from the inside surface and terminating in a large crack of super-critical size which ultimately caused the fire. The cracks were mainly unbranched, transgranular in nature and propagating from a uniform surface nitride film of around 130-350 μm in thickness. Of particular note, the cracks initiating in the hard nitride film (Figure 5) then propagated into the relatively soft underlying stainless steel metal base (Figure 6).

Figure 4. Population of ratchet cracks with inner black nitride layer

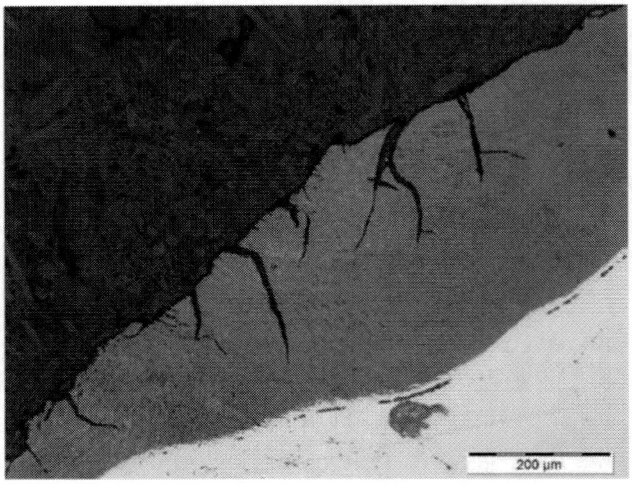

Figure 5. Crack initiating into thick nitride film (dark) above UNS 34700 stainless steel (light)

Hardness measurements of the base metal and nitride layers is presented in Table 1. In other industries commercial surface-hardening of stainless steel is carried out specifically to improve wear-resistance and usually carried out at higher temperatures in plasma [1]. While the surface properties are improved in *static* components, the component is nevertheless at risk of cracking the superficial nitride film [2, 3] under *flexion*. After perforation of the film, the crack systematically changes direction along the slip planes initially at 45° to the nitride film/base

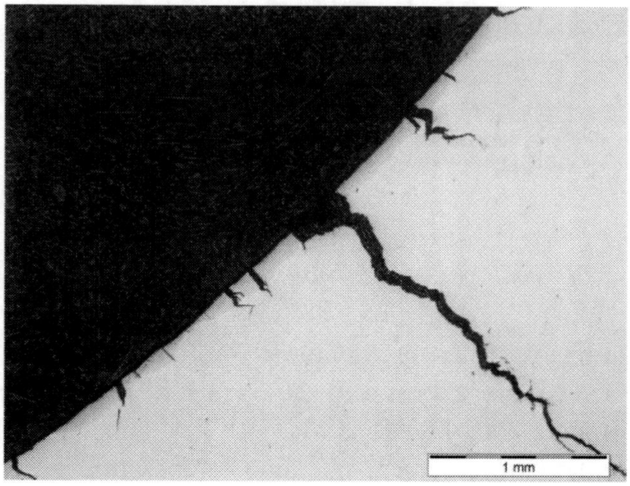

Figure 6. Crack initiating in Nitride film and propagating into UNS 34700 stainless steel base

Locus	Location	HV 0.1
1	Nitride layer edge	1116
2	Nitride layer edge	984
3	Nitride layer crack	994
4	Nitride layer crack	927
5	Nitride layer crack	1004
6	Base material	199
7	Base material	196

Table 1. Profile hardness values across inner lip-seal

metal interface in the 111 crystal plane as indicated by Varanasi et al [4]. An interesting mechanistic aspect is that the nitride film appears to *protect or reinforce* the underlying austenitic stainless steel under *compressive* load. Once the Nitride layer is cracked however, further flexing concentrates tensile loads at the gaps, producing a notch effect from which the crack propagates (Figure 6).

Effect of Nitriding

Nitriding has been a well-known degradation mechanism in ammonia plants since the 1960's [5, 6, 7]. Indeed, Redaelli and Deodato [7] identified the contribution of thermal cycling in stop/start cycles which accelerated crack propagation. The common design of many ammonia synloops employs "cold wall" pressure envelope materials from the low alloy grades (typically 1 ¼ Cr- ½ Mo; 2 ¼ Cr-1 Mo) while the relatively "hot" internals are usually fabricated in stabilized grades of stainless steels i.e. UNS S32100 (Titanium-substituted) or UNS S34700 (Niobium substituted). In 1988, Van Grieken [8] set out the first equations identifying the relationship between the process conditions, the main alloying determinant (Nickel content) and nitride film growth as:

$$LogN_t = 7.69 - \frac{2900}{T} + 0.67LogP_{[NH3]} - 2.37Log[Ni] + 0.5Logt$$

Where:

Nt =Thickness of nitride film (μm)
T= Temperature absolute (K)
$P_{[NH3]}$ = Partial pressure ammonia in bar
Ni= Nickel concentration in alloy wt%
t= Duration of the exposure in years

The Van Grieken relationship outlines the main parameters for nitride film formation, allowing some degree of predictive behavior and derivation of useful engineering properties.

Figure 7 shows the growth of such films on UNS S34700/S32100 as a function of time/temperature in a typical syngas circuit. By direct observation film *cracking and propagation* in the underlying metal is conservatively predicted for films *circa* >130 μm (flexing components only) then stainless steel grade 347 (UNS S34700) or 321 (UNS 32100) are *inapt* for service temperatures slightly above 350°C (662°F). Furthermore, given a 25-year horizon the temperature maximum of only 350°C (662°F) appears to be limiting for these alloys. In the case of the present ammonia converter WHB fire, the operating inlet process gas from the converter was *circa* 454°C (849°F).

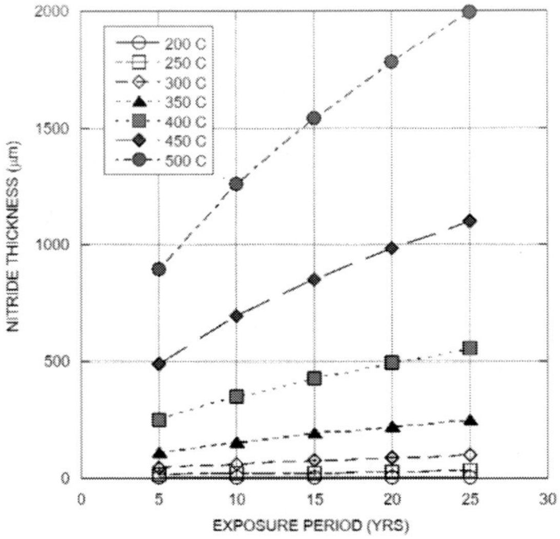

Figure 7. Van Grieken relation Nitride film thickness of UNS S34700 and S32100 as function of time

Consequently, full exposure of UNS 34700 mechanically flexing components (lip seals, bellows etc.) would be expected to rapidly crack in this service. Under normal circumstances, however the lip seal ring, the critical pressure-containing component, exist outside the main process stream and indeed are cooled by air.

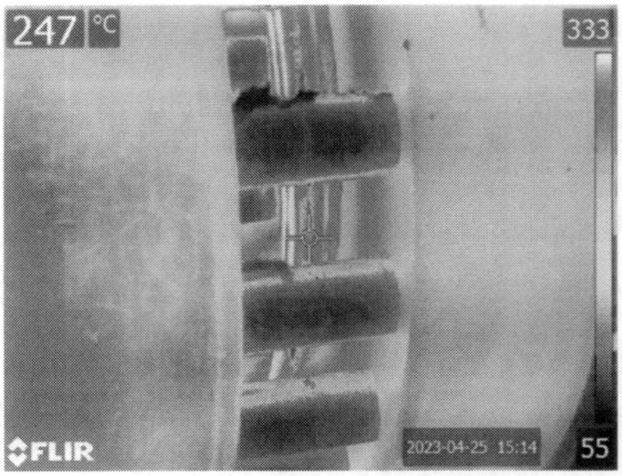

Figure 8. Infra-red inspection of lip seal joint after removal of external insulation

Infra-red measurements of the normal lip seal ring, fully exposed to the open air, were in the range 240-250°C (464-482°F), hence well outside the critical nitriding temperature range (Figure 8) and was the main reason for removing the external insulation.

Mechanical stability of lip seal joint

The mechanical stability of the joint was examined by Finite Element Analysis (FEA) specifically to confirm that the design of the joint was adequate even under external insulation. The main node mesh distribution is shown in Figure 9.

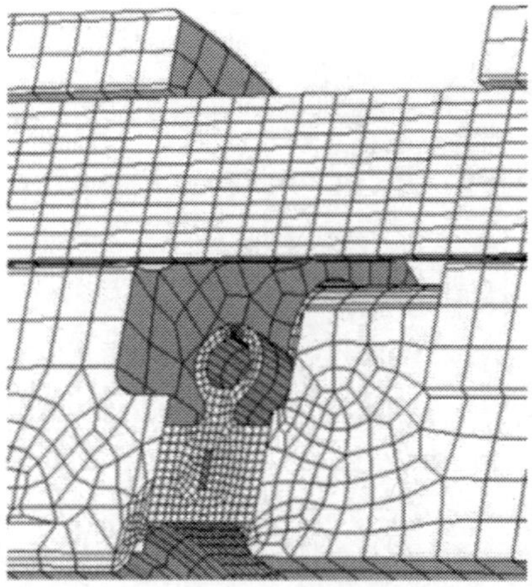

Figure 9. FEA mesh for lip seal / bolt arrangement

For this purpose, the ABAQUS FEA software package was used with input mechanical values obtained from the Materials Properties Data Base (MPDB) published by JAHM (2023). The study encompassed the B16 bolts as well as the entire flange assembly with close attention to the stress fields concentrated at the lip seal. Considerations for externally insulation as well as thermal cycling identified the effect of materials relaxation for the A193 B16 bolts as a function of time and number of cycles. Values for stud bolt stress relaxation were extracted from EN 10269-2013. While the mechanical design of the lip seal at full process temperature (subject to external insulation) permitted thermal expansion of the stud bolts, the upper toroidal part (the ring) calculated tensile stress was locally equal to only 50 MPa. The compressive stresses at the mating halves of the lip seal (Figure 10, blue lower part) predicted that even though the closure stress of the flange was much lower and decreased by stress relaxation with each thermal cycle with time, the two halves never actually separated.

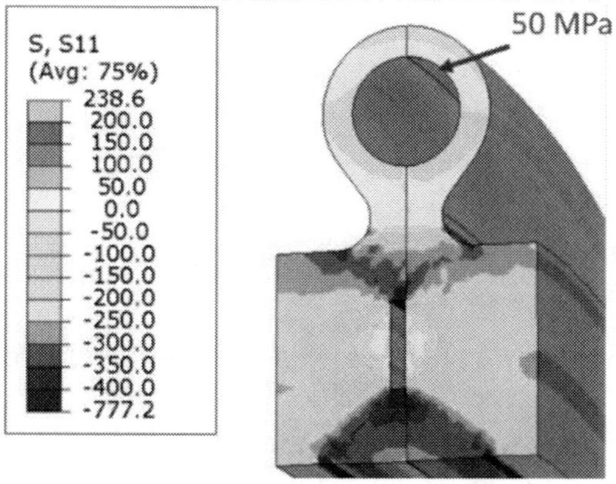

Figure 10: FEA stress fields around lip seal in service with external insulation

From a purely mechanical perspective then it was concluded that the combination of bending stresses and hoop stress *alone*, even under external insulation, combined with thermal cycling, was insufficient to cause the lip seal to rupture and was excluded as the *sole* root cause.

Contribution of *external* insulation

High temperature, high pressure joints are designed with consideration to the stress / temperature profiles such that primary containment is assured at all parts of the operational cycle (start-up, stable operation, cool down) where the components undergo expansion and contraction cycles. Flange closure calculations assume that the external environment is relatively cold and the temperature of the 16 closure bolts in ASTM SA193 B16 of the 20" flange is relatively "cool". In this plant however detailed design drawings indicating "insulation" failed to show that this referred to an *internal insulating thermal shroud* and in error was interpreted as requiring external ceramic blanketing. Consequently, during the prior shutdown, the entire lip seal joint was encased in 100 mm ceramic wool insulation plus weather shielding including the A193 B16 bolts.

The post-fire incident RCA underscored that lip seals rely on the *external cooling effect of the air* such that the hoop stress calculation for a 9 mm internal radius circumferential body may be adequately contained by a ring of only 5 mm thickness providing the ring is not continuously exposed to the hot ammonia-containing process gas. Figure 8 shows the actual temperature of the lip seal (247°C/477°F) *after* removal of the insulation to be significantly lower than the critical 300°C (572°F) indicated in Figure 7. Enclosure of the A193 B16 bolts inside the insulation resulted in thermal elongation at temperatures close to that of the process (425°C /797°F). Further to the FEA study, Sachs and Evans [9] estimated that A193 B16 bolts at this temperature retain only 43% of their residual stress after 1000 hours exposure. Values extracted from EN-10269 standard for A193 B16 stud bolts showed that after 47400 hours under external insulation, the relaxation phenomenon is assumed to be significant resulting in two major changes falling *outside* of the design assumptions as illustrated in Figure 11:

(1) Due to higher than designed flange relaxation (due to bolts incorrectly insulated), hot synthesis gas was allowed contact with the inner ring surface causing nitriding.
(2) Relaxation of the A193 B16 bolts permitted longitudinal strain of the lip seal ring.

The crucial design criterion of the lip seal ring assumed that the principal load was merely the internal hoop (circumferential) stress since the longitudinal stresses are taken up by the 16 (2 ½ inch) A193 B16 bolts. However, once the tightness of the flange was compromised through overheating and physical elongation of the bolts, some (minor) bending stresses were translated to the lip seal ring itself.

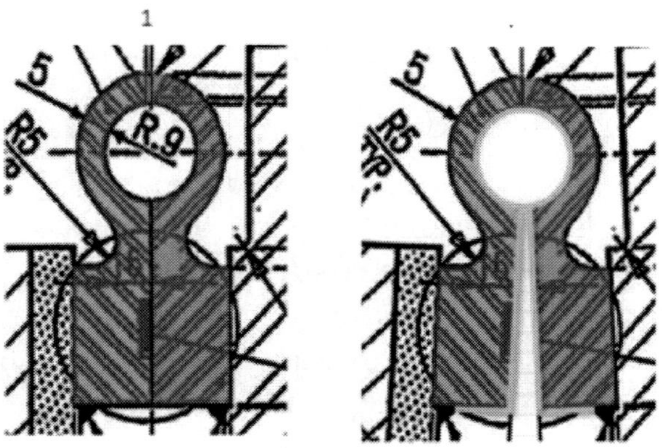

Figure 11. Effect of strain applied to the UNS S34700 lip seal resulting in internal nitriding and cyclical stress: Left: Initial state no nitriding of ring. Right: Final state: Ring forced open, process gas entry and nitriding-cracks

Discussion & conclusions

Contact with other ammonia plants in the MENA region showed that several plants with almost identical design have indeed suffered similar fires associated with incorrectly insulated lip seals. In general lip seals are considered a safe technology which avoids the high probability weld cracking mechanisms when welding materials of ISO 15608 Group 5.2 (2 ¼ Cr-1 Mo) to Group 8.1 (SS 347). Another population of plants however, incurred no such failures although they are presumed to have no external insulation around these WHB inlet nozzles or possibly use pup pieces to ensure the transition between the two dissimilar alloy categories.

General recommendations are as follows:

1) Lip seal joints in syngas service shall not be externally insulated.
2) Any pre-existing external insulation of lip seals should be carefully removed at the earliest opportunity while the plant is out of service or at least depressurized.
3) Insulated lip seals which have been in service for an extensive duration should be treated with care and barricades erected until replaced.
4) Lip seals in uninsulated UNS S34700 and where the skin temperature has been verified with calibrated infrared camera to be <300°C (572°F) are considered much *less susceptible* to nitriding-cracking and this failure mode.
5) Replace UNS S34700 lip seals with an alloy much less susceptible to nitriding (see below).

Retrofitting WHBs with completely different joint designs presents other risks (pup piece, dissimilar welds). A better option would be simple replacement of the lip seal itself with a

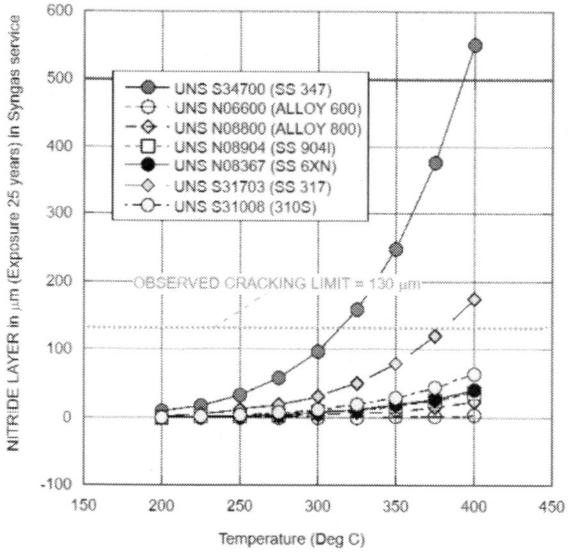

Figure 12. Material selection options for lip seals in Syngas (25 years service life)

more robust material. Lip seal manufacturers offer a range of options beyond those simply in the SS300 series. Basic elemental compositions of common lip seal alloys are presented in Table 2.

The potential replacement alloys for lip seal service are then recalculated using the Van Grieken relationship and presented in Figure 12 for UNS S34700, UNS S31703, UNS N08904 (904L), UNS N08367 (6XN), UNS S31008 (310S) and finally N08800 (Alloy 800) and UNS

		PRINCIPAL ELEMENTS				
UNS ALLOY	COMMON NAME	[Cr]	[Ni]	[Mo]	[Si]	[Ti/Nb /Cb]
S31703	SS317	19	14	3.5	0.8	0
S34700	SS 347	18	10	0	0.8	0.8
N06600	Alloy 600	16	72	0	0.5	0
N08800	Alloy 800	20	33	0	0.5	0.4
S08904	SS 904L	21	25	4	0.1	0
S08367	SS 6XN	22	24	6.3	1	0
S310008	SS 310S	25	20	0	1	0

Table 2. Elemental composition of potential replacement alloys for lip seal service in ammonia syngas service.

N06600 (Alloy 600). Direct observation of the cracked nitride layers in this study (i.e. for components requiring flexible design criteria) showed that most nitride films in excess of 130-150 μm resulted in crack propagation into the underlying alloy. Given this observation, Figure 12 predicts that over a 25 year service life, alloys UNS 32100, S34700 and S31703 (317L) are hence insufficiently alloyed. However, the intermediate Nickel grades 904L (UNS N08904) and 6XN (UNS N08367) appear to be adequate except for their high Molybdenum contents. The materials for lip seal service, showing the highest immunity against nitriding were Alloy 800 (UNS N08800) and Alloy 600 (UNS N06600) entirely as a result of their elevated Nickel contents. The optimal balance, however, of nitriding resistance to alloy content, while considering cost, appears to be UNS 31008 (stainless steel 310S) which was previously suggested by Redaelli and Deodato [7] through the Casale studies. The Van Grieken formula for alloy 310S (UNS S31008) predicts only a 63 μm nitride film over a 25 year lifetime in typical syngas environment at a surface temperature of 400°C (752°F) which is well below the critical cracking thickness of 130 μm.

In conclusion it appears that ammonia converter internals or pressure retaining components (lip seals, bellows) subject to *mechanical flexion* would be ideally served by alloy UNS S31008 (stainless steel 310S). Other ammonia syngas thicker components which are in more rigid or inflexible service (cartridge bulkheads, internal piping, supports, and catalyst bed supports) may safely be manufactured in the more conventional 300 series austenitic stainless steels (UNS S32100, S34700).

Recognition

The Belgian Welding Institute, in particular Dr Bart Verstraeten and David Kimpe as well as TWI, Cambridge are thanked for their contributions to this work. Dr Klaus Nölker is thanked for constructive improvements.

References

[1] Surface hardening of stainless steels. Materials and applications series 20. ISBN 978-2-8799-395-1

[2] Effect of nitriding on the microstructure and mechanical properties of stainless steels. P. Faltejejsek, Z. Joska, Z. Pokorny, D. Dobrocky, Z. Studeny. Manufacturing technology. Oct 2019, Vol 19, No.5 pp 745-748. ISSN-1213-2489.

[3] Cracking behaviour of AISI 316 stainless steel powder by gas nitriding and its influence on spark plasma sintering. J. Lu, Q.Wang, C. Liu, C. Zhang, D. Chen & G. Cui. J. Material Research and Technology. Vol 20. Sept-Oct 2022. Po 3796-3806.

[4] R.S. Varanasi, M. Koyama, M. Yokoi, Y. Ootani, M. Kubo, K. Tanahara, & O. Umezawa. Understanding cracking growth within γ'Fe4N in a low carbon steel during monotonic and cyclic tensile testing. 2024. Metals and Corrosion July, Vol 59, pp 14693-14652.

[5] H.D. Marsch. Nitriding of steels after 5 years in Ammonia synthesis service. Safety in Ammonia plants and related facilities symposium. 1981. Proceedings.

[6] S.Y. Sathe and T.M. O'Connor. Review of Nitriding in Ammonia plants. Safety in Ammonia plant and related facilities symposium 1987. Proceedings.

[7] L. Redaelli and G. Deodato. 2021. Ammonia nitriding, knowledge and design considerations. AICHE ammonia conference. Paper 4E.

[8] C.A. Van Grieken. Factors controlling the nitriding of stainless steel in ammonia synthesis loops. AICHE ammonia conference. 1988.

[9] K. Sachs and D.G. Evans. The relaxation of bolts at high temperature. Proceedings of Institution of mechanical Engineers 1973, pp 59-67

Failure Analysis of Syngas Piping Welds- A Case Study from an Ammonia Plant

After more than 42 years of trouble-free service, two syngas leaks were observed on welds in carbon steel syngas piping on the CO_2 absorber outlet. The CO_2 removal system uses potassium carbonate solution as the absorbent. Over time, potassium carbonate carryover in the syngas led to carbonate stress corrosion cracking in piping, which was not stress-relieved. Root cause failure analysis of two leaking syngas piping welds confirmed intergranular cracking coupled with high residual stress and weld defect (excess penetration). Leaking welds were temporarily repaired by fillet welded full encirclement sleeves until the affected piping was replaced in a scheduled turnaround.

Tayyab Ejaz
Asset & Reliability Professional (Inspection)
Fauji Fertilizer Company Limited (FFCL), Pakistan

Introduction

Fauji Fertilizer Company Limited (FFCL) is the largest urea manufacturer and operates multiple fertilizer plants in Pakistan. One Ammonia-Urea complex, using Haldor Topsoe ammonia and Saipem urea technology, (Plant-III) is located at Mirpur Mathelo, Pakistan, was acquired by FFCL in 2002. The facility was first commissioned under a government owned entity in 1980. Nameplate capacity of the ammonia plant was increased to 1250 MTPD (1378 STPD)from the original capacity of 1000 MTPD in 2008. Two incidents of syngas leakages from Ø14"-8.74 mm thick carbon steel non-stress relieved piping welds occurred within a nine month period (October 2022 and August 2023). Both leaks were found to result from cracks in the welds, accompanied by multiple fabrication defects identified in the radiographic inspection of leaking welds. There was no history of previous leakages on this syngas piping circuit. These recurring syngas leakages raised safety concerns due to the explosive nature of the gas. Leaking welds were temporarily repaired by fillet welded full encirclement sleeves until the affected piping was replaced in the next scheduled Turnaround.

Process Description

In the ammonia plant, natural gas is reformed in primary and secondary reformers to produce process gas containing hydrogen (H_2), carbon monoxide (CO), and carbon dioxide (CO_2). The process gas then passes through a CO shift conversion stage, converting CO water into additional H_2 and CO_2. In the CO_2 removal section, the CO_2 is separated from the process gas while process gas is purified into hydrogen rich syngas. This purified hydrogen-rich syngas from the CO_2 removal section, proceeds to the methanation stage and then to the ammonia converter for ammonia synthesis.

At FFCL, The CO_2 removal system uses Benfield solution, a hot potassium carbonate solution with additives to enhance reaction rates and prevent corrosion, operating in a closed loop. CO_2 from synthesis gas is absorbed in the absorption column C-302 and stripped out in the CO_2 stripper C-301. Purified syngas exits the C-302 at 28.9 kg/cm² (411.3 psi) and 79°C (174.2 °F), passing through separator vessel V-302 to remove any Benfield solution carryover before entering gas exchanger E-311. The purified syngas is then sent to methanation and fed to the ammonia synthesis reactor for ammonia production. The process flow diagram is shown in Figure 1:

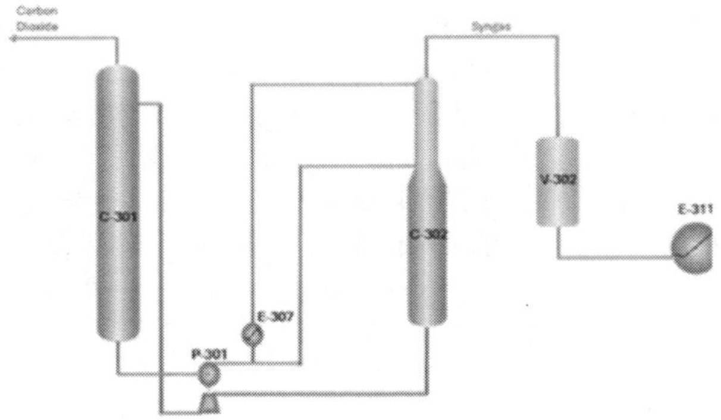

Figure 1: Process Flow Diagram of CO2 removal section

Incident Description

On October 5, 2022, a syngas leak was detected at the weld seam between a 14" × 12" carbon steel eccentric reducer and pipe (8.74 mm (0.344 inches) thick) downstream of absorption tower C-302 and separator vessel V-302 (Figure 2) Radiographic examination revealed a 28 mm long crack with multiple fabrication defects but no thickness loss was detected. A fillet-welded full encirclement temporary clamp was installed at leaky weld after implementing safety precautions.

On August 5, 2023, a second leak occurred about two meters from first leak, this time from the weld between the 14" carbon steel pipe and elbow (8.74 mm thick), releasing a 10 ft jet of syngas (Figure 2). Radiography examination confirmed fabrication defects without thickness loss, and temporary repairs were carried out using the same previously installed design of encirclement sleeve. Both leaking sections were permanently replaced with new piping during the October-2023 turnaround to ensure long-term reliability (Figure 3).

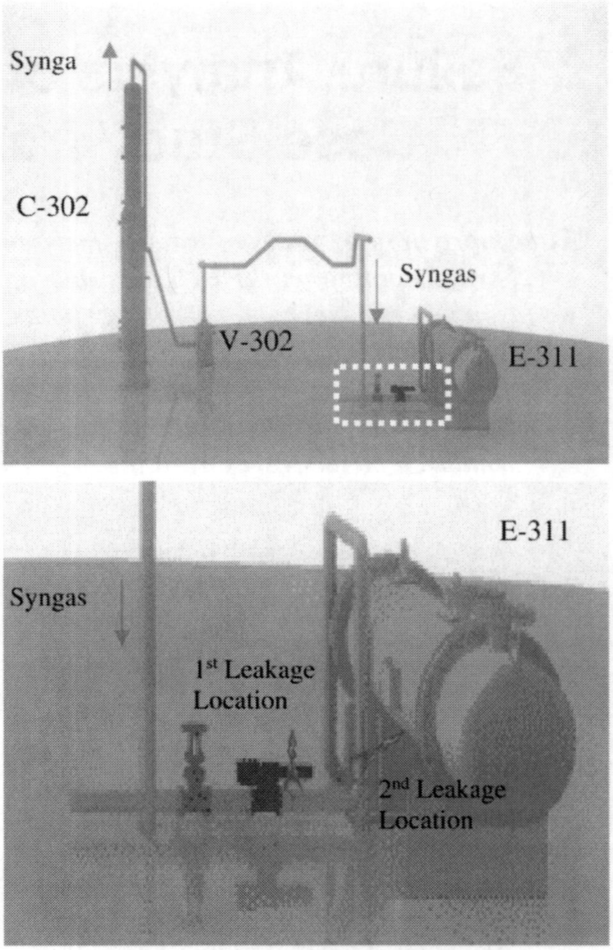

Figure 2: Syngas leakage locations

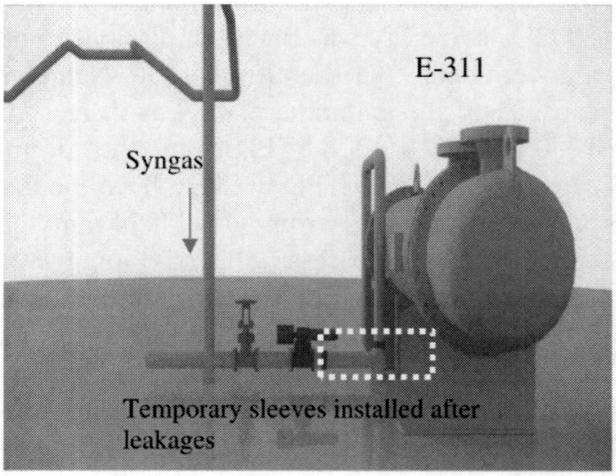

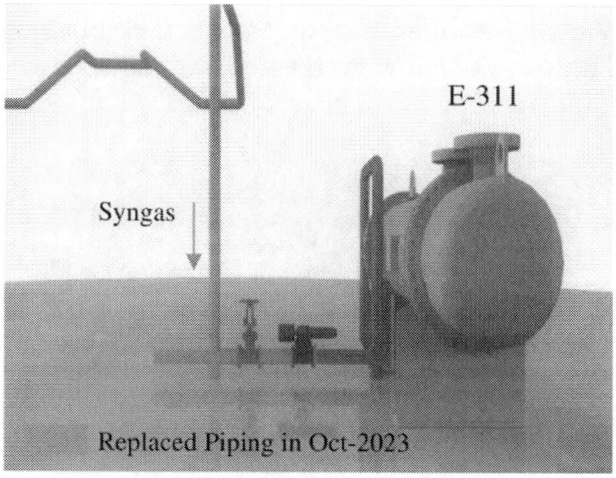

Figure 3: Temporary sleeving and replacement of leaky piping

Root Cause Failure Analysis

The hot insulated piping designed for syngas service extends approximately 230 m (754.6 feet) from the absorber C-302 to the separator vessel V-302 and then to the gas exchanger E-311 The leakage of hydrogen-rich syngas from this piping circuit raised significant concerns due to its explosive nature and the potential for production losses. Following the replacement of the leaky syngas pipe segments (approx. 12 m out of 230 m total piping) during the planned plant turnaround in October 2023, these segments became available for detailed investigation and through-thickness analysis. The relevant process and mechanical design data for the syngas piping was collected and analyzed. Additionally, previous inspection records and the leakage history of this piping circuit were reviewed, along with assessment of process conditions for any deviations or upsets. In-house non-destructive testing (NDT), metallographic testing, and scanning electron microscopy (SEM) combined with energy dispersive X-ray (EDS) techniques were utilized on the leaky welds to investigate the nature of damage. A fishbone analysis was used to systematically evaluate all possible aspects of the cause of damage and leakage. Following are the specifications of this syngas piping:

Piping Construction Code: ASME B31.3
Piping Material: API 5L Grade B (SMLS)
Size and Thickness: Ø14"-8.74 mm
Stress Relieving: No
Radiography: Spot
Operating Temperature: 79°C (174.2°F)
Operating Pressure: 28.9 kg/cm² (411.3 psi)
Service Life: 42 Years

Process Parameters Review

Although operating temperature and pressure trends of syngas piping downstream of CO_2absorber C-302 were well within integrity operating windows (IOWs), approximately 2 l (0.5 gal) of Benfield solution (mainly potassium bicarbonate) was drained from the syngas piping daily, due to carryover from the Benfield absorber C-302 and separator vessel V-302. Laboratory analysis of the carryover solution, conducted using titration methods with both a pH meter and an autotitrator, confirmed the presence of carbonate ions at a concentration of 480 ppm and a pH value of 8.5 in the leaking syngas piping. Both of these values exceed the safe limits specified in API RP 571 for prevention of carbonate stress corrosion cracking (CSCC).

Historical Data Review

There was no previous history of leakages from this circuit or its associated piping from any of the other ammonia plants being operated by FFCL. The piping material and C-302 absorber design at other plants were identical to those at this plant, where the syngas piping leak occurred. Although no syngas leakage incidents had been reported at other locations, the potassium bicarbonate drainage practice downstream of the Benfield absorber C-302 was similar to that at this location. As a result, the syngas piping at other locations operating since 1982 required inspection and analysis to assess potential damage similar to what was observed at this location.

Piping Supports Flexibility Analysis

During the failure analysis, the piping circuit supports were evaluated using CAESAR II to identify any discrepancies that might have introduced additional loads or stresses on the leaking welds. Under operating conditions, the maximum stress observed at a piping support was 38,882 KPa (5,638 psi), which is well below the allowable limit of 137,895 KPa at the operating temperature of 79°C, as per ASME B31.3 for API 5L Grade B material. Consequently, it was concluded that the piping supports did not contribute to the syngas piping failures.

In-House Non-Destructive Testing

Following the replacement of affected leaking syngas piping in October 2023 (Figure 4), these segments were available for detailed inspection and investigation. Visual inspection of both leakage location welds (1st Leakage Location: reducer to pipe weld, 2nd Leakage Location: elbow to pipe weld) were carried out from both ID & OD side. Construction imperfection / weld defects like uneven root / excess penetration, irregular weld profile at root and capping, localized lack of penetration & poor toe blend was observed in both leaking welds.

Visual examination of elbow to pipe weld revealed a linear crack both on the OD and ID of the weldment. The crack was measured to be 80 mm long on the OD (Figure 5), 110 mm long on ID and exhibited a wide opening compared to its ends. The observed crack traversed longitudinal to weld, with no extension into the base metal on either side of the weld seam. Examination of the ID exhibited that the crack followed along the weld toe region on the pipe side of the weldment. The crack did not exhibit any branching and maintained predominantly a straight trajectory. It became evident, that the entire length of the crack was within a zone where the weld exhibited excessive root protrusion of 4-5 mm compared to the rest of the weld joint (Figure 5). Similarly, through wall crack was found in the reducer to pipe weld which initiated from ID side with adjacent weld imperfection of excess root penetration.

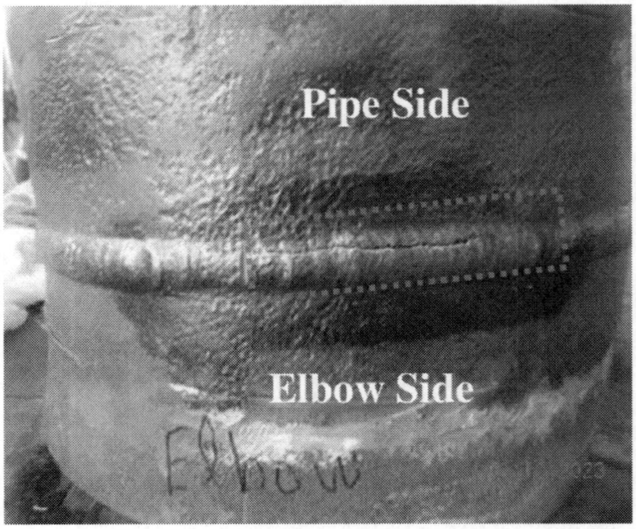

Figure 4: 80 mm long single crack on pipe OD (2nd Leakage)

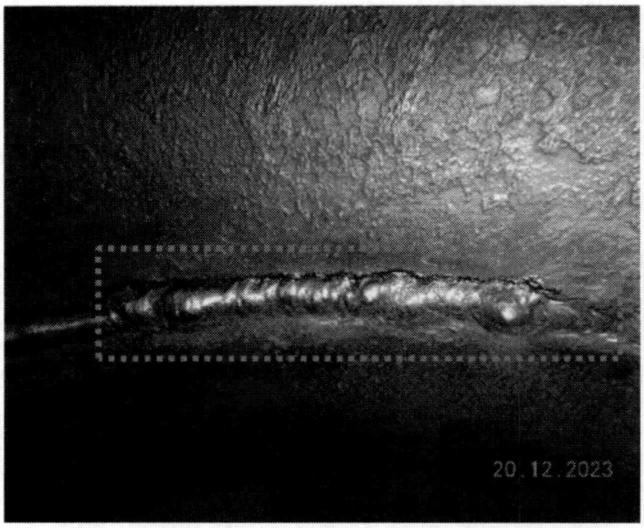

Figure 5: 110 mm long single crack on pipe ID at weld toe with excess root penetration (2nd Leakage)

Dye penetrant testing (DPT) of both leaking welds confirmed the presence of a crack initially detected during visual inspection, with no additional anomalies found. Positive Material Identification (PMI) of the base metal and welds showed no material discrepancies. Thickness measurements revealed no loss, and hardness testing on the base metal, heat affected zone

(HAZ), and welds indicated values below 200 HB across all locations.

Magnetic Particle Inspection (MPI) was carried out on both leaking weld specimens, which were halved for examination of internal (ID) and external (OD) sides. Linear transverse and longitudinal cracks were detected on the ID side, extending across the weld, HAZ, and adjacent base metal up to 12 mm from the weld (Figure 6 and Figure 7). These cracks were not visible in DPT due to oxide filling. External MPI showed no additional cracks besides those linked to the leakages.

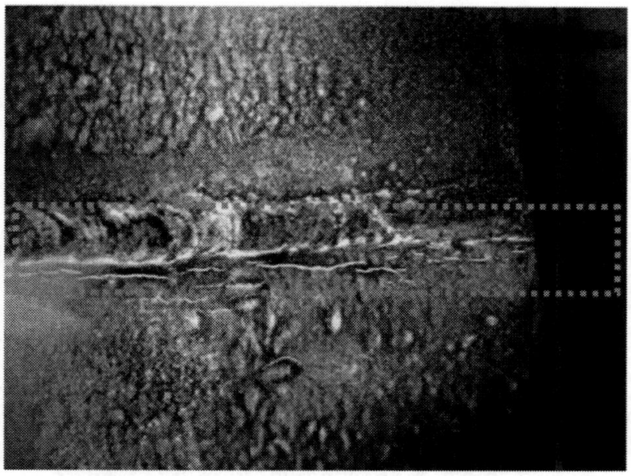

Figure 6: Longitudinal cracks on ID side of weld HAZ during MPI

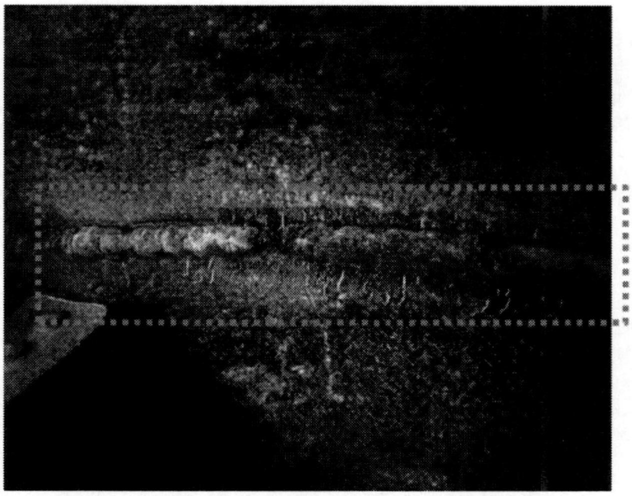

Figure 7: Transverse cracks on ID side of weld HAZ

Metallographic Testing (Macro & Micro Examination)

Macro and micro examination of leaking weld specimens were carried out by a third-party metallurgist. For macro examination, specimen was extracted and prepared to a 600-grit finish using standard metallographic techniques specified in ASTM E3. After preparation, specimen was etched using 10% Nital to delineate weld profile. Examination revealed a number of secondary fine cracks on both sides of the weld apart from the primary leaking crack (Figure 8). Moreover, the primary leaking crack exhibited branching, mostly within the HAZ region (Figure 9).

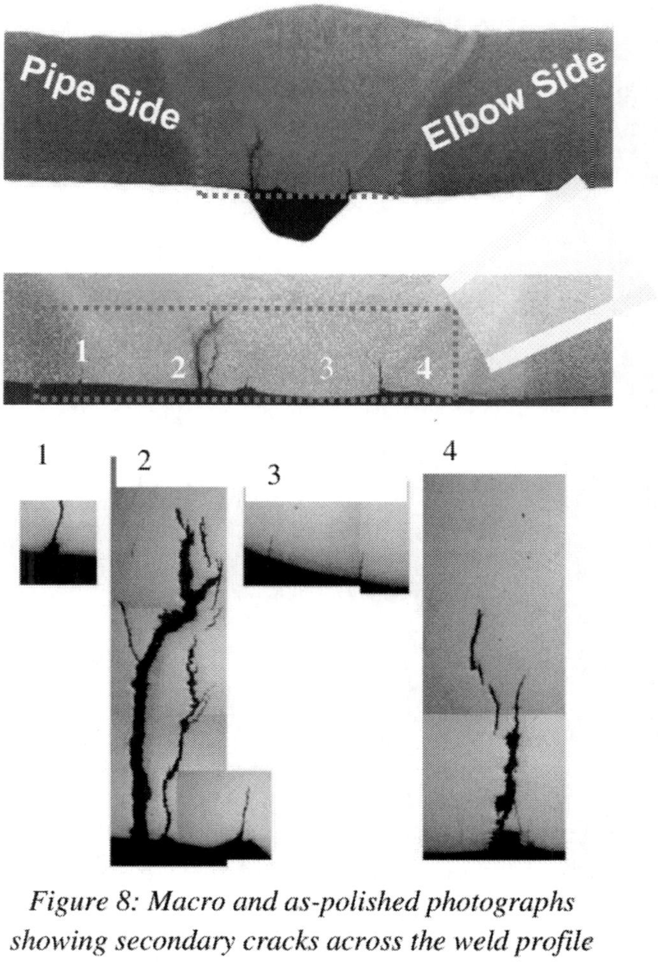

Figure 8: Macro and as-polished photographs showing secondary cracks across the weld profile

Previously prepared macro-specimens were further prepared to a one-micron finish using standard metallographic techniques. The examination in as polished condition exhibited number of cracks on the pipe ID side.

Specimens were later etched with 2% Nital to reveal the microstructure. Examination in etched condition revealed cracks that were predominantly intergranular in nature (Figure 10). The crack traversed along the grain boundary of the ferrite grains. The microstructure comprised pearlite in equiaxed ferrite grains, and no other anomalies were noted in the examined area.

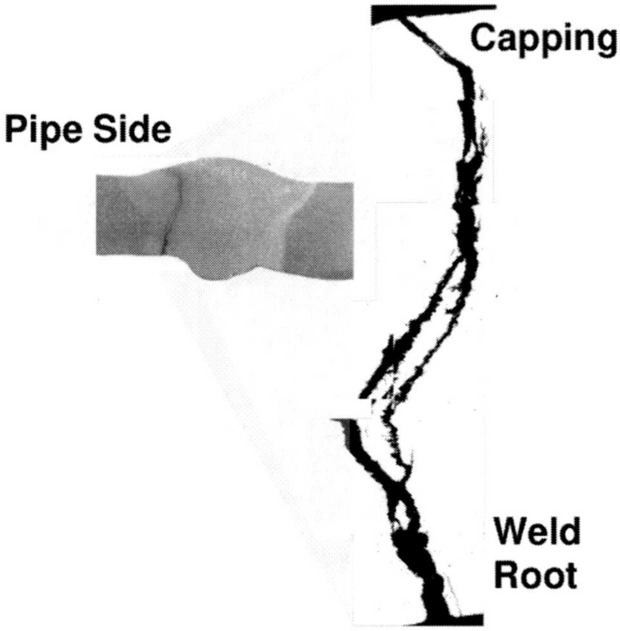

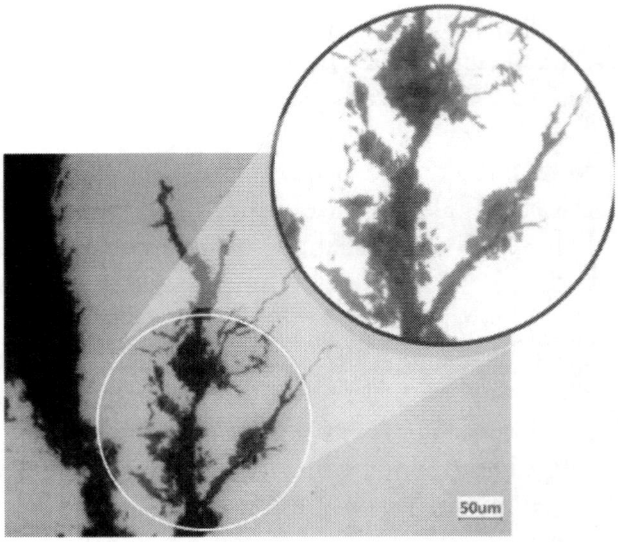

Figure 9: As-polished photograph showing main leaking crack with branching and oxide filled deposits inside crack

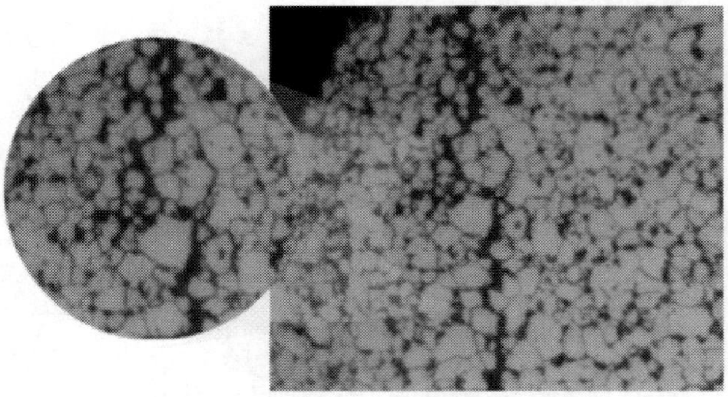

Figure 10: Etched photograph showing intergranular nature of crack

SEM & EDS Analysis:

The fracture face of the leaking weld specimen was examined with a scanning electron microscope (SEM) which revealed the intergranular fracture feature. Elemental constituents of the fracture face were analyzed using EnergyDispersive X-ray Spectroscopy (EDS) analysis. The spectra exhibited major peaks of Iron (Fe), Carbon (C) and Oxygen (O). Extraneous peaks of Potassium (K), Vanadium (V), Sodium (Na), Calcium (Ca), Magnesium (Mg), Manganese (Mn) and Sulphur (S) were also noted.

Figure 11: SEM photograph showing intergranular fracture

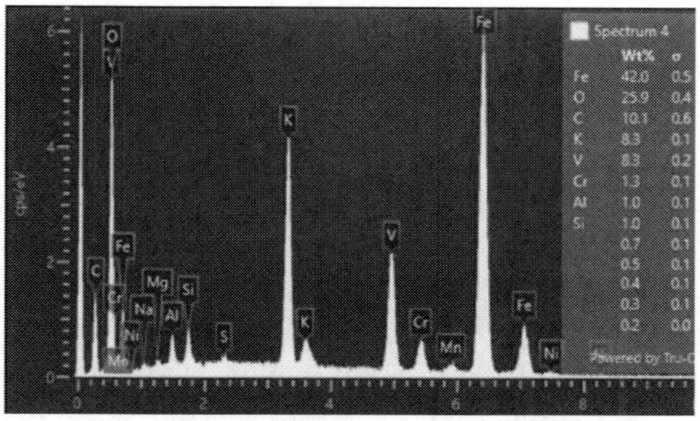

Figure 12: EDS analysis of fracture face

Discussion

Based on the above findings, it was clear that the main leaking crack originated in the weld HAZ at the inner diameter (ID) and propagated through the weld region to the outer diameter (OD) of the pipe. It initiated in the region of significant weld root penetration (4~5 mm). Visual and dye penetrant testing (DPT) did not reveal branching in the main crack; however, macro and micro-examinations identified side branching filled with oxides. Additional small cracks were detected in the weld HAZ during magnetic particle inspection (MPI). When analyzed under a scanning electron microscope (SEM), these cracks exhibited intergranular fracture features.

Excessive root penetration can create a crevice, serving as a site for corrosive fluid accumulation, leading to increased localized concentration. Energy-dispersive X-ray spectroscopy (EDS) confirmed the presence of corrosive salts, while laboratory analysis of the syngas piping drainage solution revealed carbonate ion concentrations and pH values exceeding the limits outlined in API 571 to prevent carbonate SCC. The overall fractographic and micro-examination findings, coupled with in-house non-destructive testing, indicate carbonate SCC in the syngas piping weldment during service. Non-stress-relieved carbon steel piping is typically susceptible to carbonate SCC when exposed to potassium carbonate solutions with carbonate ion concentrations greater than 100 ppm, depending on the solution's pH. Susceptibility is highest within a pH range of 7.5–11, with most failures occurring between pH 8–10.

Conclusion:

In this case, all conditions were conducive to carbonate SCC. The weld was non-stress-relieved, with localized stresses caused by excessive root penetration. Additionally, the carbonate ion concentration exceeded 400 ppm, and the pH was greater than 8. Residual stresses facilitated the initiation and propagation of carbonate SCC, ultimately leading to weld failure. The analysis concluded that the weldment failure resulted from carbonate SCC, driven by residual stresses from the poor weld profile and the absence of post-weld heat treatment (PWHT).

In-service Inspection & Repair

Since all the assessments were performed on leaking syngas piping segments (12 m) removed in October 2023 Turnaround, it was therefore necessary to monitor the remaining 218 m syngas piping from absorber C-302 to gas exchanger E-311 which was still in service since 1980.

Phased Array Ultrasonic Testing (PAUT) of all the remaining in-service welds was carried out to identify weld fabrication defects and SCC cracks. A PAUT mock-up was conducted on the leaking syngas piping segments to calibrate the testing setup. PAUT examination of the mock-up welds only detected longitudinal cracks (parallel to the weld axis) in the weld HAZ region. Transverse cracks (perpendicular to the weld axis) were more difficult to detect, as the orientation relative to the ultrasonic beam prevented the sound waves from striking the crack face at an optimal angle for reflection. This calibration was then used to examine the welds of the in-service syngas piping. Several welds in the vicinity of the previous syngas leak welds were found to have fabrication defects, likely due to the involvement of the same construction team. Additionally, nearly all of the syngas piping welds from absorber C-302 to gas exchanger E-311 examined with phased array testing exhibited cracks of different depths.

In order to avoid any future failure, an on-stream full encirclement sleeve was designed for temporary repair of welds. ASTM A516 Gr 70 plates were used for fabrication of full encirclement sleeves. The sleeve thickness was designed per ASME B31.3 to be the same as the pipe thickness. However, higher thickness was selected based on the safety factor. Temporary repair clamps were installed after monitoring cracks growth over some period to ensure that cracks will not propagate under the sleeve. Fitness for service assessment (FFS) of these temporary repairs were also conducted through a third party. Temporary repairs would remain in place until replacement of the remaining original piping in the next plant turnaround (October 2026).

Future Recommendations

The following specific actions were recommended based on this specific case study.

- Installing washing trays in the CO_2 absorber's C-302 top section just below the demister can reduce potassium carbonate carryover, which can significantly mitigate carbonate SCC in downstream syngas piping and equipment. These changes have been implemented in new designs.
- Regular monitoring of pH and carbonate ions concentration downstream of the CO_2 absorber C-302 and their levels to be maintained below the threshold limits to avoid carbonate SCC.
- All new syngas piping welds should undergo 100% radiography/ultrasonic examination to detect fabrication defects, which can serve as focal points for carbonate ion concentration and accelerate damage.
- PWHT to be performed on all carbon steel and alloy steel welds of syngas piping with chances of potassium carbonate carryover, including the HAZ, to reduce residual stresses caused by rapid heating and cooling during welding.
- Syngas piping and equipment in ammonia plant CO_2 removal sections using Benfield potassium carbonate solution in the CO_2 absorber, operating within pH and carbonate ion ranges conducive to carbonate SCC, must be inspected to prevent syngas leaks and catastrophic failures.
- Any leakage with an unidentified source or cause should be investigated through root cause failure analysis to prevent future failures and implement corrective actions.
- Any process change should be analyzed and evaluated for potential impacts or damage to the system.

References

- J. Zhang and F.J Martin, " Corrosion of steel by carbonate-bicarbonate solution for CO_2 capture" Materials and Corrosion 2012, 9999, No. 9999

- American Petroleum Institute (API). (2020). API RP 571 Damage Mechanisms Affecting Fixed Equipment in the Refining Industry (3rd ed.). API Publishing Services.

Meta-analysis of 105-D Ammonia Converter Safety

This paper presents a modern assessment of the safety performance and operational integrity of the KBR / M.W. Kellogg 105-D Ammonia Converter through a meta-analysis of over 20 published papers that examines industry experience of critical degradation mechanisms affecting both vessel and internal components. A detailed review focuses on multiple distinct degradation cases reported in the past 30 years. This is also supported by data from the successful operation of more than 112 converters, spanning vertical and horizontal configurations, over 25 years.

The review integrates KBR's root cause analysis and safety data with industry insights to identify design commonalities, and pinpoint operational risks. By systematically comparing root causes, contributing factors, and operating conditions, it formulates actionable ideas for mitigating risks and enhancing the performance of 105-D converters. These insights aim to guide current operators in preventing irreversible damage and inform design considerations for future installations.

Daragh Blennerhassett, Volodymyr Oliynyk & Sachin Kalatrakkal
KBR

Introduction

The ammonia synthesis converter is the most critical component in the ammonia facility. It is responsible for the conversion of hydrogen and nitrogen into ammonia under elevated pressures (typically 120–330 bar /1750 -4800 psi) and high temperatures (300–500 °C / 570-940°F). These demanding operating conditions presented unique mechanical and metallurgical challenges for the engineering teams involved in the first KBR (formerly M.W. Kellogg) ammonia converter design, commissioned in 1943. As plant capacities, pressures and efficiencies increase, new and interesting challenges are encountered for both the plant safety and reliability, some milestones are highlighted below;

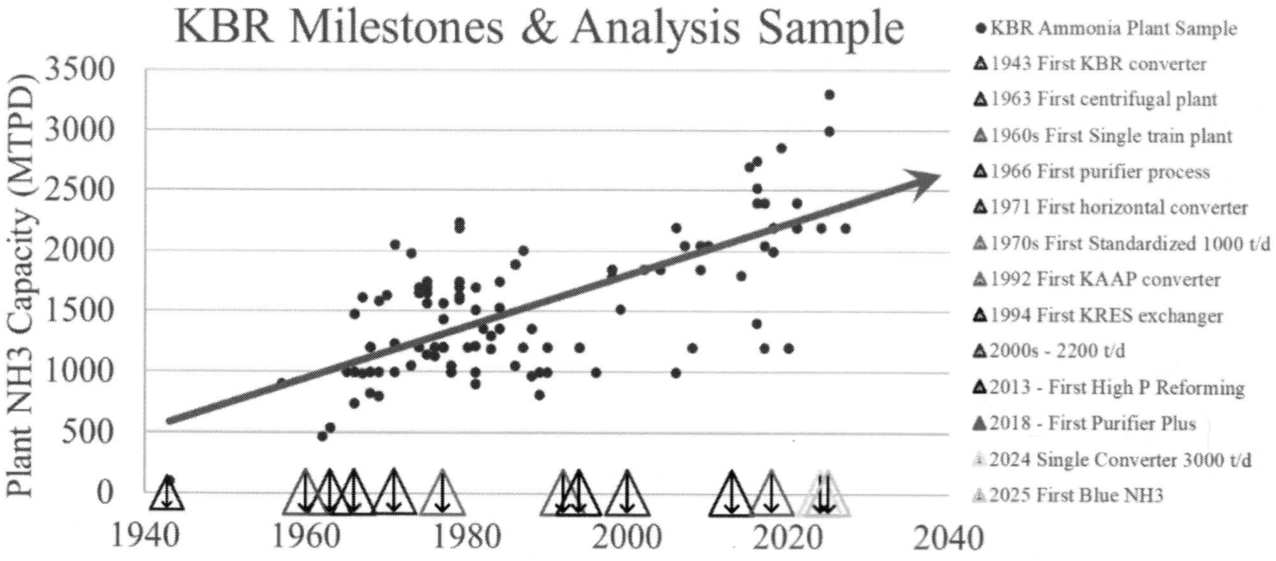

Figure 1 KBR Milestones and Ammonia Facility Analysis Sample (Malhotra, 2002) (Jain, 2008)

The installation base for KBR ammonia synthesis units is over 237 ammonia converters installed globally, 58 horizontal design and over 179 vertical. The equipment tag number 105-D is widely recognized across industry as the designation for the ammonia synthesis converter as approximately 50% of the worlds ammonia production uses KBR technology. As capacities increase so to do the opportunities and lessons learned to continue to evolve designs to industry demands. This paper aims to present a meta-analysis of mechanical failures, safety incidents, design advancements, and operational lessons learned in 105-D ammonia synthesis converters.

Hydrogen Embrittlement

Hydrogen embrittlement is a process where hydrogen reduces the ductility and tensile strength of a metal. It can lead to cracking and failure at stresses below the yield strength of the material.

Mechanism: Atomic hydrogen diffuses into the metal, often at elevated temperatures. It can then interact with the microstructure, leading to various embrittlement mechanisms.

Factors Influencing Hydrogen Embrittlement:

Material: High-strength steels and certain other alloys are particularly susceptible. Carbon steel is the most susceptible material to hydrogen damage, hydrogen can react with the carbon in the steel, leading to fissures and cracks, reducing its mechanical strength

Temperature & Stress: The effect of temperature is complex, as it influences both hydrogen diffusion and the embrittlement mechanism. Tensile stress promotes hydrogen uptake and embrittlement.

Hydrogen Assisted Cracking (HAC): a form of hydrogen embrittlement that can occur in carbon or low alloy steel in ammonia synthesis loops operating at high hydrogen partial pressure and metal temperatures above 250°C (480°F)

Hydrogen Embrittlement Industry Experience Sample		
Citation	**Description**	**Lessons Learned**
(Rugh, 2002)	*"Repair and Failure Analysis of a Crack in an Ammonia Converter"* **2002**: 1850 MTPD KBR KAAP Plant. Crack on weld of 105-D shell longitudinal weld LWL-14. Suggests poor weld repair, pre-heat duration or Post Weld Heat Treatment (PWHT) application as root cause.	The paper details the repair process, failure analysis, and eight inspection techniques to inform future operators on how to address similar situations after they occur. It does not explicitly list recommendations to avoid future issues. Adherence to welding procedures during fabrication, including preheating and post-weld heat treatment, is crucial to prevent hydrogen embrittlement.
(Sam Ternowchek, 2009)	*"Repair and online mechanical integrity monitoring of an ammonia synthesis converter"* **2007**: KBR facility commissioned 1998, no weld issues in 2002 but potential 'dirty weld' reported. 2007 TA cracks in CWL-3 discovered across the circumference ranging from 7" to 44" prompting a full weld replacement.	The root cause was deemed hydrogen embrittlement with a contribution of HTHA to crack propagation. The paper describes an interesting procedure for a full circumferential weld replacement where the whole upper portion of the vessel had to be supported during welding. The original double-vee joint could not be reproduced. After repairs, an online Acoustic Emission system was installed to monitor mechanical integrity.

Citation	Description	Lessons Learned
(Stefano Cavalleri, 2023)	*"Hydrogen Assisted Cracking in the ammonia synthesis loop"* The paper presents recent cases of HAC, analyzes contributing factors, suggests mitigation strategies during design, and discusses effective in-service inspection techniques. One case in **2019** several cracks were found at the connection of the manway and the top head.	The paper describes the methodology used for establishing Minimum Pressurizing Temperature curves with reference to API RP 934-F. In the 2019 case cracks mainly grew during start/stop cycles. There was high hardness near the welded area, possible nitriding, and internal (thermal) stresses contributed to the eventual failure. They examine other cases of HAC failures, contributing factors, mitigation strategies during design, and effective in-service inspection techniques.

Table 1 Hydrogen Embrittlement Industry Experience Citation Summaries

Preventing Hydrogen Embrittlement:

Material Selection: Using materials less susceptible to hydrogen embrittlement, such as appropriate grade of Low Alloy Steel (LAS), austenitic stainless steels or Inconel alloys

Minimum Pressurizing Temperature Curves: Consider factors like material composition, operating conditions (temperature, pressure, H2 partial pressure), and the potential for embrittlement

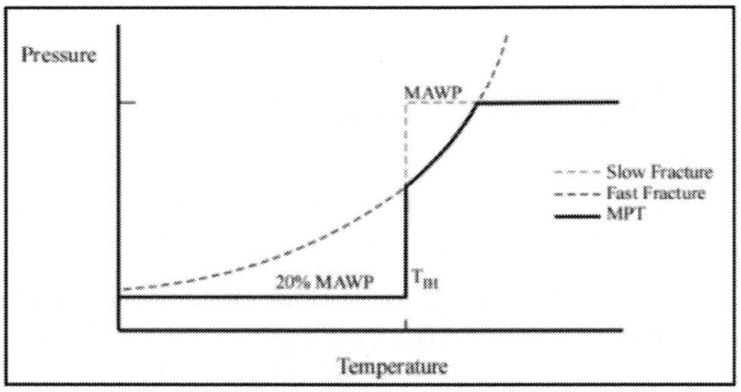

Figure 2 MPT curve example with Hydrogen Assisted Cracking limits (API RP 934-F)

PWHT: Post Weld Heat Treatment reduces residual stresses and removes diffusible hydrogen.

Maintaining Pre-heat: Maintaining pre-heat during welding to prevent cracking due to embrittlement. As described in (Michael B Rugh, 2002) a crack in an ammonia converter was attributed to hydrogen embrittlement in a poorly made weld repair.

HTHA

High-Temperature Hydrogen Attack (HTHA) is a damage mechanism that can occur in process equipment exposed to hydrogen at elevated temperatures (above 200°C / 390°F)

Mechanism: Under certain temperature and hydrogen partial pressure conditions, atomic hydrogen permeates the steel and reacts with iron carbides (Fe3C) to form methane gas (Michael W Hester, 2013). Methane gas accumulates in voids and grain boundaries, creating internal pressure, causing micro-fissures and eventually cracking.

Susceptibility: Carbon steel and low-alloy steels are susceptible to HTHA. The severity of HTHA increases with increasing temperature and hydrogen partial pressure.

Detection: HTHA is a subsurface phenomenon and not readily evident on the surface, Advanced Ultrasonic Backscatter Techniques can be used to evaluate subsurface microstructural features.

Inspection: Inspecting for HTHA can be challenging, as it can be difficult to differentiate the early stages of attack, such as fissures or small cracks. The case study paper below (D.H. Timbres, 2006) provides adequate detail to consider during inspections.

High Temperature Hydrogen Attack Industry Experience Sample		
Citation	**Description**	**Lessons Learned**
(D.H. Timbres, 2006)	*"Hydrogen Damage in a Carbon Manganese + 0.5 Moly Ammonia Converter"* **1994 – 2005**: 1320 MTPD KBR Plant, HTHA cracks developed in two closing weld seams on a 22-year-old single wall ammonia converter. Excavation of cracked weld areas, welding with filler material, preheating at 400°F (204°C), and post-weld heat treatment.	This paper details considerable inspection and repair efforts relevant to operators suspecting HTHA damage. The authors support findings of others who no longer specify the 0.5 Moly steels above the carbon steel Nelson curve due to uncertainties regarding performance after prolonged use.

Table 2 HTHA Industry Experience Citation Summary

Nelson Curves: plot temperature against hydrogen partial pressure. These curves provide guidance on the temperature and pressure combinations that carbon and low alloy steels can withstand without experiencing HTHA over a prolonged period. If operating conditions (temperature and hydrogen partial pressure) fall below the curve for a particular steel grade, the risk of HTHA is considered low. If operating conditions fall above the curve, a more resistant material should be selected.

Practical Limitations: It's important to note that Nelson curves are based on historical data and industry experience (Bernhard Geis, 2014). They provide a general guideline but may not be applicable in all situations. Factors such as steel composition, welding practices, and stress levels can influence HTHA susceptibility and may not be fully accounted for in the Nelson curves. Some sources indicate that even 0.5 Moly steels should not be used above the carbon steel Nelson curve due to performance uncertainties after prolonged use (D.H. Timbres, 2006).

Cold-Wall Design: Cold-wall designs in ammonia reactors serve to maintain low pressure vessel temperatures, irrespective of the high-temperature synthesis gas within the catalyst bed. KBR was one of the first licensors to utilize a cold-wall converter design, where the feed gas is preheated in the shell/basket annulus, keeping the converter wall relatively cool (Inderjeet Singh, 2018). This minimizes thermal stress on the pressure vessel and reduces the risk of HTHA and hydrogen embrittlement.

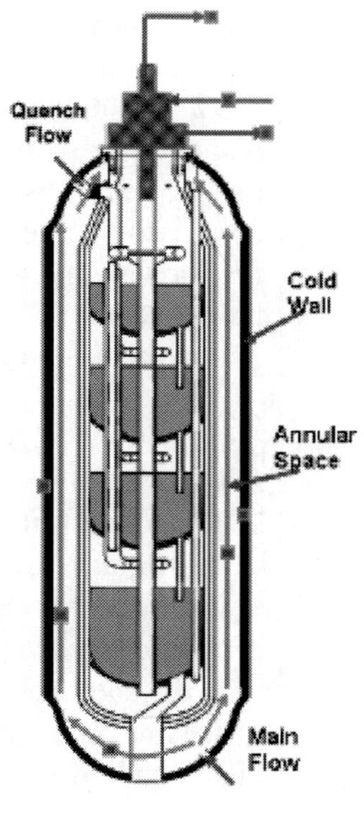

Figure 3 KBR Design Vertical 105-D Cold Wall Converter Design

Nitriding

Nitriding is a surface-hardening process that introduces nitrogen into the surface of a steel component at elevated temperatures. This forms a hard, wear-resistant layer of iron nitrides.

Mechanism: At elevated temperatures (typically above 370-380°C / 700°F-715°F), ammonia decomposes on the steel surface into nitrogen and hydrogen. The atomic nitrogen diffuses into the steel, it reacts with iron and alloying elements (like chromium) to form hard, brittle nitrides. These nitrides create a surface layer. This layer is prone to cracking, especially in areas with stress concentrations. Cracks in the nitride layer expose fresh steel to the nitriding environment, accelerating the process

Susceptibility: Nitriding can occur on both low alloy steels and stainless steels, although stainless steels have a reduced rate compared to low alloy steels. In ammonia environments, carbon steel and low alloy steels are typically not used in contact with the fluid above 370-380°C (700°F-715°F) and are replaced with austenitic stainless steel or non-ferrous alloys.

Consequences: In areas with high stress, cracks can form in the nitrided layer. The tips of these cracks can then nitride, causing the cracks to grow faster (Talarico & Scotto, 2012). Over time, the nitrated layer increases, and the ductility decreases sharply (Faraji, 2010).

Nitriding Industry Experience Sample		
Citation	**Description**	**Lessons Learned**
(Faraji, 2010)	*"High temperature Nitriding corrosion in Ammonia converters"* **1992 & 2004:** Eight cases of nitriding damage across two distinct 105-D units on two facilities showed nitriding failures across tubes, tube sheets, basket, thermowell and a thermal barrier. Samples became as brittle as glass and mainly consisted of Chromium Nitride.	When designing, be careful of low thickness material, additional corrosion allowance to be added for Nitridation. Suitable materials for a converter are AISI, 310, 321, 316 and 304 by priority. April 2005, the converter internals of ammonia plant no. 1 were replaced.
(R. Bosselaar, 2021)	*"Failure of a Hot Walled Ammonia Synthesis Converter at Yara Sluiskil"* **2019**: As part of a Risk-Based Inspection program, the second synloop converter inspection revealed a crack. Indications of Hydrogen Assisted Cracking near the shell and nozzle welds, specifically between the man-way nozzle top head and bottom inlet/outlet nozzle.	RCA determined the cracking was due to nitriding at the inner surface and hydrogen embrittlement. The nitrided inner layer acted as a crack initiation point. The usage of a high-strength material (3Cr type compared to 2.25Cr) also played a role. Using a cold wall design, adjusting inspection to use TOFD and PAUT inspections to detect cracks in all orientations. Adjusted operating procedures for MPT curves for all converters not be pressurized below 110°C (230°F)

Table 3 Nitriding Industry Experience Citation Summaries

Prevention

Inspection and Monitoring: Techniques like ultrasonic testing can be used to detect cracks and other defects associated with nitriding, these are detailed effectively by (R. Bosselaar, 2021). Regular inspection and monitoring of reactor components for signs of nitriding can help to detect and address the issue before it leads to significant damage.

Temperature: Maintaining the temperature of steel components below the threshold for significant nitriding is crucial. Generally, keeping temperatures below 370-380°C (700-715°F)

Material Selection: Austenitic stainless steels and non-ferrous alloys (e.g., Inconel 600) exhibit much lower nitriding rates compared to carbon and low alloy steels

Lining/Cladding: Applying a lining or cladding of a nitriding-resistant material (e.g., stainless steel) to the inner surface of the reactor can protect the underlying carbon or LAS from the ammonia-rich environment. However, it's important to ensure the integrity of the lining to prevent hydrogen disbanding.

Controls & Instrumentation

Poor control logic and instrumentation in ammonia converters can lead to a variety of issues, impacting safety, efficiency, and reliability. Here's a breakdown of the potential problems, drawing from the provided sources and KBR knowledge:

Temperature Excursions: Poor temperature control can damage the catalyst, reduce conversion efficiency, and lead to runaway reactions. Inadequate instrumentation might fail to detect hot spots or insufficient cooling, resulting in accelerated degradation, sintering or agglomeration of the catalyst.

Oxygenate Contamination: Ingress of oxygen containing compounds >5 ppm will reduce the activity of the synthesis catalyst at a minimum. Water ingress events can cause thermal shock to the equipment and agglomeration of the catalyst.

Pressure Control: Inadequate control logic, faulty instrumentation or passing valves can lead to rapid pressurization or depressurization, exceeding the converter's design limits and potentially causing catastrophic failure.

Control & Instrumentation Industry Experience Citation Summaries		
Citation	**Description**	**Lessons Learned**
(Gary L. Acosta, 2021)	*"Major Failure of an Ammonia Synthesis Converter Internals Due to Overpressure"* The paper details a failure in an ammonia synthesis converter during a plant start-up. The inlet and outlet expansion joints for the first bed interchanger failed due to excessive differential pressure, leading to a bypass of synthesis gas away from the catalyst beds. During start-up, the downstream pressure increased rapidly, due to the uncontrolled introduction of synthesis gas, via a manual four-inch valve used for initial pressurizing, causing a rupture disc to burst. Reverse flow led to increased differential pressure across the interchanger's expansion joints.	The operating procedure didn't account for Valve #3 sealing and preventing the pressure transmitter from accurately measuring loop pressure and there was no secondary pressure indicator. The inability to make minute changes to the four-inch valve (Valve #2) led to uncontrolled flow, causing the rupture disc (RD#1) to burst. Rupture disc RD#1, designed to protect the tubes of Exchanger #1, ruptured at 290 psi (20 bar) and the subsequent flow path through Exchanger #1 and the steam generating exchangers into Interchanger #1 caused a significant differential pressure across the inlet and outlet expansion joints of Interchanger #1. Wireless position feedback transmitters installed on manual valves, along with automated control valves, improved real-time monitoring and control of the converter.

(Arie de Bruijne, 2015)	*"Surprising Internal Damages on a Horizontal Ammonia Converter Basket"* This paper details critical issues and findings from a 2014 inspection of a 2200 MTPD nameplate KBR Purifier,, operating since 2006. Multiple rapid backflow scenarios, initiated by uncontrolled pressurizations and de-pressurizations due to faulty valves. This led to; the uplift of the catalyst beds, rupture of the bellow in the transfer pipe above bed 3B, rupture of the uplift restraint clips of the bed 2 & 3B and damage to the Johnson screen seal welds and ledge plates at the in bed 3A, resulting in a significant amount of catalyst below bed space and in between the buckled thermal barrier plates of Bed 3A. Two of the 3 thermowells showed sever deformation.	**Updated operating procedures**; Check the valve position of the Converter outlet valve HV1033 after each trip and confirm it's closed locally before restart. Set the output of HIC1101 to 0% on the 103MJ trip. Restrict the operation of the Synthesis Loop Start-up vent to avoid rapid depressurization of the loop during shutdown. **Instrumentation**: Investigate all possibilities for the replacement of HV1033 (temporary or final solution) from butterfly to ball valve. Add missing check valves to the loop. Provide a small depressurization valve in the hot synloop for controlled depressurization. Worked with KBR to design and order new basket. The paper presents very interesting images and details of damage caused by the rapid pressure and temperature swings caused by valve failure

Table 4 Control & Instrumentation Industry Experience Citation Summaries

Valve / Communication Issues: Valves can also remain open in the field despite DCS graphics indicating a closed position, which has been seen in industry to produce backflow and thermal shock events due to rapid temperature change. Figure 4 shows the temperature gradient expected during a typical shut down across the 105-D, temperature decreasing from 510°C (950°F) to 440°C (825°F), total 70°C (125°F) in about 400 seconds.

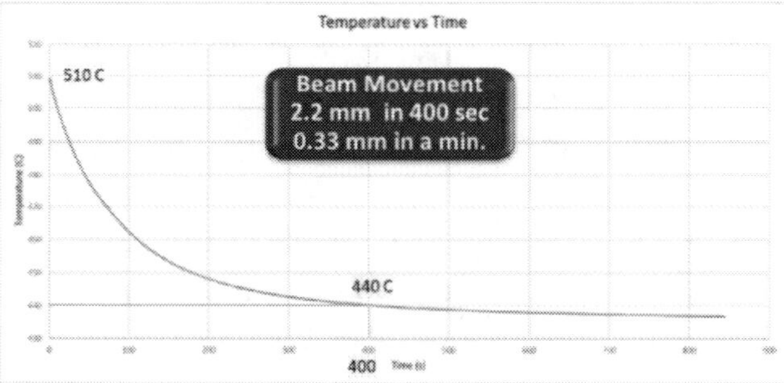

Figure 4 Synthesis Loop Isolated as Normal - Transient Case Beam Movement

Figure 5 shows the scenario where the loop is not appropriately isolated, due to a faulty isolation valve or lack of check valves and it registers a decrease of 115°C (207°F) in 15 seconds, yielding a beam movement rate of 15.2mm / minute.

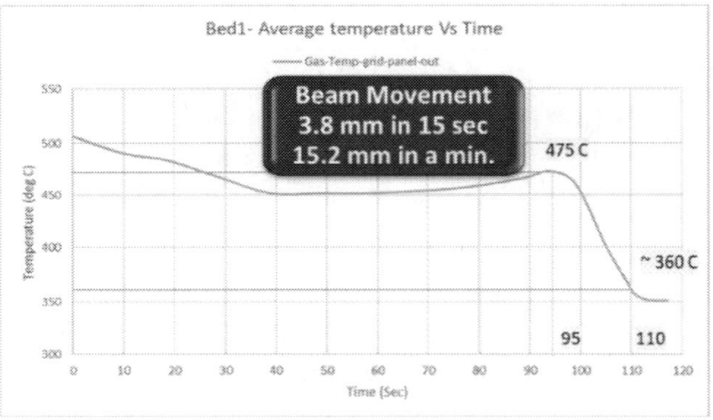

Figure 5 Synthesis Loop Not Isolated - Transient Case Beam Movement Assessment

Control Logic: In some situations, the logic implemented by compressor vendors will protect the compressor machine but can cause a high rate of depressurization in the 105-D if additional logic is not included. Synthesis gas compressor anti-surge valve tuning has been the source of pressure swings across 105-D converters.

Operator Training: Operators must be thoroughly trained on the control system and emergency procedures. Operator Training Simulators

(OTS) can be utilized, for example, to train unusual scenarios where valves are stuck open in the field but reading closed on the DCS. The distinct scenarios presented in this paper have been programmed by KBR into OTS units for training.

THERMAL / PRESSURE CYCLING

Frequent temperature fluctuations, such as during start-ups and shutdowns, cause thermal stress in the converter components. Repeated cycling leads to fatigue, which can initiate and propagate cracks, especially in welds and areas with stress concentrations.

Catalyst Damage: Wetting or damage of catalyst restricts gas flow and produces localized areas of high velocity or or temperature.

Gasket and Seal Failures: Temperature and pressure cycling can degrade gaskets and seals, leading to leaks. Such as in (Arie de Bruijne, 2015) where a loose thermowell flange and missing gasket were found after cyclical events.

Temperature / Pressure Cycling Industry Experience Citation Summaries		
Citation	**Description**	**Lessons Learned**
(Kodri, 2001)	*"Failure of Internals of Ammonia Converter"* **1999:** 105-D converter modified with Casale internals experienced catalyst migration four years after the modified basket retrofit. Experience details broken sliding expansion joints in the gas return pipes of beds No. 3 & No. 4. and a broken screen on the support thermowell of the inner collector in the fourth bed	The failure indicated that the design of the expansion joints was insufficient to absorb thermal expansion, leading to collapse. Temperature changes during startups and shutdowns caused differential growth and movement of gas return pipes and catalyst basket The support for the thermowell pipe and wire screen of the inner collector must be strong enough to withstand operational stresses and prevent catalyst spillage.
(Figuero a, 2010)	*"Troubles in radial flow reactors for ammonia synthesis"* **2006:** 105-D modified to radial flow using a Topsoe design. 2003, low flow rate through the start-up exchanger and vibration in motorized valves at the reactor's inlet and outlet were observed. June 2005, the pressure in the synthesis loop was 170 kg/cm2 then exceeding the design pressure of 160 kg/cm2 in August. In 2006 a 34" and 9" crack in the welded center screen was found.	Flow problems in the heater resulted in excessive vibration in motorized valves MOV-1 and MOV-3. The inlet and outlet of the coil were opened for inspection, but nothing was found to explain the flow problems. The paper discusses the process and procedures for removing the catalyst in inert atmosphere and repairing the cracked screen.
(Muhtad in, 2023)	*"Unprecedented Catalyst Migration Problem of 2000 MTPD Ammonia Plant"* **2021**: KBR 105-D increasing pressure drop in the synthesis loop, rising from 6.3 kg/cm2 (89.6 psi) to 8 kg/cm2 (114 psi). Catalyst migrated to synthesis-loop equipment & the Unitized Chiller tube annulus. Plant rate reduced to 75%. A 5-7 mm gap was found at the end plate-2 in bed 3B of the 105-D and leaks in the 116-C reported.	Improved control during transient conditions, such as start-ups and shutdowns, to minimize stress on the converter and prevent damage. Rapid changes in pressure and temperature in the 105-D during transient conditions and wetting of the catalyst due to 116-C leakage were presented. The paper also describes practical advice for cleaning catalyst from process equipment, particularly the downstream unitized chiller.

Table 5 Temperature / Pressure Cycling Industry Experience Citation Summaries

Summary & Combined Case

To represent an accumulation of industry experience and lessons learned, this section presents a modern converter, designed to operate safely and tolerate unforeseen operational issues and upsets, referred to as the 'Murphy Converter' for its consideration of what could go wrong and has gone wrong in real world operational experience.

Mechanical Design – Murphy Converter

1. Cold Wall & LAS material which is suitable for operations at the design temperatures without increasing the risk of HTHA.

2. Basket Connections: High temperature gas entering & exiting the converter is directly sent from the Stainless Steel (SS) external piping to the basket using SS internal pipes connecting SS/Incoloy800 extensions on the nozzles to the basket. Effectively eliminating exposure of LAS material to any high temperature area of the converter and limits probability of nitriding.

3. Grid Design Free movement of the profile wire screens eliminate stress forces in transient upsets. Rope packing and wire screen in specific areas of the support grid provide extra protection.

4. Expansion Joints: Expansion joint metallurgy and installation guidelines developed to ensure expansion joints are installed in the correct direction.

5. Thermowells: Hanger supports system redesigned to increase clearance between the hanger supports and the thermowells

6. Materials: The Murphy Converter uses LAS material based on API-941 curves and enforces the use of API-934-A or API-934-C for the manufacturing and testing. Austenitic stainless steels or Inconel alloys are utilized where applicable. Converters are manufactured as per ASME Sec VIII Div 2 and the applicable API standards.

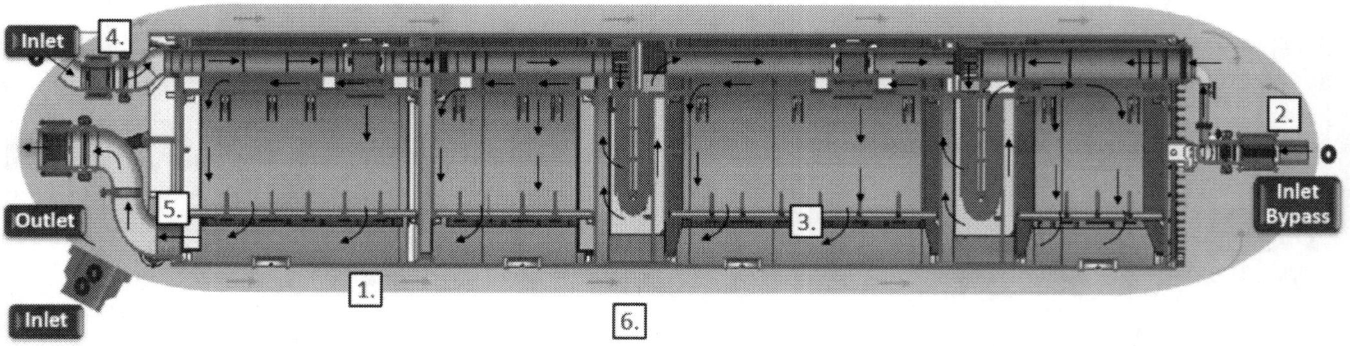

Figure 6 Hypothetical KBR 'Murphy Converter' Mechanical Design

Instrumentation and Control:

Valves: Bi-directional tight shut-off valves are specified. All 105-D valves require KBR approval of vendor drawings and datasheets.

Safety Interlocks: immediate closure of the loop isolation valves on providing manual stop command to the compressor and not wait till the machine ramps down and trips. Loop isolation valves cannot open during startup phase until the pressure on the machine side and ammonia synthesis loop are equalized.

Anti-surge Settings: Proper tuning of the compressor anti-surge system in collaboration with machine vendors is critical.

Redundant Instrumentation: Employed redundant sensors and transmitters for critical parameters like pressure, temperature, and flow to provide backup in case of failure.

Valve Monitoring: Real-time monitoring of valve positions is used via wireless position feedback transmitters on manual valves.

Integrity Operating Windows: KBR's INSITE has an Integrity Operating Window application which monitors the 105-D process conditions to track deviations over time and inform inspection and maintenance schedules.

Operator Training Simulator (OTS): A KBR high fidelity OTS system provides a risk-free environment to model, test and train failure scenarios based on plant and industry experience.

Conclusion

The KBR 105-D is the most deployed ammonia synthesis unit in the world and it presents a rich history of operational lessons learned and engineered solutions. The real-world experiences and mitigation strategies presented here aim to keep industry moving forward with iterative improvements to safety and reliability.

References

1. API RP 934-F. (n.d.). *The Effects of Hydrogen for Establishing a Minimum Pressurization Temperature (MPT) for Heavy Wall Steel Reactor Vessels.*
2. Arie de Bruijne, M. W. (2015). Surprising Internal Damages on a Horizontal Ammonia Converter Basket.
3. Avinash Malhotra, J. G. (2010). KBR's Purifier Technology and project execution options for Ammonia Plants.
4. Bernhard Geis, H.-J. B. (2014). *Experiences with 2.25Cr-1 Mo-Steel Equipment in the Ammonia Synthesis Loop after 30 Years in Service.*
5. D.H. Timbres, R. P. (2006). Hydrogen Damage in a Carbon Manganese + 0.5 Moly Ammonia Converter.
6. Faraji, A. H. (2010). High temperature Nitriding corrosion in Ammonia converters.
7. Figueroa, A. M.-H.-H. (2010). Troubles in radial flow reactors for ammonia synthesis.
8. Gary L. Acosta, O. B. (2021). Major Failure of an Ammonia Synthesis Converter Internals Due to Overpressure.
9. Inderjeet Singh, M. G. (2018). CFCL-2 Ammonia Plant Operating Experience of a KBR True Cold Wall Add-on Converter.
10. Jain, U. (2008). KBR's Purifier Technology lowers capital cost and energy consumption while improving reliability of grass-root Ammonia Plants.
11. Jing, A. (2009). KBR's New PURIFIER plus Proess for Ammonia Production. *KBR Technology Conference.* Bejing.
12. John Czerwinski, J. L. (2012). Case Study of Weld Cracking in 2 1⁄4 Cr-1Mo-1/4V Heavy Wall Ammonia Converter.
13. Kodri, H. D. (2001). Failure of Internals of Ammonia Converter
14. Malhotra, G. (2002). KBR's KRES and Purifier for energy savin and increasing capacity in an Ammonia plant.
15. Michael B Rugh, S. R. (2002). Repair and Failure Analysis of a Crack in an Ammonia Converter.
16. Michael W Hester, D. J. (2013). Elevated Temperature Failure Mechanisms in Ammonia Plant Equipment - Reducing the Risk of Failure.
17. Muhtadin, N. P. (2023). Unprecedented Catalyst Migration Problem of 2000 MTPD Ammonia Plant.
18. R. Bosselaar, S. C. (2021). Failure of a Hot Walled Ammonia Synthesis Converter at Yara Sluiskil.
19. Rugh, R. (2002). Repair and Failure Analysis of a Crack in an Ammonia Converter.
20. S.G. Choudhary, M. P. (2007). Window Repair of site joint circumferential weld seam cracks of Ammonia Converter-II.
21. Sam Ternowchek, S. D. (2009). Repair and online mechanical integrity monitoring of an ammonia synthesis converter.
22. Stefano Cavalleri, G. P. (2023). Hydrogen Assisted Cracking in the ammonia synthesis loop.
23. V. Alipour Tabrizy, B. M. (2024). *Lessons Learned from Investigations of Ammonia Incidents.*

Lessons Learned and Mitigation of Risks Associated with changeout of Internals and Catalyst in North America's Oldest Ammonia Converter

The ammonia converter reactor at the LSB Cherokee Nitrogen site has been in operation since its installation and startup in 1962. Originally fabricated by A.O. Smith in 1961, the reactor demonstrated reliable performance throughout its 63-year lifespan. In 1986 and again in 2007, its internals were upgraded with Casale components, followed by a catalyst replacement in 2013 after a rapid depressurization event in the synthesis loop. During the planned 2024 turnaround, the internals were replaced to facilitate a comprehensive inspection of the pressure vessel, and new internals from Casale and catalyst from Clariant were installed. This paper presents lessons learned from the continued operation of this aging reactor, with a focus on inspection techniques, findings, and risk mitigation strategies to ensure its ongoing reliability.

Alexis Kimpel, Gabrielle Grills
LSB Industries

Taylor Archer
Clariant

Enrico Masciarelli, Luca Redaelli
Casale

Introduction

LSB Industries' Cherokee, Alabama, facility operates a 500 STPD (450 MTPD) M.W. Kellogg ammonia plant that has been in operation since 1962. As a multi-product site, the plant produces nitric acid, ammonium nitrate, urea, and UAN in addition to ammonia, with export capabilities via rail, truck, or barge on the Tennessee River. Originally designed for 360 short tons per day (STPD) (325 MTPD), process upgrades—including the installation of a Low-Temperature Shift Converter, a larger air compressor, and modifications to the ammonia converter—have increased the plant's nameplate capacity to 500 STPD (450 MTPD).

This paper summarizes the key insights and technical findings from the ammonia converter cartridge and catalyst replacement alongside a comprehensive shell inspection in November of 2024.

Vessel Background

The ammonia converter pressure vessel, fabricated in 1961 by A.O. Smith, was commissioned in 1962. It has a Maximum Allowable Working Pressure (MAWP) of 5,170 psig (355 bar), operating at 3,400 psig (235 bar), and is fed by three 6-cylinder reciprocating compressors. Unlike most legacy Kellogg ammonia converters with a partial-opening, bottle-shaped design, the Cherokee converter features a full-opening design, as shown in Figure 1, enabling complete removal of the internal cartridge. This design choice does not impact operational performance but significantly improves maintenance accessibility and inspection capabilities.

The Cherokee ammonia converter is a multilayer construction with a forged bottom head. The shell has an internal layer of ½" (13 mm) thickness and 36 5/16" (8 mm) layers. The material of the shell is a proprietary steel grade (A.O.S 1135G Gr. A. The top flange is a carbon steel forging (ASTM A-105 Gr. 2) normalized and tempered. The bottom head is made from a single forged piece of carbon ½ Moly material normalized and tempered. The top cover is a forging made from A.O.S Grade 5002 a proprietary carbon steel enhanced with small alloys addition.

The vessel, like other legacy Kellogg designs, follows the cold wall principle. In this configuration, the pressure vessel shell is protected from high-temperature ammonia synthesis conditions by a non-pressure-bearing internal cartridge housing the catalyst beds. This design aims to mitigate degradation mechanisms such as high temperature hydrogen attack (HTHA) and nitriding by keeping the pressure vessel in a milder operating environment, while the cartridge endures harsher conditions. Nickel alloys and stainless steel are typically used for the cartridge and internals to withstand long-term exposure.

Over time, even these high-performance materials degrade, necessitating replacement. In full-opening designs, the entire cartridge can be replaced, whereas in partial-opening designs, the cartridge is permanently fixed, restricting inspection and replacement. This has led to the replacement of most partial-opening converters in the U.S. due to concerns over limited access to the cartridge and the inability to fully inspect or pressure-test the pressure vessel. However, the full-opening design of the LSB Cherokee converter has enabled multiple complete internal replacements and comprehensive pressure vessel inspections over its lifespan.

The synthesis converter was originally a five-bed axial cartridge and operated closer to the MAWP of the vessel but was replaced with a Casale three-bed radial design in 1986, among the first to incorporate Casale's axial-radial catalytic bed technology. This upgrade significantly improved performance while maintaining a straightforward design. The internals were designed for an increased nominal capacity of 660 STPD (600 MTPD), supporting future expansions.

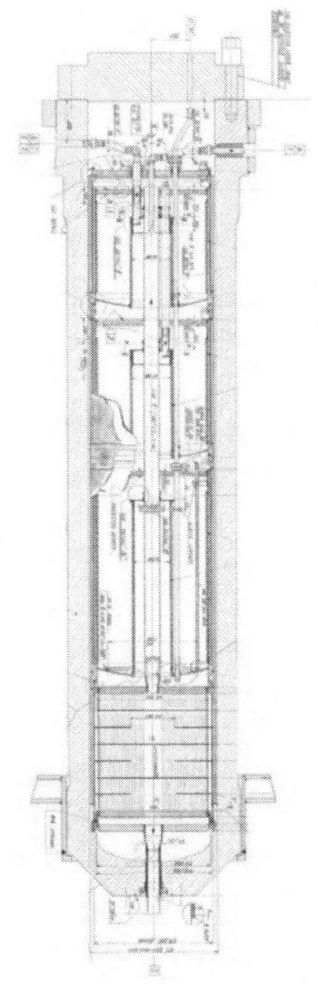

Figure 1. Ammonia converter shell featuring 1986 Casale internals

In 2007, after 20 years of operation, another complete cartridge replacement was performed, not due to cartridge degradation but because full cartridge replacement was faster than end-of-life component changes. This new design incorporated Casale's slotted plate bed collectors, a monolithic stainless-steel structure with parallel slot geometry, improving gas flow distribution while retaining catalyst integrity. Additionally, an interbed exchanger replaced the previous quench cooling system, further enhancing efficiency.

Following a rapid synthesis loop depressurization event due to HTHA in 2013, a catalyst changeout was conducted alongside a partial shell and cartridge inspection. Although no damage was observed, traces of oil contamination were detected on the cartridge shell and cover. During this outage, wet fluorescent magnetic particle (WFMP) inspections revealed longitudinal seam weld indications on the shell, though reevaluation confirmed no transverse cracking. A subsequent 100% WFMP re-inspection in 2013 found no further indications.

In the November 2024 turnaround, the cartridge and catalyst were replaced, accompanied by a detailed shell inspection and a contingent repair plan. This continues the long-standing strategy of leveraging the full-opening design to facilitate comprehensive maintenance and mitigate long-term operational risks.

Project Background

The previous ammonia synthesis catalyst operated from 2013 to 2024. During this period, the pressure drop increased to 120 psi (8 bar), causing production limitations in the recycle cylinders of the compressors. The primary suspected cause was compressor oil ingress into the converter. Additionally, the catalyst experienced ammonia carryover throughout its lifespan, along with significant exposure to CO and CO_2 due to a tube leak in the methanator feed/effluent exchanger. The catalyst exhibited slow light-off and reduced activity, further compounding operational challenges.

Decision for Cartridge & Catalyst Replacement

With the aging pressure vessel and poor historical inspection records, LSB Cherokee evaluated two options:

1. Defer full replacement of the cartridge and shell until the next planned outage in 2027.

2. Proceed with a catalyst replacement in 2024 to alleviate operational constraints.

Due to the high-pressure drop and associated production limitations, the decision was made to replace the catalyst immediately while also conducting a comprehensive pressure vessel inspection. Given that the inlet and outlet collectors and bottom exchanger (particularly the tube-to-tube sheet welds at the high-temperature interface) had reached the end of their useful life, the catalyst-containing baskets were also replaced. Following the 1987 and 2007 revamp projects, a full cartridge replacement was chosen instead of replacing only specific internal components. This approach was driven by:

- Minimized turnaround time – A complete cartridge swap is faster and simplifies maintenance operations.

- Facilitated vessel inspection – Access to the pressure shell was necessary for thorough evaluation.

Installation & Handling Considerations

Casale's special steel support stand, originally designed in 1987, was used again to pre-load the catalyst into the new cartridge before vessel insertion, significantly reducing plant downtime. This method was particularly advantageous in

2024, as it provided additional time for a thorough inspection of the pressure vessel during the turnaround.

Given the cartridge's relatively small size (~50 tons (45 MT), including catalyst), reasonably sized cranes were used for installation. However, since the previous lifting device was unavailable, a new identical lifting device was fabricated after verification of its suitability.

Internals Design & Materials Enhancements

While the general cartridge design remained unchanged—since no significant capacity increase was required—several technological advancements were incorporated to enhance reliability and extend component life. Casale implemented advanced material selection strategies based on extensive research into nitriding effects in ammonia synthesis environments.

Over the past decade, Casale has conducted a comprehensive study of nitriding in ammonia synthesis, analyzing samples from various operating conditions and performing mechanical testing. This research informed the material selection for the new LSB Cherokee converter internals, as shown in Figure 2, leading to the use of enhanced stainless steel in critical components to improve nitriding resistance and extend operational lifespan.

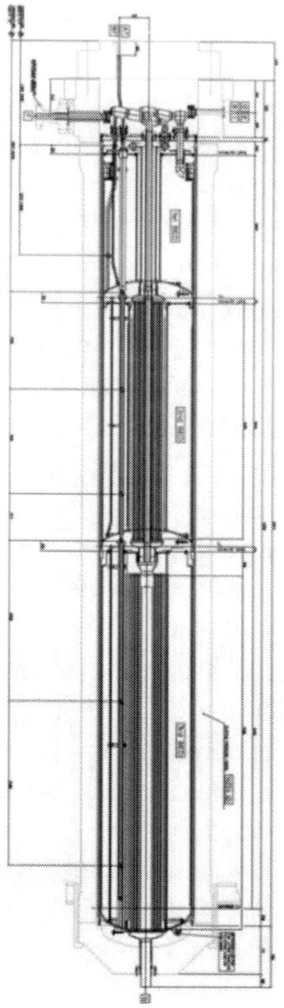

Figure 2. New Casale internals

Advantages of the New Casale Cartridge & Catalyst

Several key enhancements were realized with the new internals and catalyst selection:

1. Optimized Design for Extended Reliability
   - Casale integrated advanced durability modifications in high-temperature zones, improving the operational lifespan of critical components.

2. State-of-the-Art Catalyst Technology
   - The new catalyst selection includes AmoMax® 10 Plus and AmoMax® Casale, offering:

- Higher activity and efficiency
- Greater tolerance to poisoning and aging
- Energy savings and increased ammonia yield

  - AmoMax® Casale, developed jointly by Clariant and Casale, was introduced in 2019 and has received two major international awards:
    - 2021 Swiss Chemical Society's Sandmeyer Award
    - 2020 ICIS Innovation Award for Best Sustainable Process
  - This advanced catalyst has already been implemented in over a dozen plants worldwide.

3. Preloaded Catalyst for Faster Turnaround
    - The new cartridge and catalyst were delivered before the outage, allowing all three catalyst beds to be preloaded ex-situ.
    - This reduced the mechanical scope during the turnaround, freeing up time for a more comprehensive vessel inspection.
    - Once inspections were completed, the entire cartridge was lifted and installed as a single unit, streamlining operations.

Additional System Upgrades

Alongside the converter internals, the main inlet and quench valve control system was upgraded. Prior to 2024, these valves operated pneumatically via air pressure regulation from the control room, requiring hundreds of feet of instrument air lines.

Issues with air leaks in all five system valves led to operational inefficiencies. The converter inlet valve was typically maintained at 70-75% open to meet quench requirements, adding additional pressure drop within the system alongside the catalyst-related losses. The 2024 upgrades improved valve reliability and reduced system-wide inefficiencies.

Inspection Scope & Damage Mechanisms

The inspection scope for the converter shell included an API 510 internal inspection covering all accessible weld seams, the internal shell, top flange and cover, as well as the bottom head and nozzle. To ensure a comprehensive assessment, the following non-destructive examination (NDE) methods were employed:

- 100% Wet Fluorescent Magnetic Particle Testing (WFMT) on all internal weld seams.

- Liquid Penetrant Testing (PT) on the top and bottom flange assemblies.

- Field Metallographic Replication (FMR) at four or more locations around the bottom head.

Pre-Shutdown Inspection

Before the turnaround, several inspection activities were performed. Positive material identification (PMI) was done on the bottom head and shell to ensure that chemical analysis of the material matched what was stated in the drawings, as the vessel name plate has been removed and there was no U-1, which certifies a pressure vessel and has the materials, thickness, and pressure and temperature ratings. While this method has limitations with carbon steel, the analysis showed alloying elements that confirmed the chemical composition of the components. The weep holes of the vessel were confirmed to be free of obstructions and were tested for leaks. Each layer of the vessel contains weep holes to ensure no hydrogen gas pressure is present past the inner liner. This clever design greatly reduces the risk of HTHA in the shell and can provide early warning if the liner is breached. Inspection of the main body flange was planned, however, due to the leak present in the head, it was determined to be too great of a risk while the vessel was operational. This obstacle prompted the mechanical integrity team to develop several strategies to ensure that the flange leak could be corrected.

Contingency Repair Plans

Contingency Repair plans were created and materials were sourced in the event damage was found during inspection. Possible scenarios and their contingency plans are listed below. Scenarios that included major repairs such as head replacement or relining the vessel were considered, however, since they were unlikely and would require a significant cost to procure materials no steps beyond locating a suitable supplier were explored.

- Minor Shell Repair

In the event that minor damage was detected in the outer shell courses the following steps were planned.

  - Hydrogen bakeout at 600°F (315° C) for 4 hours, this was a precaution as the outer shell layers should not be in contact with hydrogen.
  - Preheat to be maintained at least 200°F (93° C).
  - Excavate the damaged area/s and prep for welding.
  - SA-516-70 plate was selected as a substitute for the A.O.S grade plate. Several plates were selected, and the material test reports reviewed and put on hold with suppliers before the outage, while SA-516-70 is a readily available material the original design required 26% elongation, slightly higher than required by SA-516-70 grade.
  - The filler material selected for the repairs was E-7018-H4R (SMAW) / ER70S-2, -3, or –6 (GTAW).
  - Dry Magnetic Testing (MT) to be performed as primary NDE, with WFMT on final pass.

- Internal Liner Repair

Minor cracking indications to the internal liner that did not penetrate through wall were to be ground out to prevent crack propagation, and Fitness-for-Service (FFS) was performed to ensure minimum wall thickness in localized areas.

For through wall repairs of the inner liner, the procedure follows that outlined for the outer shell courses, however, PWHT or NBIC Method 3 temper bead must be performed.

- Gasket Change/Machining

Due to the leak in the body flange, it was inaccessible before the outage. Several options were explored. Machining services were brought onsite as both a contingency and to evaluate the delta ring groove on both the cover and body flange. If significant non-conformance of the groove existed, the flange would be machined and the groove slightly enlarged to accommodate an oversized soft iron delta ring. Due to delta rings being less common than oval or octagonal ring type joints, converting the gasket type to a different style such as a Kammprofile gasket was considered. However, due to significant alterations required to accommodate this design change, it was considered a last resort.

- Stud Removal

The 24 5.5” (14 mm) studs on the main body flange of the vessel and the corresponding nuts were purchased as the age and condition of the existing hardware were unknown. A corresponding tap was also purchased to chase the threads in the vessel. There was concern that the studs would not be easily removed. Three contingencies were planned with arrangements made to ensure the necessary resources were on site or on call.

1. Mechanical stud extractor.

2. Induction heating to be used in conjunction with a stud extractor.

3. Metal Disintegration.

Inspection Methods & Objectives

Wet Fluorescent Magnetic Particle Testing (WFMT)
WFMT was selected to detect mechanical damage, cracking, and potential hydrogen-assisted cracking, including early-stage high-temperature hydrogen attack (HTHA).

Liquid Penetrant Testing (PT)
PT was utilized for mating flanges and weld seams around the top and bottom heads, specifically targeting cracking in the ring-type joint (RTJ) grooves, where stress concentrations pose an elevated risk of fatigue cracking, as well as thermal fatigue cracking in the weld.

Field Metallographic Replication (FMR)
FMR was chosen to be conducted on the bottom head of the converter to detect decarburization, an indication of near-surface HTHA damage, providing a detailed assessment of microstructural integrity. In the event that damage was found or suspected, phased array ultrasonic testing (PAUT) was to be performed to further investigate. Due to the lower risk of HTHA damage due to vessel construction and operational design, it was not deemed an appropriate primary method.

Inspection Findings

API 510 Internal Inspection Results
The overall API 510 internal inspection revealed no signs of damage or degradation that would compromise the continued operation of the converter. The delta ring groove contained grit that likely prevented proper seating of the soft iron gasket, the grit appeared to be catalyst that was not properly cleaned prior to the last assembly.

WFMT Findings
- No relevant indications were found in any internal weld seams examined within the converter.

PT Findings
- Minor surface scratches were observed on the main body delta ring joint gasket surface. These were buffed and confirmed to be shallow with no structural impact.
- Additional PT examinations revealed insignificant markings on gasket surfaces but no cracking or concerning damage.

FMR Findings
- Although FMR was conducted, difficulties in interpreting the replications led to an alternative approach using field micrographs.
- The micrographs provided a clearer view of the bottom head microstructure, showing no significant decarburization or thermal degradation of the metal.

MI & Inspection Challenges

There were several challenges during flange reassembly. Reinstallation of the cover was difficult due to the size and weight, care had to be taken to ensure that the weight of the cover was applied gently and evenly to the delta ring. While it was assembled without incident, makeshift stops were used to set the head down to make adjustments. In the future, planned tooling would be helpful to ensure success. The studs and acorn nuts specified in the drawings were too long and bottomed out before proper force was applied to the delta ring. Washers were fabricated to accommodate the extra length of the studs, and, after several torquing attempts, the vessel was leak free.

Project Execution

The converter internals upgrade, shell inspection, and potential repairs were identified as the critical path for the turnaround. To optimize the timeline, the catalyst basket was pre-loaded ex-situ before the turnaround, allowing for faster installation during the mechanical window. The loaded cartridge is shown in Figure 3. The plant shut down on November 1st, 2024, while cartridge work, catalyst screening, and loading began on October 25th.

Figure 3. New cartridge on the stand ready to be loaded with catalyst

Catalyst loading, as shown in Figure 4, was conducted using dense loading to ensure uniform bed densities across all beds. Beds 2 and 3 were loaded without issues, achieving target densities. Bed 1 catalyst, received in a pre-reduced and stabilized form, was loaded without screening and reached the required target density. Once loading was completed, the new cartridge was placed under positive nitrogen pressure for safe storage until installation into the vessel during the mechanical window.

Figure 4. Catalyst loading

During catalyst removal, Bed 1 unloading was completed on schedule, with oil contamination observed. However, Bed 2 catalyst removal proved challenging, as chipping guns were ineffective in breaking up the material. Due to anticipated delays, the team opted to remove the entire cartridge with the remaining catalyst in Beds 2 and 3, as the crane capacity allowed for the extra load. This adjustment successfully accelerated the turnaround, moving the critical path ahead of schedule. Figure 5 shows the installation of the new cartridge full of catalyst into the pressure shell.

Figure 5. Installation of the new cartridge

Following the new cartridge installation full of catalyst and converter box up, loop pressure testing at operating pressure was conducted using a nitrogen truck to verify proper head installation and torque values before introducing syngas into the synthesis loop.

Startup & Reduction Process

Before introducing syngas, the team established a structured startup sequence to ensure a smooth and controlled startup. The process began with loop pressurization and startup heater ignition, followed by temperature ramp-up in the converter according to Clariant's specifications. Reduction of Bed 1 was initiated first, with Beds 2 and 3 following in sequence. Once the exothermic reactions were self-sustaining, the startup heater was shut down, and the synthesis loop pressure increased to operational levels. At this stage, the second and third syngas compressors were started, and final converter optimizations were carried out to achieve full plant capacity.

The CO and CO_2 slip after the methanator was monitored to ensure it was within acceptable limits before introducing syngas into the synthesis loop. One of the three syngas compressors was started, allowing gradual pressurization to protect the converter internals. To monitor water evolution during reduction, Clariant's ActSafE® non-dispersive infrared (NDIR) measurement system was deployed, providing continuous real-time readings of water and ammonia at the converter exit. Initially installed in an analyzer shelter, the system can also be integrated into the DCS system for long-term monitoring.

Historically, the main inlet valve was kept closed until the exotherm began, likely exceeding the design metal temperature of the pressure vessel. Moving forward, additional loop circulation was introduced manually to ensure proper heater function while maintaining thermal protection for the shell. This adjustment aligned the startup process with industry's best practices for cold wall designs.

During initial bulk reduction, minimal water was removed due to the inability to commission the refrigeration chillers (115-Cs). The Cherokee site refrigeration loop operates at 25 psig (2 bar), with an ammonia saturation temperature of ~11°F (-12° C). At this pressure, condensed water can freeze, as the submerged tube bundle in liquid ammonia may reach temperatures low enough to cause ice formation. To mitigate this, cooling water was manually introduced into the shell side of the refrigeration chillers, and the oil separator drain was opened to allow continuous water removal.

The bulk reduction of Bed 1 lasted until 09:00 on December 7th, with an inlet water concentration of 562 ppm before reaching the oil and secondary separators. Once Bed 1 completed bulk reduction, the startup heater firing rate was increased to bring Bed 2 to reduction temperature. By 01:00 on December 8th, the desired temperature was reached, and the heating rate was slowed to control water production.

Once the freezing risk of aqueous ammonia was mitigated and converter inlet ammonia concentration reached 8%, the 115-Cs refrigeration chillers were commissioned, significantly improving water and ammonia knockout. The refrigeration compressor was initially started on total kickback, bypassing the chillers while drawing ammonia vapor from the receiver. Once stable compressor operation was established, liquid ammonia refrigerant was gradually introduced into the 115-Cs, ensuring controlled cooling of the syngas stream.

By 18:45 on December 8th, Bed 2 bulk reduction was completed, and flow rates were adjusted to increase differential pressure (ΔP) and improve flow uniformity. The final Bed 3 reduction proceeded smoothly, completing at 05:00 on December 10th. The converter was then handed over to operations for normal startup. Minor compressor and control system issues were quickly addressed, and the plant was stabilized at design capacity.

Results & Performance Improvements

Following the converter internals and catalyst replacement, the plant achieved record-high production rates and significantly improved process gas efficiency as reflected in Table 1 below.

Time Period	Production Rate, STPD (MTPD)	Process Gas Efficiency, MMBTU/ton (GJ/MT)
Post Turnaround (First Full Month)	541 (491)	20.63 (23.99)
Pre-Turnaround (October '24)	507 (460)	21.35 (24.82)
Winter '23	525 (476)	21.15 (24.59)

Table 1. Process gas efficiency improvements

Additionally, the pressure drop across the converter was reduced from 120 psig (8 bar) to 60 psi (4 bar), and the synthesis loop pressure decreased to ~3,350 psig (231 bar). The converter inlet valve operation was improved, allowing for additional pressure drop relief with upgraded positioners and DCS controls. These modifications have led to enhanced process stability and efficiency, positioning the Cherokee ammonia plant for sustained high performance until the next planned outage.

Acknowledgements

We would like to express our sincere gratitude to Seth Storie and Scott Curl for their contributions to this report. Their expertise in mechanical integrity and inspection methodologies has played a crucial role in the development of this paper.

Hidden Threat - The Weld Failure Behind a Reverse Flow and Major Unplanned Shutdown

A major mechanical failure at an Ammonia Plant in Trinidad led to a Tier 1 Loss of Primary Containment (LOPC) and an emergency shutdown. The incident was traced to a compromised weld on the Primary Reformer, causing reverse flow and significant damage to various components.

This paper analyses the incident, exploring its consequences and failure mechanism. It also outlines the inspection techniques and repair methods employed to restore serviceability. Furthermore, best practices for preventing similar incidents in the future are discussed, such as enhancing equipment specifications, conducting asset integrity inspections, and implementing engineering controls.

Rajindra Benny, Dane Hosein
Proman Limited, Trinidad and Tobago, W.I.

Introduction

In April 2024, a major mechanical failure occurred causing a loud noise and subsequent emergency plant shutdown. Further investigation revealed that the outlet header to end cap weld on the Mixed Feed Coil had failed, resulting in a Loss of Primary Containment (LOPC) with the end cap separating from the header and being projected to grade.

This failure resulted in a reverse flow which exposed other components of the reforming section to temperatures above their design limits. These were subsequently inspected with repairs being performed as required.

This paper analyzes the incident, consequential damage, the failure mechanism and outlines the inspection techniques and repair methods used to restore serviceability. Best practices for preventing similar incidents in the future are discussed, including improved equipment specifications, asset integrity inspections, and engineering controls.

Process Description

The ammonia plant was commissioned in 2002 and built with a capacity of 1850MTPD.

Incoming natural or feed gas undergoes a reforming process in the presence of steam, at elevated temperatures to produce hydrogen.

The feed gas is compressed to 732psi (5047kPa) using a single-case, centrifugal Feed Gas Compressor, heated to 700°F (371°C) in the primary reformer convection section and sulfur is removed in the Desulfurization Vessels. After desulfurization, the feed is mixed with Medium-Pressure (MP) steam to form the Mixed Feed (MF) required for reforming. The mixed feed is preheated to 1150°F (621°C) in the MF coil in the convection section of the Reformer before being distributed to the catalyst tubes.

Nitrogen, the other essential component for Ammonia production, is introduced from the Process Air Compressor to the Secondary Reformer.

The Secondary Reformer effluent flows directly to the Waste Heat Boiler where high-pressure (HP) steam is generated. The effluent then moves to the HP Steam Superheater before entering the High Temperature Shift Converter at 700°F (371°C). A simplified Process Flow Diagram (PFD) for the reforming and downstream shift system is shown in Figure 1.

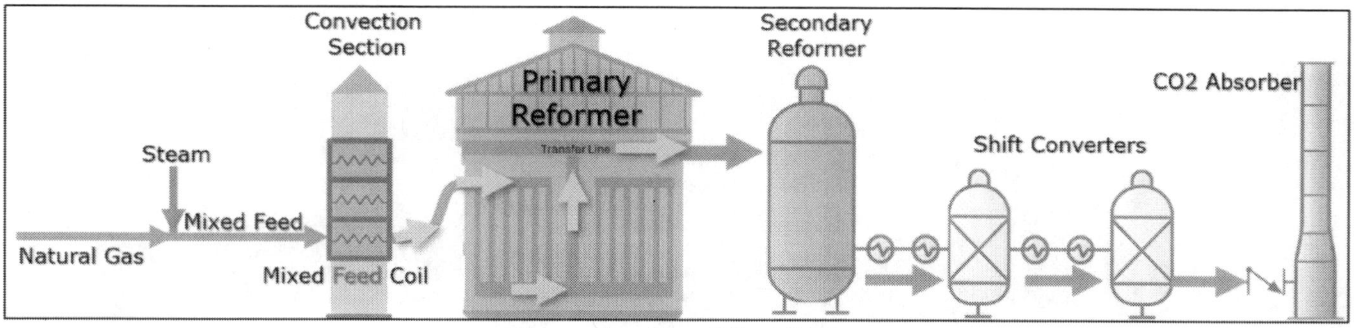

Figure 1 – Simplified PFD

Mixed Feed Coil (MFC) Data

The MFC is an important heat exchange component in the primary reformer and essential for optimal Plant performance. MFC is the first of five coils in the convection section and as the lowest coil it is the one which is first to be contacted with hot flue gas from the reformer.

Figure 2 shows a 3-Dimensional view of the headers, branches and mixed-feed flow direction. By design, there is a total of 180 tubes, with 72 bare tubes on the lower section and 108 finned tubes on the upper section of the assembly. The bare tubes are made from 4-inch Schedule 80S piping and finned tubes are made from 4-inch Schedule 120 piping.

The Inlet and Outlet Headers connected to the tubes are constructed from 16-inch Schedule 120 piping. The inlet header is made from ASTM A335 P11 (Low Alloy Carbon Steel), while all tubes and Outlet headers are constructed from ASTM A312 TP304H (Stainless Steel) material. On the opposite end, each pair of tubes is connected by a 180° short radius return bend, constructed from ASTM A403 WPS 304H material.

The End Cap on the Outlet Header is also constructed from ASTM A403 WPS 304H. The tubes are designed to withstand temperatures up to 1290°F (699℃), with the outlet header (including the End Cap) designed to withstand temperatures up to 1225°F (663℃) at a pressure of 860psi (5929kPa), in accordance with API RP 530 (1988).

The mixed feed coil is housed within the convection section and encased with carbon steel panel inner lined with refractory insulation.

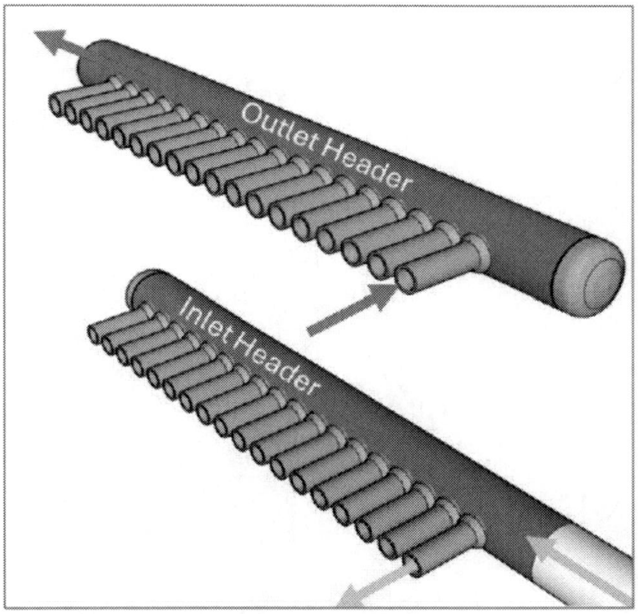

Figure 2 – View of Headers and branch connections

To gain a better understanding of the location of the mixed feed coil, Figure 3 shows a sectional view of the primary reformer, highlighting the mixed feed coil and associated outlet piping. An isometric view of the mixed feed coil assembly is shown in Figure 4, identifying the inlet and outlet header.

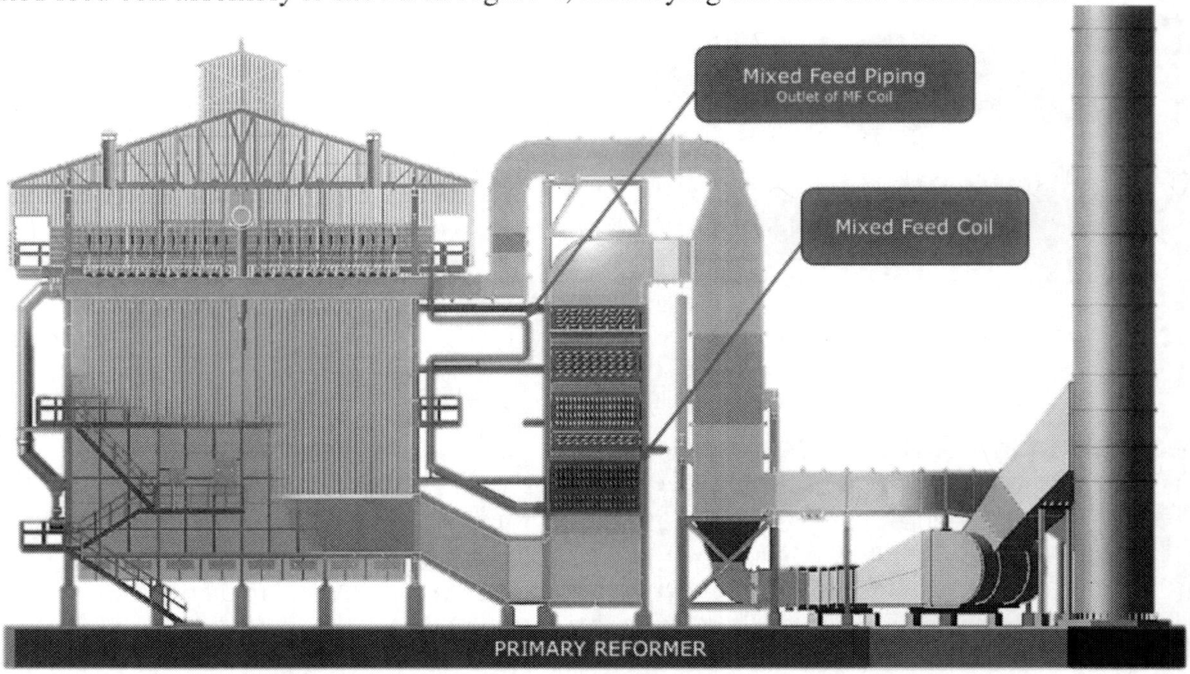

Figure 3 – Sectional View of Primary Reformer

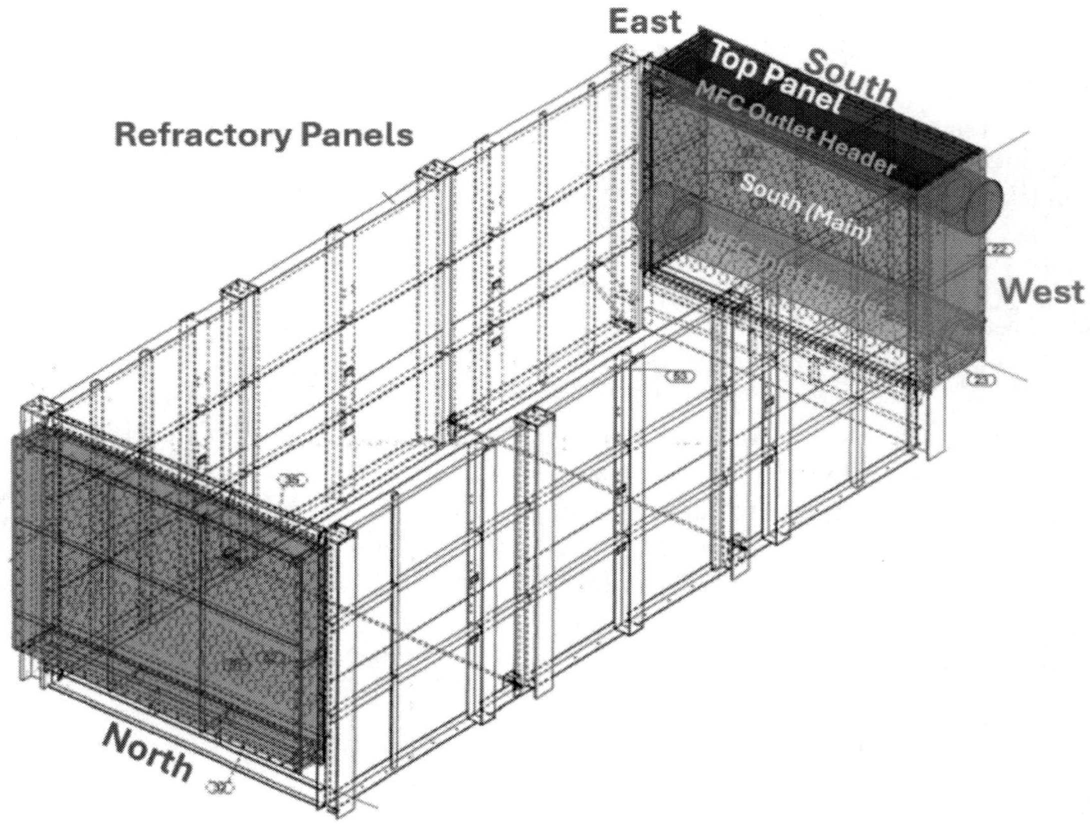

Figure 4 – Detailed view of MFC

Major Inspection Findings

Figure 5 shows a view of the mixed feed coil from the southern side, where the failure occurred.

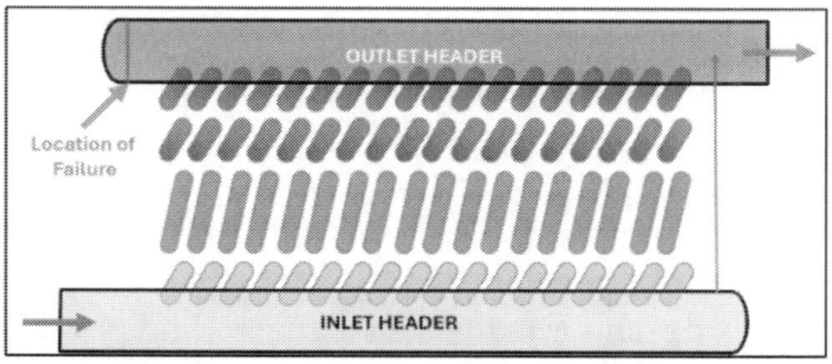

Figure 5 – MFC View from Southern Side

An initial inspection revealed a failure at the outlet header to end cap weld. The 16inch (40cm) diameter end cap [weighing approximately 167.5lb (76kg)] was found at grade, having been projected from an elevation of 27ft (8.2m) west and away from the convection section. See Figure 6.

Figure 6 – Dislodged end cap (left), outlet header (right)

This sudden mechanical failure and release of process gas resulted in significant damage to the south and western refractory covers and the following consequential damage to neighboring components.

a) Debris consisting of mineral wool insulation, insulation sheeting, refractory and structural members were found at grade encompassing an area with a footprint of approximately 240 x146ft (73 x 44m).
b) A diagonal brace (structure support) from the platform on the western side of the MFC coil was observed on the ground. See Figure 7.
c) Insulation Jacketing/Cladding were hanging from elevated areas.
d) Instrumentation Cables were hanging and the armouring/insulation covering for same were noted to be damaged and exposed. Approximately 50ft (15m). of instrumentation cabling was found dislodged with evidence of exposure to high temperature.
e) Deformed structural beam, broken handrail, mid rail and ladder cage observed on the western side of the MFC coil.
f) Structural beams were bent, and springs cans damaged.

Figure 7 – Consequential Damages

Consequence of Reverse Flow

This LOPC event, which occurred on 2024 April 20, was a result of a complete separation of the end cap from the MFC Outlet Header due to a failure of the Outlet Header-to-End Cap Weld. This failure created an area of low pressure (atmospheric pressure) within the referenced circuit and hence, a reverse flow condition was created (flow path seen in Figure 8). Due to the presence of a check valve installed at the Process Gas Inlet to the CO2 Absorber, the reverse flow occurred from the CO2 Absorber, through the breached outlet header and to atmosphere as identified in Figure 9. Analysis of the temperature trends revealed values greater than design, with internal inspection finding significant damage due to reverse flow from Secondary Reformer to MFC. The subsequent maximum temperature experienced in the Transfer Line was 2064°F (1129ºC) when compared to the design of 1557°F (847ºC). The Primary Reformer risers and harps experienced temperatures above 2000°F (1093ºC) when compared to the design of 1675°F (913ºC) for approximately four (4) minutes. The maximum temperature at the MFC outlet was 1200°F (649ºC) when compared the design of 1200°F (649ºC)).

Note that prior to the loud continuous noise heard at approximately 6:06 am, the main Front End process parameters were steady at a rate of 95% with no abnormalities or evidence of pressure, temperature or flow excursions. The estimated quantity of flammable gas released during the reverse flow episode was circa 2240lbs (1016Kg). in 4 minutes, categorizing it as an API Tier 1 Process Safety Incident with no recordable injuries.

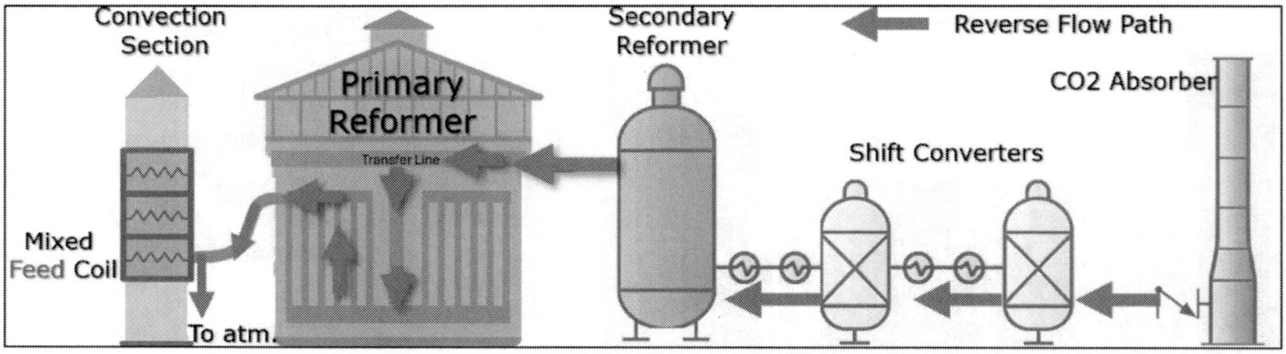

Figure 8 – PFD showing reverse flow path.

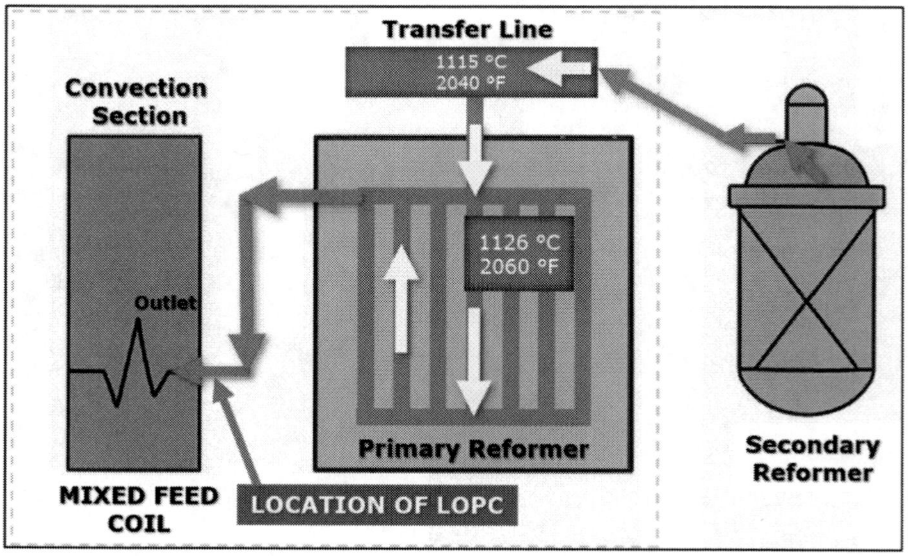

Figure 9 –Detailed PFD showing the affected Circuit

Damage to Transfer Line

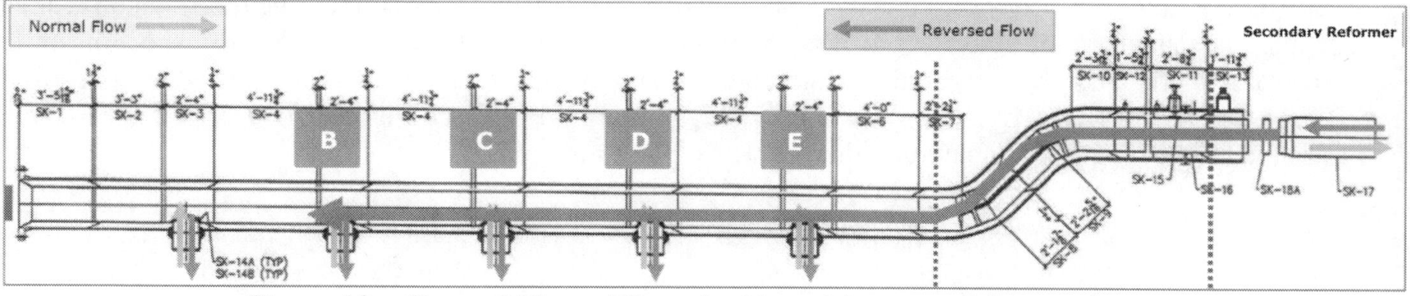

Figure 10 – General View of Transfer Line with reversed flow path

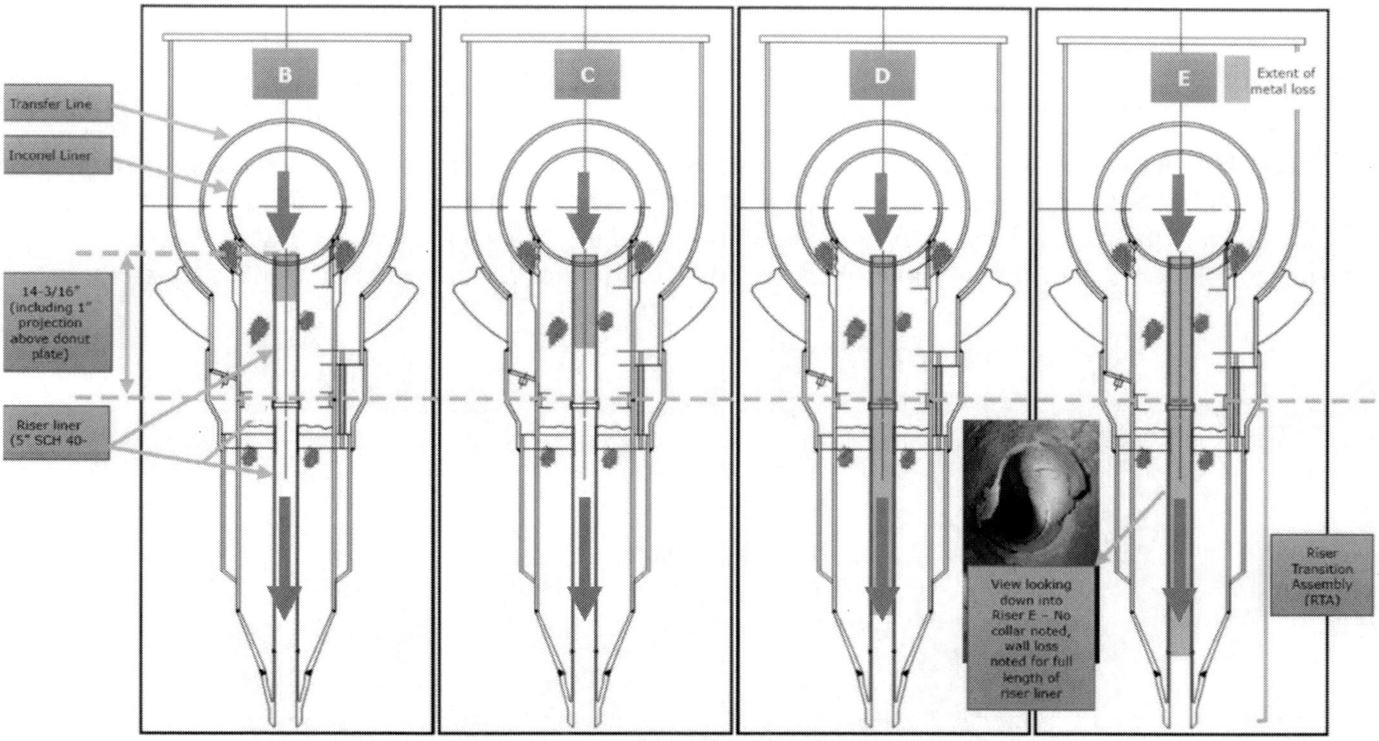

Figure 11 – Detailed cross-sectional view at affected Rows along transfer line

During the reverse flow, temperature indicators recorded a maximum temperature of 2064°F (1129ºC) which was significantly above the design limit of 1557°F (847ºC) within the transfer line. Based on the findings, there were varying degrees of riser liner (Incoloy) loss observed in Risers B to E (see Figure 10 and Figure 11). Figure 12 shows damage sustained by the inconel liner within the transfer line. All damaged sections of the transfer line liner were replaced and/or weld repaired via removal of the inclined section to grade. The riser liner at B and C were repaired in-situ with pup pieces. The riser at D was lowered to the radiant section to facilitate replacement of the thinned liner section. The Riser Transition Assembly (RTA) at E was replaced in kind. Note that the damage observed was a direct consequence of the temperature excursion and reversed flow path of the process gas.

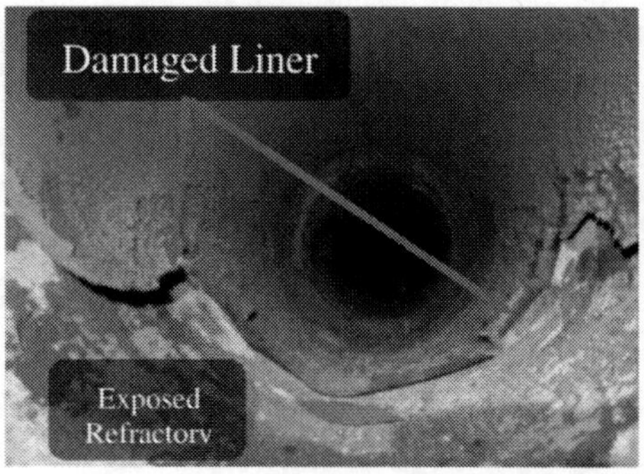

Figure 12 – Damaged Liner on Transfer Line (Secondary Reformer End)

Damage to Primary Reformer

Outlet Header – Row E – Based on normal flow path, the outlet header collects the reformed mixed feed gas and channels it into the riser at the centre and into the Transfer Line. As-found inspection on the internals of the riser tee revealed a significant amount of molten liner material, carried over from the transfer line and Riser Transition Assembly (RTA) internal liner. See Figure 13. Due to this row being affected by the highest temperature, this tee was replaced in kind.

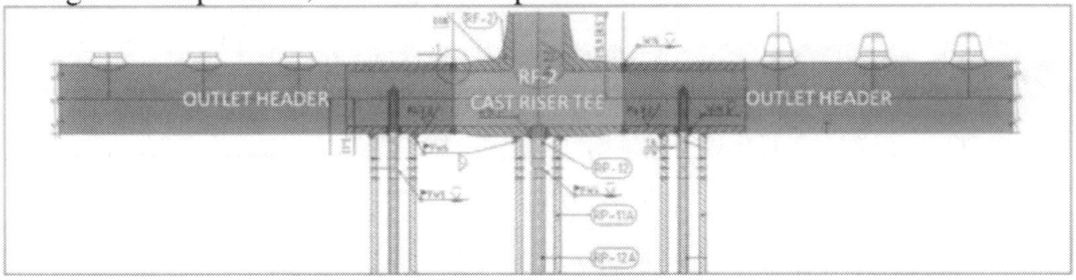

Figure 13 – Sectional View of Row E Outlet header assembly

Outlet Header – Row D – Remote visual inspection (via the transfer line) revealed a significant amount of molten liner material and loose refractory internally. The outlet header to tee weld was cut to facilitate cleaning and the tee was reused. A large piece of molten liner was removed. See Figure 14 for a sectional view and Figure 15 for debris and molten liner material found within the outlet header and riser tee.

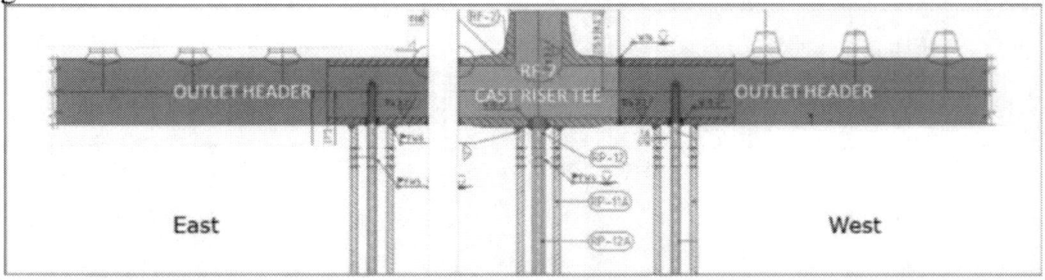

Figure 14 – Sectional View of Row D Outlet header assembly

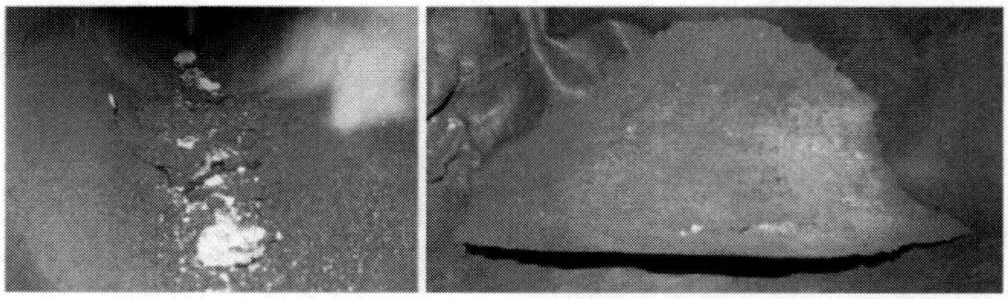

Figure 15 – Typical debris/ molten liner material found within Rows D and E outlet header.

Primary Reformer Catalyst

Based on temperature measurements recorded on the reformer harps, the catalyst sustained temperatures exceeding 2000°F (1093°C), as indicated in Table 1. The catalyst was inspected and verified to have nickel spinel formation. This formation was an indication of catalyst deactivation resulting from exposure to temperatures exceeding its design value for an extended duration. Consequently, the catalyst was replaced.

Table 1 – Temperatures on Harp

Equipment	Temperature Design	Temperature Maximum
101-B Harp Pass 1	1675°F	2062°F
101-B Harp Pass 2	1675°F	2056°F
101-B Harp Pass 3	1675°F	2053°F
101-B Harp Pass 4	1675°F	2063°F
101-B Harp Pass 5	1675°F	2068°F

Detailed Assessment of MFC

Visual Inspection of the failed cross section revealed two distinct regions as seen in Figure 16.

a) A light grey, relatively smooth texture which originated at the toe of the root weld
b) A darker and significantly textured region which showed evidence of tensile fracture. Upon measurement, the smooth region was found to vary between 0.315-0.550inch (8-14mm) in width. Note that the nominal thickness of this cross section is 1.221inch (31.01mm).

Figure 16 – Failed Cross Section of end cap weld

Replica Metallography was performed on various components along the mixed feed coil assembly to assess impact of elevated temperature and reverse flow on serviceability. These included various points along the inner and outer diameter of the outlet header, end cap and accessible coils. Based on the findings, there were no anomalies noted on the outlet header parent material; however, sigma phase was evident within the grains of the end cap weld. Sigma phase is an intermetallic phase which is formed and can serve as nucleation sites for creep damage. Figure 17 shows a magnified replica of the failed end cap weld with sigma regions highlighted in red.

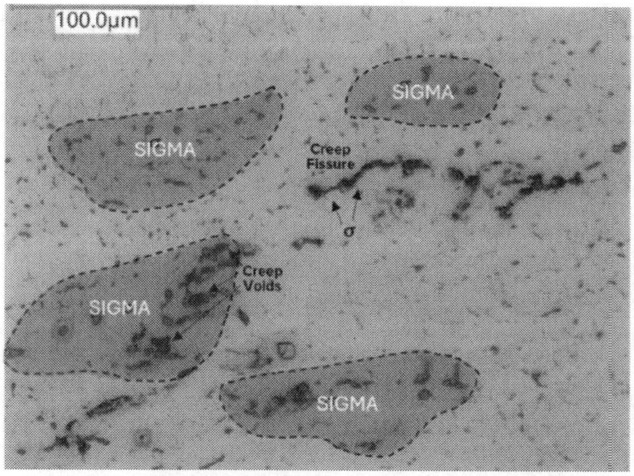

Figure 17 – Identification of Sigma Phase

Dimensional Verification and **Ultrasonic Thickness testing** confirmed the thickness of the outlet header and return bends to be acceptable. A combination of surface techniques such as **Dye Penetrant Testing (PT)** and **Ultrasonic Thickness Testing (UT)** was performed on all accessible welds on the outlet header, weld-olet, branch pipes, inlet headers. See Figure 18 for welds.

Figure 18 – Identification of welds Inspected

A combination of volumetric techniques such as **Phased Array Ultrasonic Testing (PAUT) and Balanced Field Electromagnetic Testing (BFET)** was done on the following welds, revealing acceptable results.

- 100% Outlet Header Welds.
- 100% Outlet Header to Weldolet
- 100% Outlet Header Weldolet to Branch Pipe/ Tubes
- Same for Inlet Header.

Destructive Testing

To understand the damage mechanism, two (2) sample rings shown in Figure 19 were cut from the failed end cap and outlet header which includes the remnants of the failed weld cross section and 0.16inch (4mm) of the parent material measured from the toe and sent for destructive failure analysis.

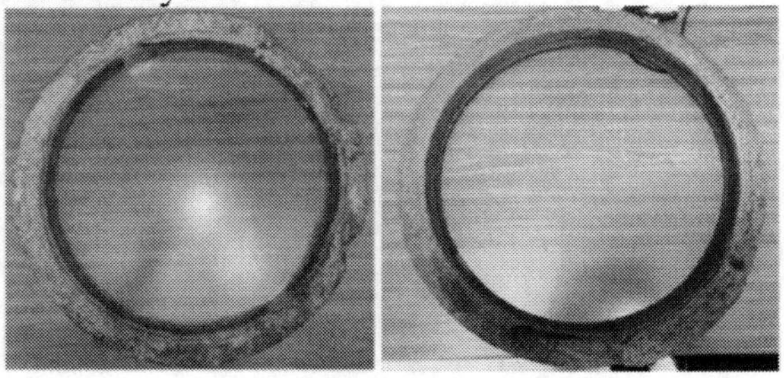

Figure 19 - Sample Ring from End Cap (left), outlet header (right)

Visual Examination – Two (2) pieces were taken from both the cap side and pipe side which included the representative base metal, failed weld metal, and fracture surface. The end cap ring was selected for detailed metallographic examination.

Metallographic Examination - The base metal microstructure was as expected for 304H. There were creep fissures and cracks visible throughout the weld metal, while the base metal showed no cracking or other damage (See Figure 20). A modified Murakami's etchant was used (containing KOH and NaOH) and identified sigma in the end cap base metal and weld metal.

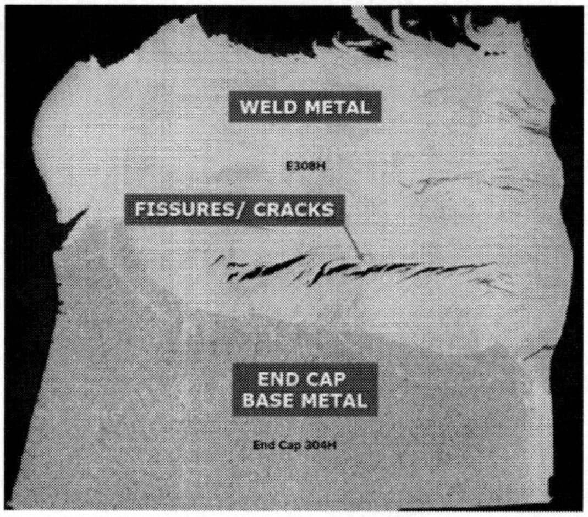

Figure 20 – Metallographic Sectional View

Scanning Electron Microscope (SEM) Analysis - There were three (3) distinct regions present as identified below (See Figure 12).

1. Thermal Fatigue in the base metal and heat affected zone, near the root.
2. Fast Fracture zone where the cracks connected and resulted in ductile overload (fast fracture).
3. Creep damage or creep cracks in the weld metal.

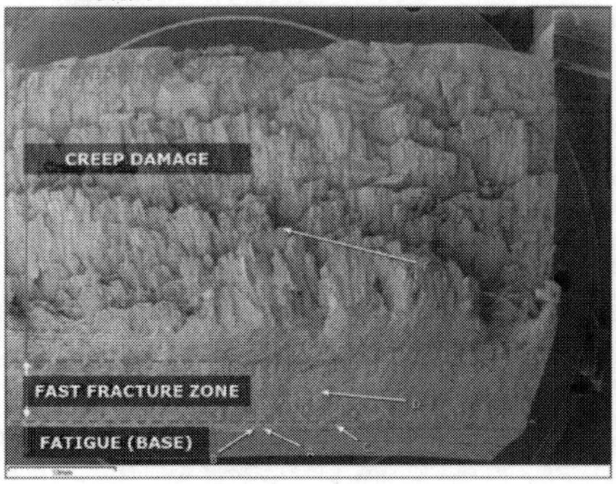

Figure 21 - Zones

Chemical Composition with Optical Emission Spectroscopy (OES) revealed no issues.

Energy-Dispersive X-Ray spectroscopy (EDS) maps (Figure 22) confirmed the presence of sigma in the end cap weld through analysis of the elemental composition. Sigma typically has elevated Chromium and moderate amount of Iron. The EDS map shows the high chromium sigma phase present (in red) in the weld metal, including the creep fissure.

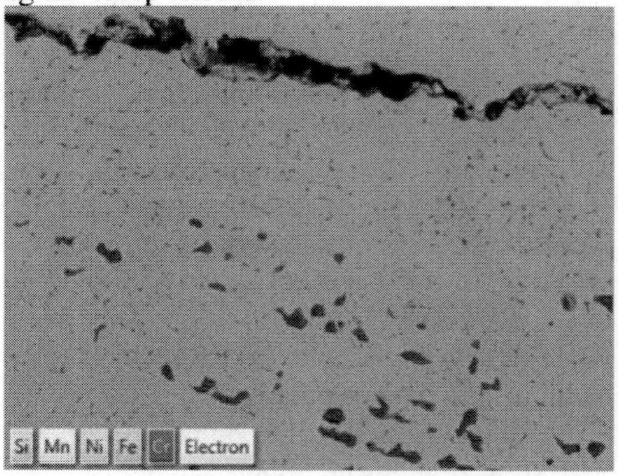

Figure 22 – EDS Map showing sigma in red

Root Cause Analysis

Based on equipment design, inspection history and latest corrosion study, the following possible damage mechanisms could have been experienced on the MFC:

- Dissimilar Metal Weld (DMW) Cracking due to temperature and stress
- Thermal Fatigue due to temperature
- High Temperature Creep due to temperature and stress
- Short-Term Overheating due to temperature

Based on these damage mechanisms, visual and detailed inspection is typically carried out during a turnaround (TAR) opportunity. Visual inspection would identify any sign of distortion, mechanical damage, fouling and assess the overall condition. Detailed inspection comprising of PT and UT is then conducted, with a primary focus on components experiencing the highest levels of stress, such as the tube to weldolet and weldolet to header pressure retaining welds as well as monitor thickness of the return bends.

Based on the failure analysis results, DMW cracking, thermal fatigue and short-term overheating was ruled out and the primary damage mechanism for the weld failure was attributed to creep cracking.

Creep Cracking is defined as material deformation that occurs under prolonged exposure to elevated temperature and stresses. This phenomenon ultimately results in the material becoming weaker as micro cracks form along the grain boundaries. Specifically, with 304H stainless steel, creep damage typically becomes a concern above 950°F (510°C) according to API 579. The mixed feed coil outlet header operates around 1100-1150°F (593-621°C). Creep cracking often occurs in areas of high restraint and stress concentrations, but can also happen independently, such as along solidification grain boundaries. Two (2) common contributors for reduced creep strength are detailed below.

1. Insufficient or excess carbide formation: Carbides (e.g. chromium carbide $Cr_{23}C_6$) form distinct phases within the microstructure of metals and alloys and often emerge during solidification or subsequent heat treatments. These carbides strengthen the material by pinning dislocations and stabilizing grain boundaries, which prevent creep deformation at elevated temperatures. Too low carbon content can lead to insufficient carbide formation. However, too high carbon content can lead to excessive carbide precipitation (leading to the coarsening of the carbides, stress concentrations, brittleness, and chromium depletion), which can reduce toughness and corrosion resistance, and indirectly reduce creep strength.

High temperature creep resistance is enhanced by the carbon content in the metal. Based on chemical analysis, the carbon content in the weld metal was found to be at the minimum specification of 0.04 wt.% out of a range of 0.04 to 0.10 wt.% suggesting that the creep resistance would have been on the lower end of the scale.

2. Sigma phase, which is an intermetallic phase formed in service that can serve as nucleation sites for creep damage. This phase forms due to high-temperature exposure [above 1000°F (538°C) per API 579] and affects the mechanical properties of the material by causing a loss of ductility (increasing hardness and brittleness). Continuous formation of sigma phase forms a network which ultimately weakens the material's ability to withstand creep. Furthermore, sigma formation can deplete the matrix of chromium and molybdenum, elements that are essential for the formation of creep-resistant carbides (like chromium carbides mentioned above).

Sigma phase was confirmed present in the weld as per the replica testing at R1 and R2, i.e. 12 and 6 o'clock positions. Detailed microstructural analysis also confirmed a significant presence of sigma throughout the endcap to header weld as seen in Figure 17 which likely contributed to the creep cracking observed.

Both carbides and sigma are forms of precipitates. Carbides are considered good if properly managed through optimised carbon content and controlled heat treatments and cooling. However, sigma is almost always considered bad due to its negative effect on mechanical properties. Sigma phase in welds can be minimized by controlling the ferrite content, i.e. <5 % for 304 SS (as per API 571). Lean electrodes such as 16-8-2 (vs. 308H) can also be used where creep resistance is required due to its lower alloy content.

Through the study of the failed sample's microstructure (metallography), the presence of intergranular fissures and cracks in the weld metal was observed, thus confirming the mechanism of creep cracking. These fissures in the weld metal would have started as voids, but due to the low creep ductility of the weld metal, the voids grew and easily linked with other voids, leading to the failure.

Note that intergranular refers to defects formed along the grain boundaries of the microstructure. Additionally, there were some trans-granular, dagger-shaped cracks found at the 5 o'clock position, consistent with thermal fatigue, however, it is suspected that creep cracking remains the primary contributor to the failure. It is relevant to note that this damage mechanism was limited to the weld metal and no damage occurred in the base metal of either the end cap or the header except near the weld root. The lower carbon content played a role in the creep cracking, but the damage was more pronounced near the sigma phase, making it the primary contributing factor.

Based on the smooth cross section observed near the ID, detailed analysis of this area via etching was done and confirmed that there was no lack of fusion at the root or groove faces. See Figure 23.

Figure 23 – SEM image

Detecting Sigma Phase

Sigma phase is a metallurgical change which is very difficult to detect without relying on destructive and laboratory testing methods. Damage due to sigma phase embrittlement and creep initially forms as microcracks, particularly at the welds or areas of high stresses which eventually grow to macro cracks, connecting to cause failure. For non-destructive testing to evaluate the condition and in lieu of taking destructive samples, the following techniques suggested below would primarily focus on identifying cracks.

- Phased Array Ultrasonic Testing (PAUT)
- Time of Flight Diffraction (TOFD). PAUT and TOFD are the most reliable high resolution volumetric ultrasonic testing methods for detecting micro cracks.
- Dye Penetrant testing (PT)
- Field Metallurgical Replication (FMR), Hardness testing and PT would evaluate the surface condition.
- Eddy Current Testing (ECT)

Repairs to MFC Outlet Header

The following outlines the steps taken to repair the failed weld.

a) Both cross sectional joints for the replacement end cap and existing outlet header were bevelled to the OEM joint design. PT was performed on both bevels, revealing acceptable results.
b) Following welding the root and hot pass, Radiographic Testing (RT) and PAUT were performed with acceptable results.
c) Following completion of welding the fill and cap, PT and PAUT were completed on full circumferential weld, revealing acceptable results.
d) Pickling of the completed weld was completed to restore corrosion resistance and remove any unnecessary contaminants.

Proactive Management

To proactively manage the consequence of this failure and reverse flow condition, the following recommendations were proposed.

1) Safety Instrumented System (SIS)

Implemented a modification to the Safety Instrumented System (SIS) logic by utilizing existing instrumentation. This modification involved adjusting the SIS logic to trigger the primary reformer fuel valves and the Process Air Compressor. An alarm was also configured to alert the Distributed Control System (DCS) via Mod-Bus in the event of high flows of either steam or gas.

2) Additional Instrumentation

There is currently planning ongoing for the installation of additional instrumentation on the Mixed Feed Crossover piping to detect sudden loss of pressure (indicative of rupture / loss of containment) and initiate the relevant automatic actions to mitigate the severity of the rupture. This will be executed during the next available opportunity.

3) Conducting Inspections and modifying the Inspection and Test Plan (ITP)

Based on the root cause, the inspection techniques used on this occasion were shared with other plants and inspections were performed. This was also institutionalized as part of the ITP, documenting the inspection techniques and locations for inspecting.

Additionally, information was also shared with other business units having similar plants or materials of construction within the susceptible temperature range.

Conclusion

This weld failure resulting in an LOPC serves as a stark reminder of the importance of robust inspection and maintenance practices in industrial facilities. The incident, which was ultimately traced back to a compromised weld on the outlet header of the MFC, caused reverse flow and extensive damage to various components on the primary reformer. The root cause of the weld failure was creep cracking, exacerbated by the presence of sigma phase in the weld metal which negatively affected the mechanical properties.

Extensive inspection was performed during this occasion which confirmed serviceability of the outlet header and coils. Repairs to the MFC involved replacing the end cap in kind, welding to the existing header and inspecting to verify acceptance. Repairs associated with the reverse flow were also performed and serviceability was restored.

To mitigate the risk of such failures moving forward, key recommendations were implemented. A proactive approach was taken with inspecting and testing of similar plant components at susceptible locations to identify potential vulnerabilities.

References

American Petroleum Institute. API RP 571: *Damage Mechanisms Affecting Fixed Equipment in the Refining Industry.* 3rd ed. Washington, DC: American Petroleum Institute, 2020. Section 3.56 Sigma Phase Embrittlement.

Use of Risk Based Inspection (RBI) to demonstrate the industrial robustness of the long neck single nozzle ATR design, providing extended operating cycles

Through experimental work at Johnson Matthey (JM)'s Advanced Reforming Development Unit and shared experience of operators who have accessed lifetimes of over 4 years from their O_2 driven ATR burners, the longevity of this ATR design to provide at least 4 years continuous operation in the first operations cycle is demonstrated. A Risk Based Inspection (RBI) approach can further be used to extend the operating cycles of ATR operations and reinstate the same burner without need for repair or replacement. The paper concludes by detailing the safety and financial benefits of longer operating cycles. and proposes that there is no need to reduce operating cycles current ammonia operators have achieved, where 4-year operating cycles are the norm, and 6 or more years are now anticipated.

Matt J Cousins
Licensing Manager – Johnson Matthey Low Carbon Hydrogen

Abdulaziz N Al-Sakran
Section Head, Process Engineering - SIPCHEM (Sahara International Petrochemical Company)

Introduction

Ammonia and methanol flowsheets utilizing steam methane reformers (SMRs) have realized improvements to operating cycles over past decades. Some reformer designs and catalyst combinations have achieved cycles of 12 or more years operation before tubes and catalyst have required replacement.

This is a step change from syngas flowsheets from 1950s-1960s, where the achieved cycle times of months to a few years were typical [(1)].

This step change has required investment in more robust technology, better materials from which tubes, pigtails and manifolds are manufactured as described in an AIChE paper presented in 2022 [(2)] and better catalyst solutions, such as the KATALCO 57-4XQ a catalyst that has provided cycle-times of over 12 years.

Additionally, monitoring of the process to provide confidence to increase the operating cycle is key. This includes operating conditions to indicate catalyst performance while the plant is operating. However, this does not inform the operator of the health of their process equipment. This is often only measured periodically when the plant is in turnaround and inspections can be made. Sweet et al[(3)] discusses the application of Risk Based Inspections (RBI) to an ammonia plant, considering high temperature hydrogen attack in the reforming and water gas shift (WGS) sections, and amine stress corrosion in the CO_2 removal section [(3)].

Risk Based Inspections

RBI considers (i) the likelihood of failure and (ii) the consequences of a failure.

To draw conclusions, it is required to consider the:

- Process conditions
- Process fluids
- Materials of construction and
- Operating history

The aim is to estimate the present risk and forecast this into the future, providing justification when setting (i) inspection frequencies and (ii) making modifications to equipment to extend life.

The formalization of this approach was led by the American Petroleum Institute (API), from the inception of RBI in the 1990s their use has spread across the chemicals sector. The methodology most referred to is API 580 & 581[(4)].

API methodology uses qualitative and quantitative approaches. Qualitative methods are generally used to prioritize areas where more detailed quantitative assessment might be needed.

An overview of the elements that are considered is shown below (Figure 1).

Once the consequence and probability are understood they can be applied to a risk-grid. JM uses a risk grid, such as the one detailed in Figure 2.

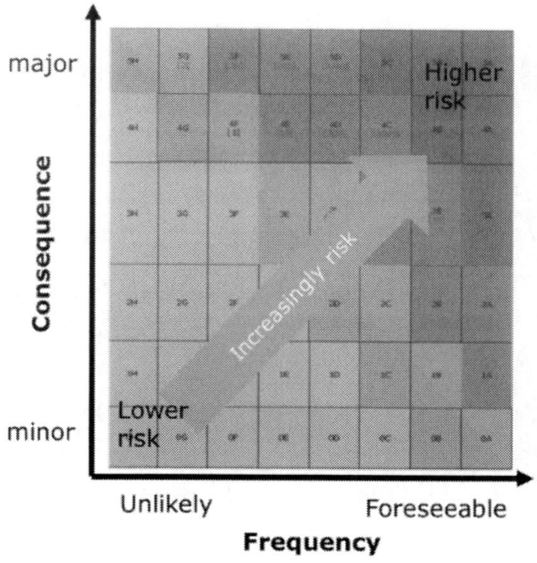

Figure 2. Simplified risk grid, example of full risk grid shown in appendix A.

This can be used to assess the risk today, and crucially to estimate the risk in [X] years' time, where [X] is the period of the planned operating cycle.

This is where RBI has been used to justify why it is possible to extend operating cycles, coupling knowledge of the materials and process alongside real operating experiences to manage risk.

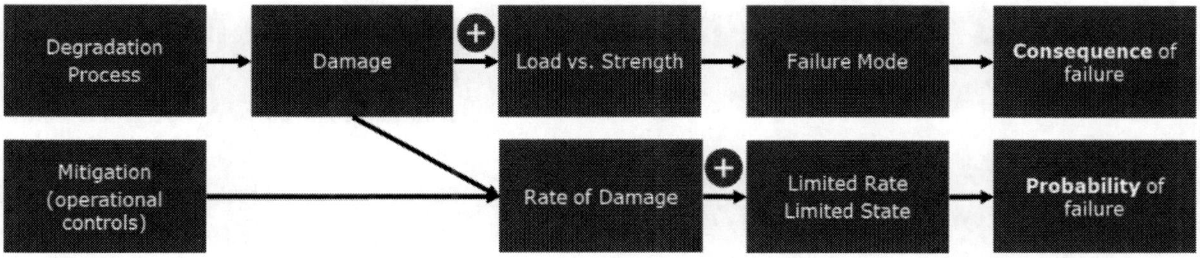

Figure 1. Details the elements considered when applying API based RBI.

The advantages of longer cycles are:

- Plants spend more time producing products, and less time in turn around.
- In turn this increases the profitability from the plant, by minimizing operating costs, and avoiding the capital cost of asset repair and replacement,
- While additionally lowering the risks for the operations team. As they can spend more time in the safety of the control room running the plant, and less time is spent repairing or replacing items in the field.

The value of this is shown by the graphic below (Figure 3), which measures decreasing impact to gross margin due to extending operating cycles from 2 to 12 years, by calculating the cost per year attributable to the outage period from the different operating cycles. The cash cost of ammonia is calculated using the online tool from Yara [(5)], for a 3500MTPD plant, with gas price of \$4/MMbtu. The cash price for ammonia is from a S&P Global report [(6)], and outages were set at 30 days duration from stop to restart. The gross margin was determined by subtracting direct costs of production from the product price.

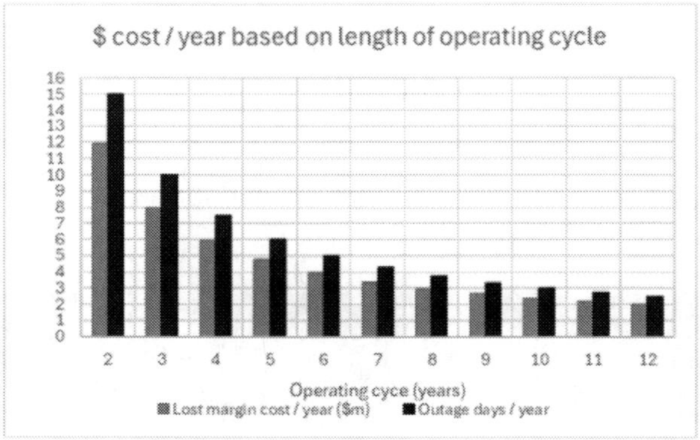

Figure 3. Cost avoidance by achieving longer operating cycles for a 3500MTPD ammonia plant.

This explains why SMR based flowsheets, producing grey ammonia, now target 4-6 years cycles, and some seek further extension to this.

Future ATR flowsheets.

As we look at future flowsheets, that will allow our industry to increase ammonia production while simultaneously reducing CO_2 emissions key parts of the flowsheet will change. The SMR will be replaced by an oxygen blown ATR (Auto Thermal Reformer). This offers a pathway to the lowest carbon intensity products when utilising a natural gas or fuel gas feedstock.

The same principles as described above on longer operating cycles apply to ATR flowsheets, longer cycles have value for the same reasons. This value was shown to contribute a 30% OPEX improvement in the value obtainable from an ATR flowsheet when an optimized S:C (1.3 vs 0.6) and longer cycle time, (6 vs. 2 years), were considered for a JM ATR design, operating conditions and catalyst [(7)].

This paper will review the mechanical robustness of the ATR burner, a key component of the ATR. This is a critical piece of propriety equipment. The burner, and more specifically the nozzle, is at the interface of the pure oxidant jet (leaving the nozzle) and the oxidant mixing with process gas to produce a gas mixture that ignites. At this point heat is released that:

A. Drives the reforming reaction
B. Thermally stresses equipment

The ATR reforming process produces hydrogen and CO_X at the process conditions. In combination with Water Gas Shift (WGS) this enables low carbon hydrogen and ammonia production through capture of the CO_2 for storage. JM's LCH™ process utilising a standalone oxygen-blown ATR, when fully integrated with an ammonia loop, allows for tailoring of the carbon capture rate between 90% and >99%. The carbon intensity of the product will vary based on that of feeds to the process. Typically, it is possible to provide Scope 1 values of 0.1 – 0.7 kg CO_2 / kg H_2, where Scope 1 are the direct CO_2 emissions from the flowsheet.

The temperatures at this position in the flowsheet are the hottest we see, with combustion temperatures >2500°C (>4532°F), noting the adiabatic flame temperature of O_2 in H_2 is 3200°C (5792°F) [(8)]. The hottest area is associated with the stoichiometric combustion with the flame front, shown red within Figure 4 below. This starts at the interface of the pure O_2 mixing with the process gas close to the nozzle tip, towards the left of Figure 4, and extends with the flame-front away from the burner, carried by the flow of gases moving forwards (to the right of the image).

The temperature profile for a standalone ATR, i.e. an ATR processing natural gas, not downstream of an SMR, are shown below (Figure 4). This is an important differentiation, as it is standalone ATRs that do all the reforming, so for a given H_2 output these must introduce oxidant to make all the production capacity.

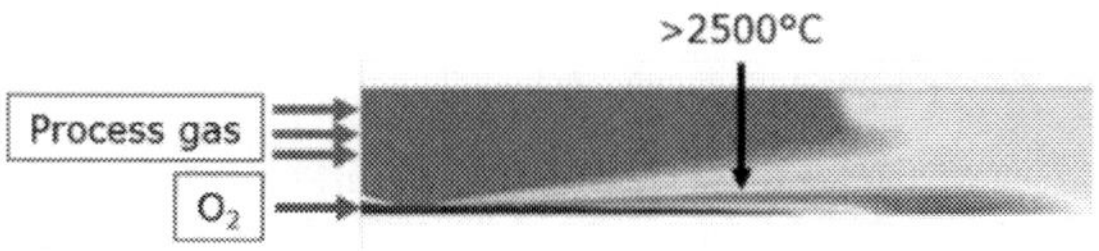

Figure 4. Temperatures in and around the flame in a JM Oxygen blown ATR.

The description of the process thus far is common to all ATR designs, as all ATRs must introduce an oxidant and support a combustion process to provide heat.

Not all ATR burners are the same.

The design of burners is based on proprietary information, and so the details are not publicly known. At a high level we can distinguish some differences in the approach different licensors take in providing a solution.

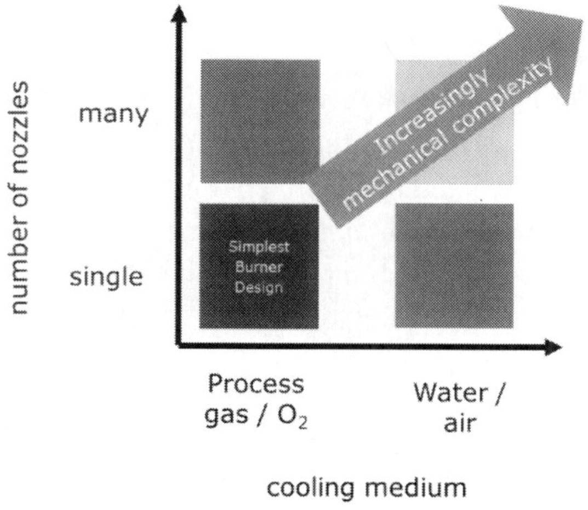

Figure 5. Categorisation of burners deployed in ATR license technology

There is different operating experience of the ATR burners, across the range of alternatives described in figure 5. The difference between the burner designs when used in an ATR is immediately noticeable shown by Figure 6.

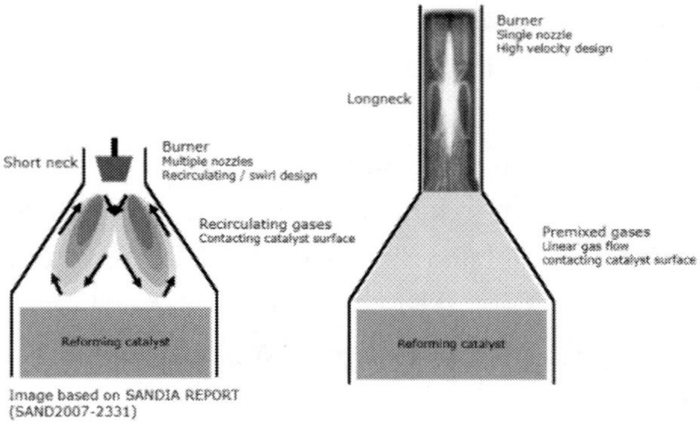

Figure 6. Characteristics of different ATR burner designs. (on the left: ATR burner with multiple nozzles and on the right: JM's ATR burner with single nozzle)

Burners with multiple nozzles closer to the catalyst surface (lefthand image) employ technology to intensely mix the gas as it leaves the burner tips [(9)]. This is something ICI/JM experimented with and chose to not progress prior to developing the single nozzle design in

1969. The multiple nozzle burners seek to moderate the gas velocity, so the hottest gases do not hit the catalyst bed and rely on mixing the gas directly above the catalyst volume to prevent hot jets hitting the catalyst surface.

Comparatively the JM single nozzle design (righthand image) provides a high velocity O_2 jet, that effectively mixes with the process gas and re-entrains some of the combustion products into a recirculating gas-flow, part way down the long-neck. The burner and neck are both designed so the recirculation, and therefore the mixing, are complete before the gas exits the neck. This means the gas exiting the neck is (i) fully mixed, (ii) ensuring a uniform temperature and composition, so there are no hotter gas jets that contact the catalyst surface and (iii) provides time for gas-phase reforming to occur in the neck and headspace above the catalyst bed, lowering the gas temperature contacting the catalyst.

The high velocity of the O_2 pulls the flame away from the burner nozzle, so the flame is detached from the nozzle. Because the mixing regime does not cycle the gases all the way back to the tip, this allows for the JM single nozzle design to be cooled by the gases passing through and around it, counteracting the radiated heat from the flame and neck wall. This design is well proven with over 25 references deployed globally.

There are also hybrids of these two designs. These elect to use a single nozzle in a short neck design. Because of this difference the gas velocities likely need to be adjusted to limit the impact of the gas flow and temperature at the top of the catalyst bed, and this anecdotally appears to be the case as they need to employ water cooling so the nozzle temperatures can be managed.

Application of RBI methodology to the JM ATR burner.

Following Figure 1 to think about the Consequence and Probability of a failure.

Consequence:

The **degradation process** we need to consider is high temperature phenomena; (oxidation and thermal fatigue)

Damage, that operators seek to mitigate is burn-back of the nozzle tip, this alters the nozzle diameter, lowering the gas velocity and changes the intended design.

Load is defined by the temperature of the gas at the nozzle tip. Key to the JM design is (i) the flame is detached from the nozzle tip, lowering the temperature the tip is exposed to, (ii) the high velocity O_2 and process gas flows provide cooling, pulling heat away from the nozzle.

Strength is the proprietary materials of construction and design rules used to fabricate the JM burner for each ATR design.

Failure mode, of the ATR burner, JM's design avoids is oxidation of the burner, causing burn-back of the nozzle tip. In turn if the mixing of the O_2 and process gas is poor, the combustion process will happen in an uncontrolled manner. This will induce a horizontal temperature profile across the catalyst bed, reducing the performance of the ATR / ATR catalyst. Ultimately the worst-case scenario is hot gas jets impact the refractory of the ATR and damage this, leading to increased stress on the ATR vessel.

Consequences of the failure mode, if captured by the plant operations team early would be to shorten the operating cycle, causing a loss in production, and the cost to repair and replace the ATR burner, and/or catalysts, and/or refractory. Alternatively, if not spotted, the problem may escalate potentially effecting people, equipment and the environment.

Mitigation through operational controls. JM have operated long-neck, single burner designs since 1969, so have amassed a wealth of operating experience. The operating conditions are developed for each flowsheet type. JM's ATR offer is a product of rigorous testing and development. JM have invested in an Advanced Reforming Development Unit (ARDU), located at our UK research facility in Billingham and a Materials Demonstrations Unit (MDU).

ARDU replicates industrial conditions at a semi-tech scale and offers a way for JM to stretch operating conditions. This provides a way to test boundary conditions, as well as verification of the performance our technology will provide when operated at these conditions.

Figure 7. The ARDU located at the JM research facility Billingham, UK.

The MDU has allowed JM to test different materials at industrial conditions for longer periods. This facility has been operated at customer sites globally, exposing materials of interest to realistic process conditions, through startup, shutdown and normal operation.

Rate of damage, this can be estimated through JM's work, using facilities like ARDU.

Nevertheless, the proof statement, that really matters is what happens in the field. Indeed, RBI would look for experience at the specific plant conditions.

SIPCHEM (Sahara International Petrochemical Company) has operated JM's standalone ATR since 2018 at their site in Al Jubail, in the Kingdom of Saudi Aribia.

Here they have operated a JM methanol process since 2004 and driven by a desire to expand capacity they added JM's standalone ATR to process additional natural gas in parallel to the SMR.

Once the plant had finished commissioning a few months later, in 2019, ATR operations have been stable and there was no need to inspect the burner until a planned outage in 2023. At this point the JM ATR burner was removed for inspection, supporting SIPCHEMs own RBI process.

After 5 years of operation the burner was in good condition, the photograph below (Figure 8) shows the burner tip, where O_2 is fed into the process, and process gas is fed around this into the mixing zone below it. The burner still presents a clean edge, with no change to the diameter, meaning there is no change to the O_2 velocity it provides to the process, at the selected operating conditions JM have designed and agreed with SIPCHEM for this ATR.

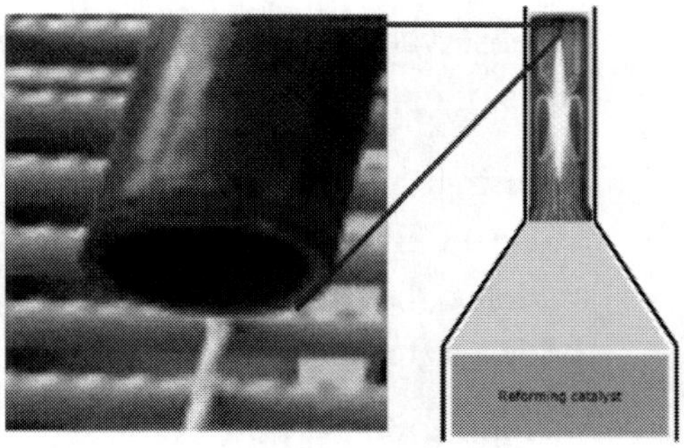

Figure 8. JM ATR burner nozzle tip after operation by SIPCHEM, Al Jubail.

The only difference to the burner from when it was installed 5 years earlier was a thin layer of black deposit on the outer surface, this colouration can be seen in Figure 8.

The inner surface, all the way to the connecting flange was in good condition.

Minor "coating damage" was noted on the external face of the flange that connects the burner to the ATR vessel, shown in Figure 9.

Figure 9. JM ATR burner, external flange face, after operation by SIPCHEM, Al Jubail.

The inspection report SIPCHEM commissioned concluded the burner remained in good condition. This did not surprise JM. Like other JM ATR reference plants SIPCHEM were able to reinstate the same burner back into service without need for any repair.

To date the maximum demonstrated operating life of JM's single nozzle burners are

- >25 year in air fired service
- >12 years in O_2 fired service

Limited Rate/ State, the SIPCHEM inspection report we have referred to in this paper shows no observable difference to the burner after 5 years of operation.

By using the operating conditions JM advises, and SIPCHEM has proven, we can reference evidence that the burner is expected to provide at least a further 5 years' service.

Probability of failure

JM would expect a burner to last multiple 5-year cycles, where operation is in accordance with our recommendations. With a low risk that the burner would be a cause of failure.

With routine inspections every 5 – 10 years achieving over 20 years life before any repair is required appears possible based on the data we are aware of in the field.

The value of extended operating cycles in O_2 driven ATRs

An ATR can be integrated to provide a highly efficient and reliable ammonia flow sheet.

The drivers are to convert the gas feed stock to H_2 (to provide makeup gas to the loop) and CO_2 (for storage).

Because the product molecules, H_2 and NH_3, contain no carbon there is a driver to fully convert the methane feedstock by the reforming and WGS reactions.

The reforming reaction:

$CH_4 + H_2O \Leftrightarrow CO + 3H_2$ (reaction 1)

The Water Gas Shift (WGS) reaction:

$CO + H_2O \Leftrightarrow CO_2 + H_2$ (reaction 2)

The reforming reaction including WGS:

$CH_4 + 2H_2O \Leftrightarrow CO_2 + 4H_2$ (reaction 3)

Conditions should be selected to maximise both reaction 1 and 2. This is not different to the optimisation Max Appl sets out for the current ammonia process, noting the theoretical optimal stoichiometry of our process [(10)].

Additional steam can be added downstream of the ATR, this limits the mass flow through the ATR to fit the reaction-pathway, reducing the size of the ATR and the pre-heat of the ATR feed gases. JM's ATR process, referenced to capacities equivalent to >5000 MTPD ammonia, has been fully integrated with Uhde's proven 3500 MTPD ammonia synthesis loop.

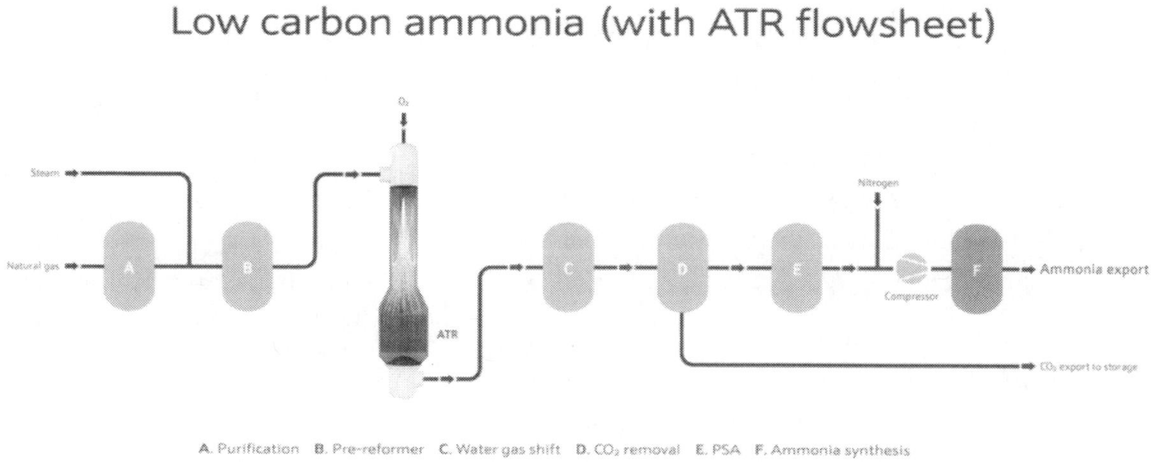

Figure 10. JM-uhde® decarbonised ammonia process

JM-Uhde are the only provider offering technology and catalyst solutions proven at >3500MTPD. We provide references of the 5 largest ammonia plants, and large standalone ATR deployment. The JM and uhde® process (Figure 10) offers leading efficiency in low carbon ammonia production, by combining JM and Uhde referenced technology.

This paper next assesses the value of extending the operating cycles for the decarbonised flow sheet and compares this to those described in Figure 3, determining if the same value remains to seek longer operating cycles.

The same basis to calculate the cash cost was used, while recognizing that the energy import is now a mix of gas and imported electric power. This is because it is no longer attractive to burn marginal gas / waste streams to produce power, as this would impact the carbon intensity of the product. The electrical price was set at \$30/MWh. The cash price for ammonia was taken from the same S&P Global report[6], this noted a premium for blue ammonia, 470 \$/T compared to a 400 \$/T value for grey ammonia. Outages were set for the same duration, 30 days from stop-restart. The result from conducting the same assessment detailed in Figure 3 is shown in Figure 11.

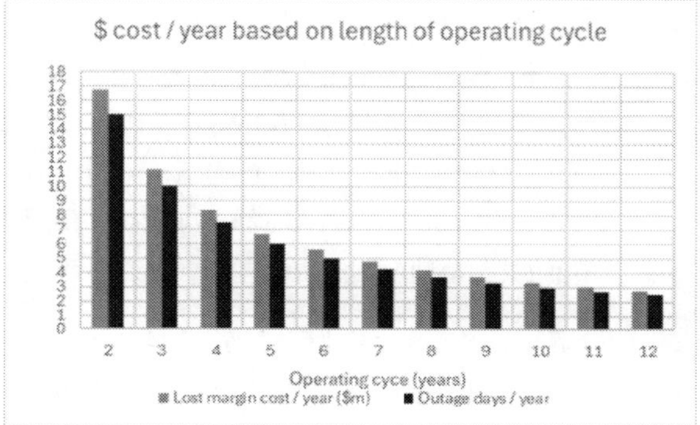

Figure 11. Cost avoidance by achieving longer operating cycles for a 3500 MTPD grey and blue ammonia plant.

The results show the same trend, but with a greater impact calculated for shorter plant cycles in blue ammonia plants. Comparing the costs for a 2-year cycle shows an impact of \$11.9m vs \$16.7m for the grey vs. the blue ammonia plant.

Calculating the cost differential over a 25-year basis. 2-year operating cycles requires 15 days of outage per year. Figure 11 shows the cost = \$16.7m/year, so over 25 years = \$418.2m. 4-year operating cycles requires 7.5 days of outage per year. Figure 11 shows the cost = \$8.3m/year, so over 25 years = \$209.1m.

Making the differential worth a cost saving >\$200m, showing why operators seek longer operating cycles, noting further savings can be gained by pushing from 4 to 6 years cycles.

Conclusions

The driver for longer operating cycles, is based on the same principles operators have already validated, and the value that is there to be gained is at least the same, if not greater, as our industry moves from grey to blue ammonia production.

Through the operational experience of SIPCHEM, this paper has shown the JM ATR burner can be operated for periods of over 4 years without any damage to the nozzle tip.

The same ATR burner can be re-instated for the next operating cycle, without any repair. After the inspection and reinstallation, the same ATR burner is still installed and has additionally achieved more than 600 days of uninterrupted operation, which is a new record for SIPCHEM methanol plant.

This, alongside JM's other operational experience with our proven ATR design shows us that operating cycles of 4 years can be achieved from the designs we provide in their first operating cycle.

Using RBI operators can push for even longer cycles, by establishing their own experience to inform their own RBI process.

This paper shows operators that the drive to extend operating cycles in past decades need not be stopped. The JM ATR process provides a technology solution to continue this journey.

References

1) M. J. Cousins and P. J. Carnell, Johnson Matthey. Cleaner for Longer, meeting the reliability and purification demands of the Ammonia Process. AIChE proceedings 2015.
2) M. P. Brongers, J. K. Hong, Self Employed, C E. Jaske, HIS Group Inc., J. B. Nelson Koch Ag & Energy Solutions LLC, and J. E. Ramirez, Air Products. Fitness for service assessment and materials properties for Alloy 20Cr32Ni1Nb in outlet manifolds, AIChE proceedings 2022
3) Risk Based Inspection Prioritization applied to an ammonia plant. Lily Sweet, Lynne C. Kaley, Riccardo R. Valbuena and Alan Warnock. Det Norkse Veritas. AIChE. Chemical Engineering Progress 1957. Vol 55.
4) American Petroleum Institute (API) 580 &581.
5) Yara. Ammonia and urea cash cost. https://www.yara.com/investor-relations/analyst-information/calculators/ammonia-and-urea-cash-cost/, accessed February 2025
6) James Burgess and Mollie Gorman, S&P Global. US Gulf becomes the lowest-price blue ammonia delivery region. June 2024. https://www.spglobal.com/commodity-insights/en/news-research/latest-news/energy-transition/062724-us-gulf-becomes-lowest-price-blue-ammonia-delivery-region#:~:text=Platts%2C%20part%20of%20Commodity%20Insights%2C%20assessed%20blue%20ammonia,Europe%20remained%20the%20most%20expensive%20region%2C%20averaging%20%24508%2Fmt.
7) M. J. Cousins Johnson Matthey. Reliability and safety of oxygen blown ATRs to enable Low Carbon Ammonia. AIChE proceeding 2024.
8) https://www.engineeringtoolbox.com/adiabatic-flame-temperature-d_996.html
9) S. F. Rice ad D. P . Mann. Autothermal Reforming of Natural Gas to Synthesis Gas. SANDIA REPORT SAND 2007-2331. Published 2007, Sandia National Laboratories
10) Max Appl. Ammonia: Principles and Industrial Practice, Wiley VCH, 1999.

Appendix A, example Risk Grid

CONSEQUENCE CATEGORY

CATEGORY A	(5) CATASTROPHIC	Environment	Major / readily identifiable environmental damage many kilometres from the site boundary.
		Health	Exposure to workplace health hazards rated as high and there is a high incidence of occupational illness cases with one or more fatality.
		Safety	Multiple onsite fatalities (greater than 1). Any number of offsite fatalities or major injuries.
		Authorities	Fines affecting profitability or significant custodial sentences.
		Fires	Whole plant or office significantly affected. Over one month of business interruption. Significant damage to multiple plant items. Significant loss of business information.
		Reaction	International attention / outcry threatens business.
		Financial	Replacement cost / loss > £10M.
	(4) MAJOR	Environment	Major / readily identifiable environmental damage within a few kilometres of the site boundary.
		Health	Exposure to workplace health hazards rated as high and there is a high incidence of occupational illness cases with some resulting in permanent disability.
		Safety	Fatality, life threatening or life changing injury, e.g. amputation of limb or injury resulting in permanent loss of sight or hearing. Any number of offsite severe injuries.
		Authorities	Severe fines (>£20k) or custodial sentences.
		Fires	Whole plant or office significantly affected. Up to 4 weeks of business interruption. Significant damage to multiple plant items. Significant loss of business information.
		Reaction	National headlines and continuing local attention.
		Financial	Replacement cost / loss of £1M to £10M.
CATEGORY B	(3) SEVERE	Environment	Observable many kilometres from the site and measurable off-site environmental impact within a kilometre of the site boundary.
		Health	Exposure to workplace health hazards rated as high and/or there is a high incidence of occupational illness cases.
		Safety	Fractures, other than fingers or toes. Amputation of fingers or toes. Significant injury to eyes requiring corrective surgery. Crush injury to head or torso. Burns that cover more than 10% of the body or cause significant injury to eyes or respiratory tract. Loss of consciousness due to head injury or asphyxiation. Any accident involving injury in a confined space. Any injury or illness involving resuscitation or hospitalisation for more than 24 hours. Any number of offsite serious injuries.
		Authorities	Prosecution with potential for fines up to £20k.
		Fires	Multiple areas affected. Up to one week of business interruption. Significant heat damage to equipment. Minor loss of business information.
		Reaction	Public concern and some national press attention.
		Financial	Replacement cost / loss of £100k to £1M.
CATEGORY C	(2) SERIOUS	Environment	Measurable on-site environmental impact. Major breach of regulatory consent and/or observable off site.
		Health	Exposure to workplace health hazards rated as significant and/or there is a high incidence of occupational illness cases.
		Safety	An incident involving medical treatment beyond defined first aid. A lost time accident without severe consequences. Any number of offsite first aid injuries.
		Authorities	Notifiable to regulator with possibility of minor notice of violation.
		Fires	One plant area or office significantly affected. Up to two days of business interruption. Some heat damage to equipment. Minor loss of business information.
		Reaction	Some local attention, for example local media coverage.
		Financial	Replacement cost / loss of £10k to £100k.
	(1) SIGNIFICANT	Environment	Minor breach of regulatory consent with no measurable environmental impact.
		Health	Exposure to workplace health hazards rated as significant and/or occupational illness cases may have occurred.
		Safety	A first aid incident.
		Authorities	No expected enforcement action by regulator.
		Fires	Only one plant area or office affected. Only one day of business interruption. Minor heat damage to equipment. No loss of business information.
		Reaction	Complaint from local resident.
		Financial	Replacement cost / loss of £1k to £10k.
	(0) SITE ISSUE	Environment	No environmental damage or pollution, readily controlled and contained on company property.
		Health	Exposure to workplace health hazards rated as low and no cases of occupational illness have occurred.
		Safety	EHS Learning Event.
		Authorities	Not applicable.
		Fires	Not applicable.
		Reaction	Site issue only.
		Financial	Replacement cost / loss up to £1k.

[Numbers in brackets for prioritisation of Cat A/B high risk events]

Consequence								
(5) CATASTROPHIC	5H	5G [2]	5F [20]	5E [200]	5D [2000]	5C	5B	5A
(4) MAJOR	4H	4G	4F [1]	4E [10]	4D [100]	4C [1000]	4B	4A
(3) SEVERE	3H	3G	3F	3E	3D [0.5]	3C [3]	3B [30]	3A
(2) SERIOUS	2H	2G	2F	2E	2D	2C	2B	2A
(1) SIGNIFICANT	1H	1G	1F	1E	1D	1C	1B	1A
(0) SITE ISSUE	0H	0G	0F	0E	0D	0C	0B	0A
EVENT FREQUENCY (YR-1)	10^{-7} (H)	10^{-6} (G)	10^{-5} (F)	10^{-4} (E)	10^{-3} (D)	10^{-2} (C)	10^{-1} (B)	1 (A)
	EXTREMELY UNLIKELY	VERY UNLIKELY		UNLIKELY		POSSIBLE	PROBABLE	REGULAR
	Theoretically possible but extremely remote chance of occurrence	Foreseeable event but very remote chance of occurrence during the lifetime of the plant / facility / process. Requires multiple independent protective layers to fail on demand.		Incidents known in industry. Unlikely events not expected during the lifetime of the plant / facility / process. Requires 1 or 2 independent protective layers to fail on demand.		Possible during the lifetime of the plant / facility / process. Root cause likely to have occurred.	Event or near miss that has occurred during the plant / facility / process life time.	Event occurs regularly on the plant / facility / process.

Extreme Risk: Terminate activity/process immediately.

High Risk: Operational or Maintenance tasks, do not begin task until additional controls have been implemented.
New or modified processes, add safeguards to reduce risk before starting process.
Existing processes, a Business/Group EHS approved plan to reduce risk is required with approval for continued operation from Business Operations.

Medium Risk: Manage existing safeguards/barriers to ensure continuous risk reduction and demonstrate that relevant good practice has been applied and risk is 'as low as reasonably practicable'.

Low Risk: Manage existing safeguards/barriers and take opportunities for further risk reduction when appropriate.

Catastrophic Tube Failures While Starting up a Reforming Furnace

This paper describes multiple tube failures and furnace damage which occurred during start-up of a large hydrogen plant. In it we will share the sequence of events leading up to the incident and the extent of damage caused by the failures. But most importantly, we will review the causes identified during the incident investigation, and the key lessons learned from this incident to prevent recurrence.

Ken Wohlgeschaffen
Chevron Products Company (a division of Chevron U.S.A., Inc.)

Introduction

This incident occurred in a 2004 vintage top-fired reforming furnace in a hydrogen plant designed to produce 85 MMSCFD (94,000 nm^3/hr) of 99.9% hydrogen. The furnace has six rows of tubes. Each row contains fifty tubes for a total of three hundred tubes. There are seven rows of burners, fifteen burners per row, for a total of one hundred and five burners.

Incident Summary

As with most reforming furnace tube failures, this incident occurred during transient (non-steady state conditions). In this case, during a plant start-up. The feed gas compressor was out of service to clean the steam turbine driver. The plant was on "hot standby", feeding only steam and using imported refinery fuel gas to fire the furnace.

Once the maintenance work was complete, the feed gas compressor was handed over to Operations, and they initiated normal start-up of the plant.

The Outside Operator began lighting the furnace's pilot burners. After heating up on steam feed, natural gas feed was re-introduced into the plant. Pressure Swing Adsorption (PSA) tail gas was then introduced as fuel to those burners which were in service. The Board Operator then increased plant feed rate.

After three hours, the Outside Operator left the furnace to take a break and was relieved by another Operator from the control room. The Relief Outside Operator left the control room and approached the furnace. Standing at ground level, he noticed large areas of unusual discoloration of the paint in three locations on an external wall of the furnace (Figure 1).

Figure 1. Paint discoloration and Damage to the Furnace Wall (circled in red)

The Relief Outside Operator immediately notified the Board Operator and Shift Supervisor via radio. He then proceeded to check the condition inside the furnace box. Looking through the peep doors, the Relief Outside Operator noticed several broken tubes and tubes glowing very brightly (Figure 2).

The Relief Outside Operator immediately blocked in the PSA tail gas which was the main source of fuel to the burners. Six minutes later the furnace tripped when the flue gas leaving the radiant section of the furnace reached the high-high crossover temperature trip point of 2000°F (1093°C).

Figure 2. Photo Taken at the Time of the Incident Through a Peep Door Showing Ruptured Tubes in Two Tube Rows (on the left and right)

Extent of the Damage

After the furnace had cooled down, a confined space entry was made to conduct an internal inspection. Forty-nine of the tubes (16% of the total) in the radiant section had failed. Some had split open longitudinally in the classic "fish mouth", creep-induced failure style (Figure 3).

Figure 3. Longitudinal Crack in a Furnace Tube

Other tubes had cracked radially and the sections which had not fallen down to the floor were hanging from the roof (Figures 4 and 5). Catalyst from inside the radially cracked tubes had fallen onto the furnace floor beneath the damaged tubes.

Figures 4. Radially Failed Tubes Hanging from the Roof

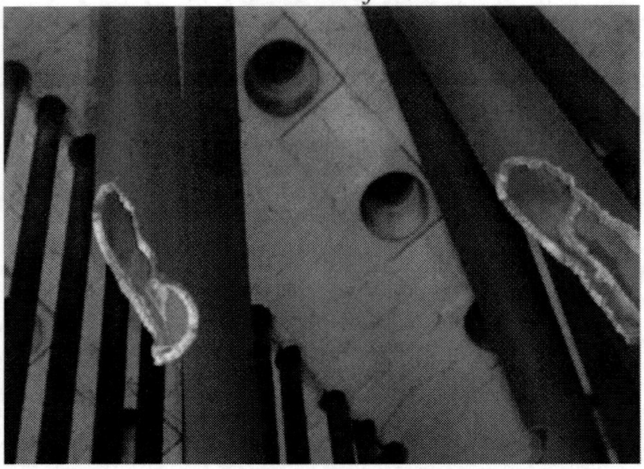

Figure 5. Radially Failed Tubes Hanging from the Roof

Some of the tubes which split open were also bowed. This was most likely due to excessive overheating on the side exposed to lit burners than the side exposed to unlit burners, due to the very high differential temperature on either side of the tubes (Figure 6).

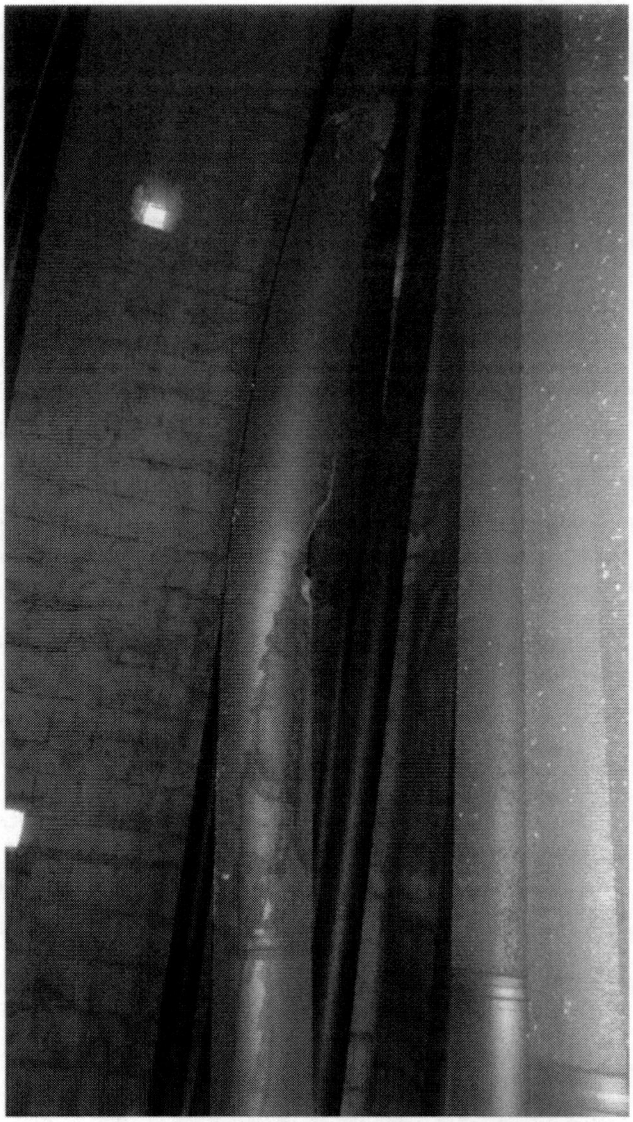

Figure 6. A Bowed and Cracked Tube

An external tube crawler eddy current inspection was conducted. A remaining life assessment utilizing this data revealed that an additional thirty-three tubes were no longer fit for service due to exposure to high temperature. These tubes had to be removed.

In total, eighty-two of the tubes (27% of the total) had to be replaced. Of the remaining tubes, it was determined that 11% had less than 3 years of life left. The rest varied in life expectancy from 3-6 years to 14-16 years.

There was significant damage to the furnace wall in the locations where the paint discoloration or "hot spots" on the walls had been observed. Refractory and insulation had been stripped away from the furnace wall, likely due to direct flame impingement from adjacent leaking tubes. One of these hot spots is shown in Figures 7 and 8.

Figure 7. Damaged Wall Refractory Looking Directly at the Wall

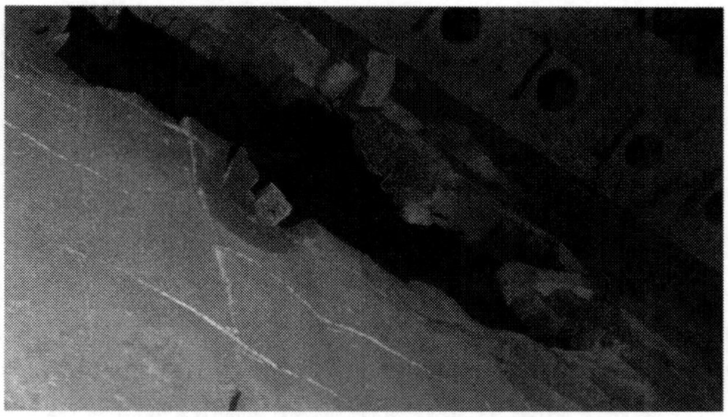

Figure 8. Damaged Wall Refractory Looking Up into the Furnace

Investigation Findings

It was clear from the location of the damaged tubes (Figure 9) that the heat distribution in the furnace during start-up was extremely uneven. The three tube rows which suffered immediate damage were all on one side of the furnace. This side had been severely overheated.

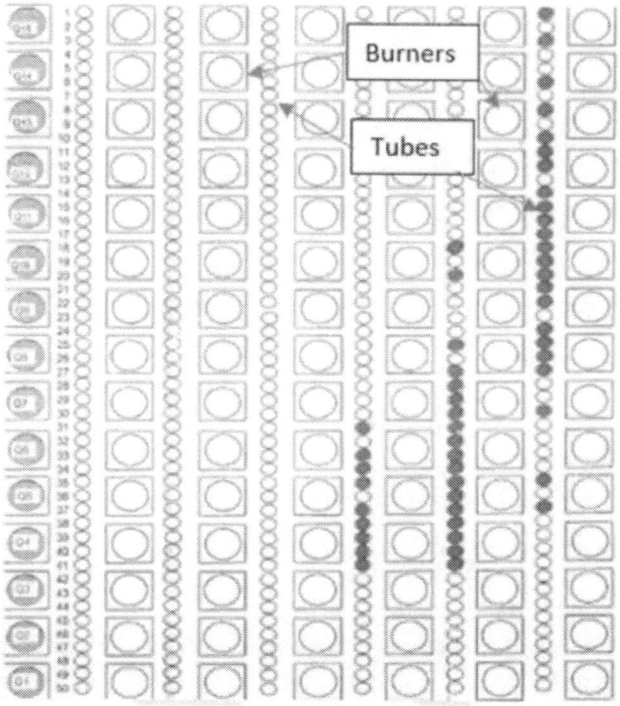

Figure 9. Plan View Showing Location of Damaged Tubes in the Furnace Fire Box (in red)

A review of historical process data indicated the furnace heat-up rate was excessive. The flue gas temperature increased at a rate of 500-700°F/hr (260-370°C/hr). The acceptable heat-up rate is 100°F/hr (38°C/hr).

A normal plant startup takes about two days. Operations was under pressure to start the plant up as quickly as possible. As a result, the start-up task was completed in just one day. Thus, the maximum heat-up rate had been exceeded.

Trends highlighted that the process gas temperature exiting the outer row of tubes (row #6 on the far-right hand side of Figure 6) had been over 1500°F (816°C) for 30 minutes. This was the target exit temperature. Since the temperature was at the target, it had been assumed that the start-up was proceeding rapidly but normally.

The investigation also identified that the Outside Operator performing the burner light-off task was not familiar with the task. He had never performed this task in the past and had not been trained on how to properly do it.

Furthermore, a startup procedure did not exist for how to start up the plant when it is on hot standby and there was no procedure describing the correct burner light-off sequence to ensure even heat distribution in the furnace. The Board Operator had simply instructed the Outside Operator to go out and quickly light the burners. Additionally, there had been no communication between the Board Operator, Outside Operator, and Shift Supervisor on which burners were being lit during lighting of the burners. As a result, all the burners on just one side of the furnace had been lit.

It should be noted that the burner lighting sequence provided by the furnace designer had not been followed in the past, with no adverse consequences. Thus, Operations felt a low sense of vulnerability based on previous experience. In other words, they had had been caught in the classic trap of "normalizing deviation".

The Outside Operator, who was focused on lighting burners, was not looking inside the furnace box frequently. The Outside Operator was also not aware that when furnace tubes are too hot they can be detected by visual inspection.

There are temperature indicators on several of the flue gas tunnels in the bottom of the furnace. However, only one of them was configured with an alarm set at 1900°F (1038°C). This temperature indicator was damaged. The alarm did not activate when the flue gas temperature got too high.

There is also a skin temperature indicator on one of the radiant section tubes which failed. However, it too was not functioning at the time of the incident. The stack oxygen analyzer was also not functional.

Key Lessons Learned

Like most incidents in the industry, there were multiple failures of the layers of protection which were designed to prevent such an incident from occurring.

This incident reinforces that:

- The furnace should only be started up by skilled Operators familiar with the task. Also, more than one Field Operator should start up a furnace of this size.
- Roles and responsibilities during startups should be clearly defined.
- It is essential to have clear communication between the Board / Panel Operator and Field Operator during furnace startups.
- The burner light off sequence is critical to ensuring the furnace is started up safely and should be included in the startup procedure.
- As noted in previous papers [1][2], it is important to visually inspect the furnace box frequently during startup. Instrumentation alone is a poor indication of the condition inside the firebox.
- Furnace instruments designed to protect the plant from overheating incidents – in this case the tube skin temperature and tunnel flue gas temperature indicators, and the stack oxygen analyzer – need to be properly configured and maintained.
- Alarm and shutdown setpoints which are in place to protect the plant from overheating should be evaluated. Specifically, the temperature alarm and trip points should be examined to ensure they are not set too high.
- Always follow procedures. A clear expectation should be set that "There is always time to do it right". Production goals should not be reached at the expense of safe work practices.
- A risk assessment should always be conducted whenever there is a deviation from normal start-up procedures.

It is our hope that by sharing this incident and its learnings with other operators, we can help prevent future incidents of this nature from occurring in the industry.

References

[1] D.Timbres, "Primary Reformer Failure", Ammonia Technical Manual ©AIChE 2002

[2] B.Rani, "Catastrophic Failure of Reformer Tubes at Courtright Ammonia Plant", Ammonia Technical Manual ©AIChE 2006

Addressing Reverse Flow Hazards in High Pressure Boiler Feed Water Pumps

Reverse flow and rotation induced failures of high-pressure boiler feed water pumps were experienced at three Koch Ag & Energy Solutions (KAES) ammonia production locations during the time period from 2018 to 2020, causing significant equipment damage and plant downtime. This paper summarizes how KAES analyzed and addressed the risk across the fleet.

Mark Schroeder, James McMurphy
Koch Ag & Energy Solutions, LLC

Introduction

Reverse flow and rotation of high pressure (HP) boiler feed water (BFW) pumps caused by failures of various check valve styles occurred at AES ammonia production facilities in Brandon MB, Enid OK, and Ft Dodge IA over the time period from 2018 to 2020. These failures led to significant equipment damage and production downtime, with the potential for injury to personnel. This paper summarizes the failures that were experienced, the fleetwide hazard analysis conducted, and the additional protection layers applied by KAES to reduce this risk.

Description of the Facilities

KAES owns and operates 6 fertilizer and one methanol production locations in North America, along with a network of 21 distribution terminals. Plant ammonia capacities range from 900 to 3,500 STPD, along with a single plant methanol capacity of 5,000 STPD. All of the production facilities generate and utilize high pressure steam at 1,550 psi/(10,700 kPa) to drive the syngas make-up/recycle turbines.

Incidents

Brandon, MB 8/9/2018

The plant was operating normally and was in the process of performing a routine BFW pump swap from the electric driven BFW pump to the turbine driven BFW pump. While the turbine was being brought up to offset flow supplied from the electric BFW pump, it exceeded its rated governor speed of 2,500 rpm and the steam supply automatic shut-off valve was closed. It was observed that the turbine was still rotating, so the emergency shutdown was activated in the field and the manual steam isolation valve upstream of the automatic shut-off valve was closed.

At this point in time, BFW flow to the steam drum dropped suddenly from 1,550 gpm to 355 gpm (352m^3/hr-80m^3/hr). The operator increased the speed of the electric BFW pump attempting to restore the forward flow to the drum. It was noticed at this point that there was a 15 second gap in DCS data for the turbine speed, presumably when it went into reverse rotation. Subsequently when the turbine speed began to read a value again, it showed a reading of 3,400 rpm and proceeded to rise to a maximum speed of 3,476 rpm, exceeding the overspeed trip setting of 3,465 rpm. Approximately five minutes later, the field operator closed the manual discharge and suction valves on the turbine driven BFW pump and the turbine speed began gradually reducing. The entire plant tripped on low steam drum level shortly thereafter, followed by the electric BFW pump tripping on motor overload.

During this event, the synthesis gas feed/effluent heat exchanger 121-C and methanator outlet trim cooler 115-C also experienced temperature excursions above their maximum allowable working temperature (MAWT) due to loss of BFW flow through the boiler feed water pre-heaters 123-C on the converter outlet and 114-C on the methanator outlet. The waste heat boilers 101-C and 102-C also experienced temperature excursions above MAWT following loss of level in the steam drum. Rubbing damage was sustained ohe turbine 3rd stage disc.

The root cause of the incident was determined to be the failure of the ARC valve on the discharge of the turbine driven BFW pump. The turbine driven BFW pump was induced into reverse rotation by the backwards flow of BFW through the pump suction, setting up a circulation pattern that prevented the admission of BFW from the de-aerator (see Figure 1).

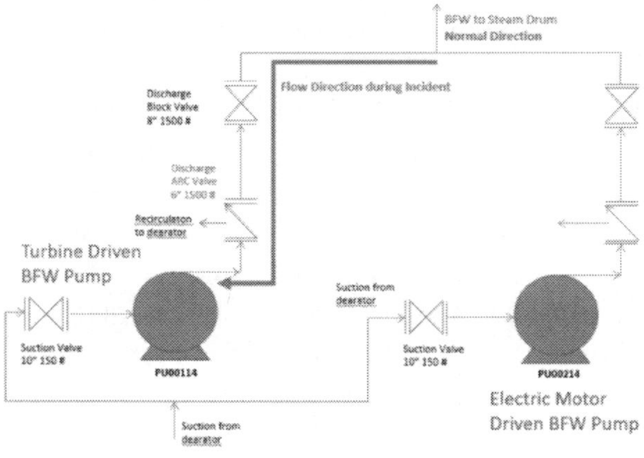

Figure 1. Brandon BFW Reverse Flow Path

It was found that the top web was damaged with the bore excessively increased, the disc/piston was tilted due to the plunger rod being broken, and the lower web support was also damaged (Figure 2). The suction and discharge block valves were hydrotested and found to be leaking through due to excessive erosion in the valve disc.

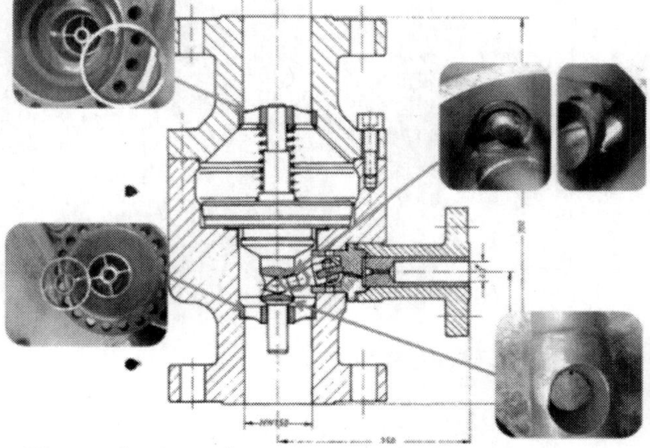

Figure 2. Brandon ARC Valve Failure Pictures

The ARC valves were added to the critical check valve program in 2014 and the first recorded inspection occurred in the subsequent 2017 turnaround. The 2017 inspection was performed in the field by removing piping above the valve (see Figure 3) to reveal and inspect the top of the disc, then leak testing by filling it with water above the top disc to the upper flange and confirming that the water level remained steady for one hour. The bottom half was not visible, which prevented inspection of the seat, bypass head plunger mechanism, and bottom support web.

Figure 3. Brandon ARC Valve Field Installation

Enid, OK 1/26/2020

The plant was in the process of start-up and were swapping from one turbine driven HP BFW pump to the other (104-J to the 104-JA). As the swap proceeded and inlet steam was closed to the 104-J turbine, the outside operator noticed oil spraying from the internal oil pump on the 104-J. The control panel operator notified the outside operators that the 104-J had not come to a complete stop and outside operators confirmed it was still rolling. After closing the inlet steam manual valve, they observed that 104-J was still spinning and its rpms were increasing as 104-JA was being brought up to a higher speed. At this point the operators concluded that 104-J was spinning backwards and attempted to close its discharge valve, but it would not move. Loss of level in the steam drum caused the reformer and Aux boiler to trip, aborting the start-up.

It was found that a small steel bolt about 1" long was wedged under the disc of the ARC valve as depicted in Figure 4 and prevented full closure of the valve, so that there was reverse flow from the discharge side of 104-JA through 104-J to the deaerator.

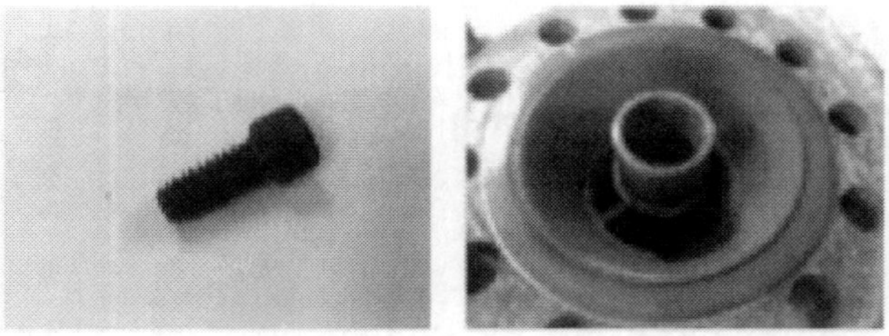

Figure 4. Enid ARC Valve Failure Cause

Ft Dodge, IA 3/26/2020

The plant was operating at normal production rates and in the process of swapping between the two turbine driven HP BFW pumps (from south pump 2002-JA to north pump 2002-J). The procedure called for operating both pump turbine speed controllers in manual while the console operator slowly raised the speed of the on-coming pump and lowered the speed of the off-going pump while simultaneously monitoring the steam drum level. The off-going pump would switch to its minimum flow recirculation orifice line once its discharge swing check valve seated. There were operations and maintenance personnel observing the swap in the field because of a loss of BFW flow experienced when attempting the swap a day earlier that was suspected to have been caused by an issue with the north pump.

As the speed of the two turbines was equalized, a decrease in BFW flow occurred and the steam drum level began to fall. The console operator responded by raising the speed to both BFW pump turbines, restoring flow and stabilizing the steam drum level. he operator resumed gradually lowering the speed of the off-going (south) pump in small increments a few minutes later and placed the on-coming (north) pump speed controller in automatic. Another sudden drop in BFW flow occurred and the oncoming (north) pump sped up to its maximum to compensate, invoking an overspeed trip. Over the next several minutes, the oncoming (north) pump was attempted to re-start twice, but ramped up too quickly and tripped on overspeed (set at 4150 rpm).

During this 6 minute period, the operator attempted to gain control of the steam drum level by speeding up the off-going (south) pump, but could not establish stable forward BFW feed flow and it eventually tripped on overspeed as well, but continued to spin. The south pump speed indicator temporarily went off-scale high, exceeding 5,300 rpm four times. The duration of each time spent above the maximum transmitter range increased from 2 to18 seconds. The fifth and final time the pump turbine speed transmitter went above range, it remained over range for 49 seconds, then rapidly descended to to 0 rpm over 8 seconds once it resumed reading a value. It is presumed that the turbine seized up at this point in time due to disintegration of the outboard bearing.

Catastrophic damage was sustained to the south HP BFW pump turbine outbound bearing housing during this episode. Figure5 shows where the outbound bearing housing came apart, with its integral lube oil pump separated from the shaft and dangling in the air. Figure 6 shows where the turbine foundation cracked. Figure 7 shows internal damage incurred by the separated turbine blades inside the case. Ball bearings and fragments from the disintegrated outbound bearing were found scattered on the floor in the vicinity of the turbine as seen in Figure 8. The vibration also tore a hole through the the turbine's steam exhaust expansion joint as shown in Figure 9. The turbine case fortunately remained intact, minimizing the amount of shrapnel that could have otherwise been caused by the separation of the turbine blades, but did sustain cracks.

Figure 5. Outbound Bearing Damage

Figure 6. Cracked Turbine Foundations

Figure 7. Turbine Blade Damage

Figure 8. Bearing Fragments Found on Floor

Figure 9. Hole in Outlet Steam Expansion Joint

During this 6 minutes, period, there were also alarms received on low lube oil pressure, which is driven by the pump attached to the turbine. It went into alarm, suggesting that the south pump slowed, recovered, then slowed again. he temperature on the BFW preheater at the converter outlet (123-C) syngas increased from 325° F (163°C) to 607° F(354°C) while the temperature on the BFW pre-heater at the methanator outlet (114-C) increased from 270°F(132°C) to 527°F(275°C).

Once the south turbine was at 0 rpm, its manual suction valve was closed in the field. This allowed the north pump turbine to be reset and stabilize its speed, but the plant subsequently experienced a full front end trip on low level in the steam drum, stopping the process gas side flows while maintaining steam flow to the reformer tubes. From that point forward, BFW flow was restored on the north pump and the temperatures at the outlet of the 123-C and 114-C cooled back to within normal limits within the next half hour.

As mentioned previously, there were several mechanics and operators observing the pump swap in the field because of the failed attempt to start up the north pump on the prior day. At the moment that the BFW flow was lost the first time, an experienced mechanic detected a change in the operating sound coming from the pump and turbine and the group quickly evacuated the building in which the BFW pumps are housed.

After the south turbine damage was sustained, the group re-entered the building and extinguished a small fire on the south HP BFW pump end. The location of the fire and residual charring is shown in Figure 10. The pump itself was still rotating (backwards) until its suction isolation valve from the deaerator was closed. After closing the south pump suction valve, the north turbine was reset and able to maintain speed and forward BFW flow.

Figure 10. Location of Fire on Pump

Figure 11. Failed Check Valve As Found

The cause of the event was determined to be a failure of the swing check valve on the discharge of the south HP BFW pump (Figures 11 to 13). It is believed that this was done from chattering of the flapper which caused parts to wear out not allowing it to close. A secondary check valve further down the line going to the BFW coil by the Reformer, also suspected of leaking, showed no visible damage.

Figure 12. Wear on Check Valve Pin

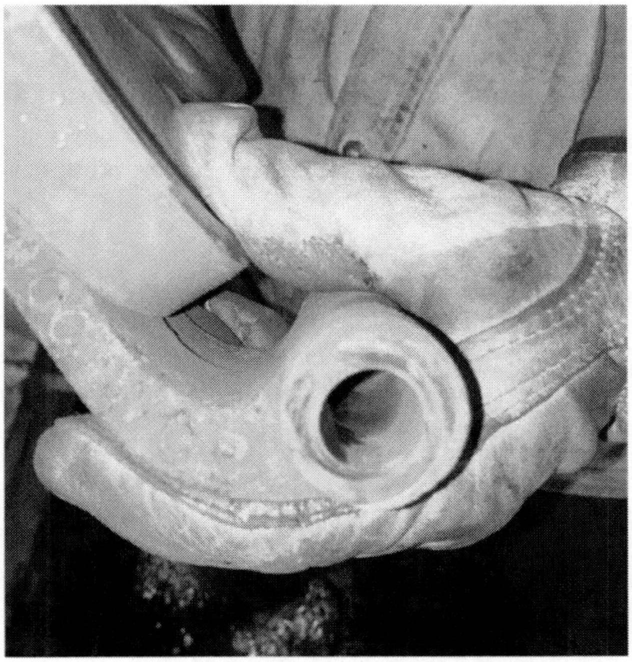

Figure 13. Wear on Check Valve Flapper Hinge

Short Term Actions

Following this series of fleet incidents, a cross-site team was assmbled to evaluate the risk posed by reverse HP BFW pump failures and mitigate any determind gap in protection layers. Because any additional equipment solutions recommended would require engineering and installation during a plant turnaround, interim actions were taken to reduce the risk across the fleet:

1) Operational awareness of the risk: A meeting was conducted with all of the Operations managers across the fleet to share the learnings from the event and raise awareness that a demand is placed on the discharge check valve any time a pump shuts off or is taken out of service. Loss of BFW flow when bringing a new pump on-line in parallel with another operating pump could be caused by failure of the off-going pump's discharge check valve. Planned pump swaps should be performed with all personnel removed from the vicinity of the pump and turbine planned to be taken down.

2) Check Valve Inspections and PM effectiveness: Plants verified that the BFW pump discharge check valves were part of the critical check valve preventive maintenance (PM) program and shared best practices for inspecting and maintaining them. n the case of ARC valves, the disc must be removed, preferably in a shop environment, in order to allow inspection and mainatenance of all the pistons and moving parts underneath.

With these recommendations in place, each plant scheduled the addition of secondary check valves and isolation valves at future turnarounds. Figure 14 shows the Check Valve Inspection Checklist completed at the Bradon facility following the failure.

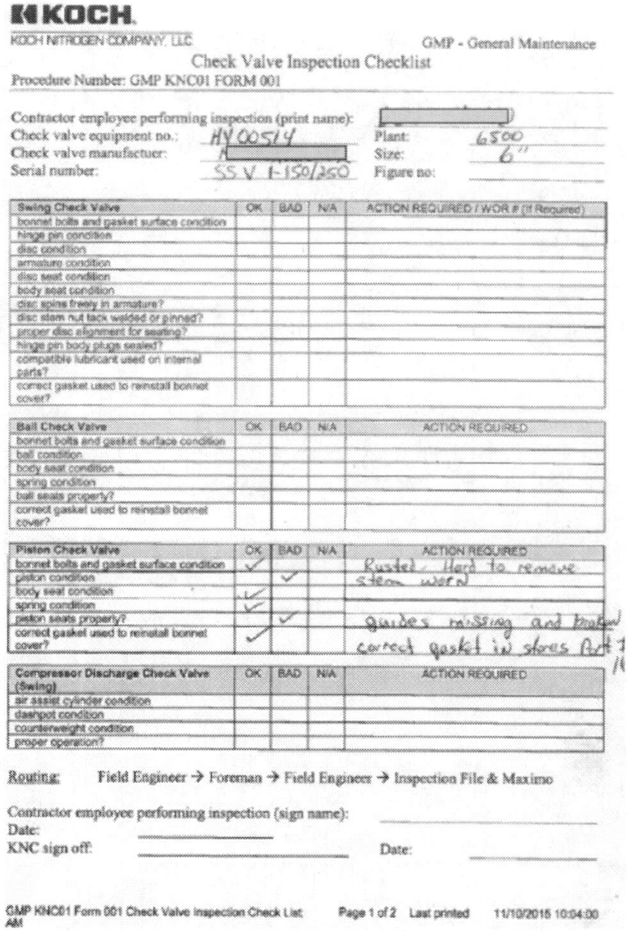
KOCH
KOCH NITROGEN COMPANY, LLC
GMP - General Maintenance
Check Valve Inspection Checklist
Procedure Number: GMP KNC01 FORM 001

Contractor employee performing inspection (print name): [redacted]
Check valve equipment no.: HY00514 Plant: 6500
Check valve manufactuer: [redacted] Size: 6"
Serial number: SSV 1-150/250 Figure no:

Swing Check Valve	OK	BAD	N/A	ACTION REQUIRED / WOR # (If Required)
bonnet bolts and gasket surface condition				
hinge pin condition				
disc condition				
armature condition				
disc seat condition				
body seat condition				
disc spins freely in armature?				
disc stem nut tack welded or pinned?				
proper disc alignment for seating?				
hinge pin body plugs sealed?				
compatible lubricant used on internal parts?				
correct gasket used to reinstall bonnet cover?				

Ball Check Valve	OK	BAD	N/A	ACTION REQUIRED
bonnet bolts and gasket surface condition				
ball condition				
body seat condition				
spring condition				
ball seats properly?				
correct gasket used to reinstall bonnet cover?				

Piston Check Valve	OK	BAD	N/A	ACTION REQUIRED
bonnet bolts and gasket surface condition	✓			Rusted, Hard to remove
piston condition		✓		stem worn
body seat condition	✓			
spring condition	✓			
piston seats properly?		✓		guides missing and broken
correct gasket used to reinstall bonnet cover?	✓			correct gasket in stores Part # [illegible]

Compressor Discharge Check Valve (Swing)	OK	BAD	N/A	ACTION REQUIRED
air assist cylinder condition				
dashpot condition				
counterweight condition				
proper operation?				

Routing: Field Engineer → Foreman → Field Engineer → Inspection File & Maximo

Contractor employee performing inspection (sign name):
Date:
KNC sign off: Date:

GMP KNC01 Form 001 Check Valve Inspection Check List Page 1 of 2 Last printed 11/10/2015 10:04:00 AM

Figure 14. Check Valve Inspection Checklist

Long Term Actions: Close Independent Protective Layer (IPL) Gaps with Plant Modifications

Recognizing the need to re-evaluate hazard scenarios pertaining to these BFW events, a new Process Hazard Analysis (PHA)was performed to evaluate risk profiles at each plant and close any gaps with capital projects. While these reverse flow scenarios had been previously evaluated in PHA with no risk gaps, several important differences impacting the new evaluation included the following:

- The observed frequency of equipment failure as an initiating event was higher than the probability assumed in the original PHA.

- The probability of personnel in close proximity to equipment at the time of failure was different from the original PHA. When possible, a conditional modifier for presence of personnel was used after administrative and procedural controls were implemented to prevent personnel from occupying certain areas during pump swaps.
- The probability of shrapnel ejection from equipment based on previous incidents, pump design, pump operating procedures, and manufacturer consultation.

One goal of this holistic, fleetwide evaluation of risk gaps was to implement a common design framework that would yield a shorter timeline to implement solutions. The team performing this evaluation produced two recommendations centered around installing additional check valves and an instrumented function to detect and isolate reverse flow.

IPL #1: Dissimilar Check Valves

Additional check valves were specified based on each plant's existing configuration, with the basic requirement of installing an additional check valve that was of dissimilar design than the existing type. The basis of this recommendation was to eliminate a common failure mode among check valves. Most existing plant configurations featured typical swing-check valves with separate minimum flow lines, with Automatic Recirculation Check (ARC) valves being the only other type in use.

Based on observed failure modes of these two types of valves, nozzle check valves were recommended as the preferred dissimilar design. Furthermore, an evaluation of check valve installation details and maintenance practices was also performed to ensure existing check valves conformed to the manufacturer's recommendations. Figure 15 shows the completed installation at the Ft Dodge, IA plant.

Figure 15. Dissimilar Check Valves installed at BFW pump discharge in Ft Dodge, IA

IPL #2: Instrumented Protective Functions

To achieve an additional layer of protection against reverse flow, automatic discharge isolation valves were recommended on each individual BFW pump, upstream of their common discharge line, as shown in Figure 16.

Figure 16. BFW Pump Automatic Discharge Isolation Valves installed in Ft Dodge, IA

Requirements for instrumented protective functions were specified for the sensor group, logic solver, and final elements based on the required Risk Reduction Factor (RRF). A reverse rotation event would be detected in its incipient stage by sensors, and the logic solver would actuate a final element to achieve a safe state of the process.

When permitted by a lower required RRF, the instrumented function could be implemented in the plant's Basic Process Control System (BPCS). Higher RRFs, which necessitated a SIL rated system to close the remaining risk gap, would require implementation of these functions in the plant's Safety Instrumented System (SIS). This approach allowed sites to effectively manage their individual risk profile and avoid the complexity of additional SIS instrumentation when the site's risk profile did not warrant it.

Several sensor types were considered including pump discharge pressure, discharge flow, suction flow, suction pressure, differential pressure across the pump, and pump shaft speed with direction. A focus was placed on the ability to detect the hazard in its incipient state, i.e. before reverse rotation would result in damage to the equipment or catastrophic failure.

Discussion among internal subject matter experts (SMEs) concluded that suction or discharge pressure alone would not reliably correlate with flow reversal, as downstream check valve failures would not necessarily produce a low discharge or high suction pressure at the pump. Similarly, SMEs concluded that pump d/p would also not easily correlate with flow reversal. Further consultation with pump vendors reinforced this assessment.

Ultimately, the team narrowed sensor recommendations to pump discharge flow measurement and pump shaft rotational speed with direction. Recommended sensors included HART-enabled d/p flow transmitters and specialized reverse-rotation relays.

While reverse rotation relays would still be capable of isolating flow before a runaway pump condition occurs, flow transmitters provided the additional benefit of isolating reverse flow before the pump shaft reaches the manufacturer's stated "breakaway torque" in which the pump begins to spin in reverse. Accordingly, choosing flow transmitters for the sensor group would provide added protection to mechanical components, such as shaft-driven oil pumps, that are sensitive to reverse rotation. Furthermore, use of HART-enabled flow transmitters facilitated use of the control system's diagnostic capabilities in determining degrading sensor performance prior to an otherwise undetected failure of the sensor. Reverse rotation relays were considered an acceptable alternative when piping configuration would not facilitate installation of flow transmitters.

When reverse flow conditions are detected on multiple flow transmitters, voting logic in the BPCS or SIS triggers the closure of a pneumatic block valve. Recommended logic design featured a straightforward two-out-of-three (2oo3) voting of sensors to actuate a single pneumatic block valve. Flow transmitters were configured for bi-directional flow measurement, with forward (positive) flow being the normal process condition. Setpoints for reverse flow detection were set between no flow (zero flow) and the reverse (negative) flow needed to generate the pump manufacturer's stated breakaway torque to initiate reverse rotation. After evaluating the Process Safety Time of this scenario, a trip setpoint was selected with enough conservatism to prevent the reverse flow from developing the pump's breakaway torque. A non-zero trip setpoint was also desirable to prevent spurious trips when securing a pump.

This logic design allowed the automatic block valve to remain open under normal operating conditions, including normal pump shutdowns, and only actuate when actual reverse flow conditions are detected. A major benefit of this arrangement was that standby pumps could quickly start and develop forward flow without having to wait for the discharge block valve to open. Reset logic was specified based on each plant's operating procedure, but automatic resets were not recommended. Existing manual block valves were retained for Lockout/Tagout purposes.

Based on the principle of fail-safe design, the pneumatic block valve was specified with an air-to-open/fail-closed actuator with a trunnion ball valve or butterfly valve specifically designed for high pressure boiler feed water applications. The minimal risk of spurious closure inherent to this valve design was acknowledged by plant leadership and considered a small tradeoff for the substantial risk reduction it provided to personnel safety.

Project Management

Implementation of these solutions considered plant turnaround schedules and procurement times for long-lead items. These considerations necessitated a staged approach to implementation of these risk reduction measures. Installation of check valves was typically the easiest modification and could usually be performed in less time than the engineering required for instrumented protective functions. Block valves and pneumatic actuators often had longer lead times.

When turnaround schedule did not allow sufficient time to design instrumented protective functions, several plants still accomplished an order of magnitude risk reduction with check valve installation. Installation of the remaining instrumented protective function occurred in the subsequent turnaround.

One plant was undergoing a major capacity expansion project, and this scope was identified early enough to include reverse flow protection for new equipment in the expansion project. A separate project to address existing equipment was managed.

Conclusion:

The need for reliable reverse flow protection in high head pumping systems is important to prevent potential catastrophic damage, particulary with turbine driven equipment. Check valves are a critical component that are called upon every time a pump is shut down, and must be inspected and maintainted properly and at appropriate time intervals.

Where the risk justifies additional protection, redundant check valves and/or automatic discharge isolation valves that fail closed to prevent reverse flow/rotation should be considered. When multiple pumps can operate in parallel, the additional discharge protection should be added to each pump to prevent one pump's dischage from pushing backwards through another's that could be operating with a compromised discharge check valve.

Thrust Collar Failure on Integrally Geared Process Air Compressor in Ammonia Plant

ESSA PT Panca Amara Utama, Banggai Ammonia Plant (BAP) operates KBR Purifier plus unit to produce 2,167 MTPD of cold ammonia. After 238 continuous days of safe and reliable operations, the Process Air Compressor (101-J) tripped on high vibration at Stage-1 of Compressor. After the compressor tripped, an inspection was conducted and found that the thrust collar at stage-1 of the compressor and turbine connection was damaged which resulted in the full replacement of the compressor rotors with new spare rotors and 30 days plant shut down for maintenance. Further analysis was performed by the OEM, using fractography inspection and finite element method analysis, to find the root cause of failure with the result that the thrust collar was detached and damaged due to reduced contact leading to fatigue fracture. Improvement was performed by changing the installation method for the thrust collar to the shaft and installation of additional axial vibration sensors for online continuous monitoring.

Mukesh Agrawal, Novi Ariharnomo
ESSA – PT Panca Amara Utama

Introduction

Operating since 2018, ESSA PT Panca Amara Utama (PAU) is the largest private Ammonia producer in Indonesia. The PAU Banggai Ammonia Plant (BAP) is located at Banggai, Central of Sulawesi, Indonesia, produced 734,818 metric tonne (MT) (809,990 short tonne (ST)) of ammonia in 2024, and runs on the KBR Reforming Exchanger System (KRES) and Purifier Technology. This represents its first application in the world, which puts the BAP at the forefront of Ammonia production worldwide.

On April 30, 2021 after 238 days of continuous plant operation, the 101-J Process Air Compressor tripped due to high vibration on first stage, with two-out-of-two (2oo2) voting logic. Actual vibration of VI-10132X and VI-10132Y are 95 and 104 micrometer peak-to-peak (μmPP), respectively. The trip set point vibration is 88 μmPP (3,465 μinPP).

The intent of this paper is to describe PAU's challenges related to finding the root cause of failure and improvements applied to prevent future failure reoccurrence.

Process Description

BAP is single train ammonia plant based on the KBR Purifier plus process technology; a low energy natural gas reforming process. BAP production process consists of three main processes; Synthesis gas reforming (Front End), Purification of synthesis gas (Middle End), and Ammonia synthesis (Back End).

Natural gas is available at BAP battery limit at normal pressure of 50 kg/cm^{2}g (712 psig). The natural gas supplied is clean and dry, with usual methane (CH_4) content of 91.8%mole and carbon dioxide (CO_2) content 0.01%mole.

In the Front-end process of BAP (Figure 1. Process Flow Diagram of Synthesis Gas Production (Front End Process)), natural gas is fed to the Mercury Guard Removal Bed (102-D), which removes mercury.

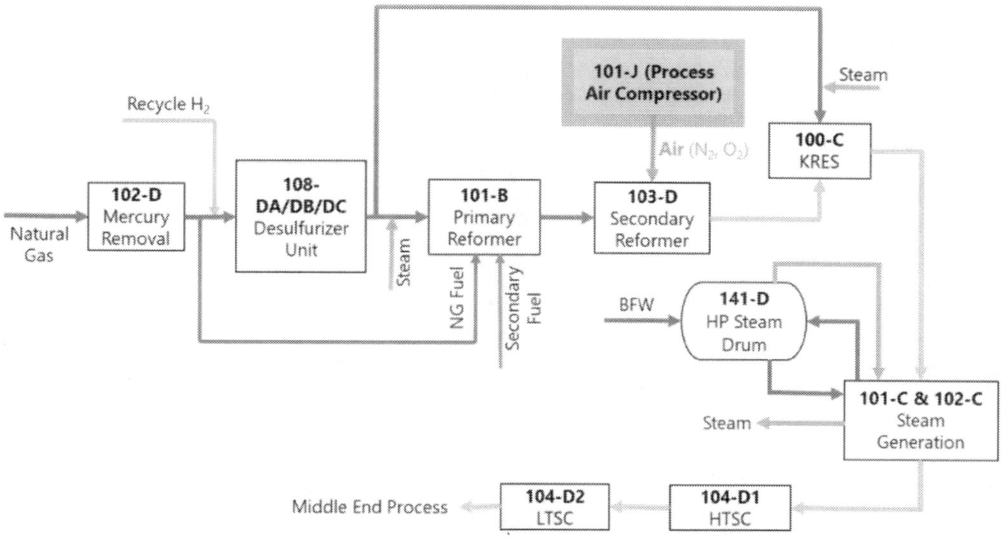

Figure 1. Process Flow Diagram of Synthesis Gas Production (Front End Process)

After mercury removal, the natural gas is sent to a Hydrotreater (108-DA) to convert organic sulphur compounds to hydrogen sulphide as the first step of desulfurization process. In the second step, the process gas passes through the Desulfurizer (108-DB/DC) which is running in series "lead-lag" configuration and produces effluent stream having less than 0.1 ppmv (0.139 mg/m^3) sulphur.

After the desulphurization process, the natural gas is mixed with steam and fed to Reforming Section in Primary Reformer (101-B), Secondary Reformer (103-D) and Reforming Exchanger (100-C).

The de-sulphurised feed is split to 83.7% gas to Primary Reformer (101-B) and 16.3% gas to KRES (100-C). After splitting, medium pressure steam is added to achieve designed steam-to-carbon molar ratio to 101-B (2.7 S/C) and 100-C (3.0 S/C).

The process air is filtered and compressed in a six-stage centrifugal integrally geared Process Air Compressor (101-J) to a pressure of 44.5 kg/cm^{2}a (633 psia). The process air is then heated in the convection section of the primary reformer to about 492°C (917 ^{o}F) and introduced to the secondary reformer coil.

Process Air Compressor (101-J)

The Process Air Compressor (101-J) is a six-stage centrifugal, integrally geared compressor manufactured by Mitsubishi Heavy Industries Compressor (MCO). Its specifications are listed in Table 1. The 101-J has a bull gear and 3-pinion shafts with impellers attached to each end. The bull gear has thrust bearing and each of the pinions use thrust collars as shown in Figure 2.

Description	Specification
Type	6-Stages Centrifugal Integrally Geared Compressor
Model	12C6-BG20
Original Equipment Manufacturer (OEM)	Mitsubishi Heavy Industries Compressor (MCO)
Capacity (Rated)	125,185 Nm^3/hr (4.4x10^6 Nft^3/hr)
Monitoring System	System-1 Bently Nevada-Baker Hughes (BN-BH)
Driver Type	Steam Turbine
Operating Pressure - Suction - Discharge	0.95 kg/cm^{2}a (13.5 psia) 44.5 kg.cm^{2}a (633 psia)
Speed	5,498 rpm (Turbine) 1,324 rpm (Bull Gear) 8,182 rpm (Pinion-1) 12,272 rpm (Pinion-2) 15,194 rpm (Pinion-3)

Table 1. Process Air Compressor Detail Specification

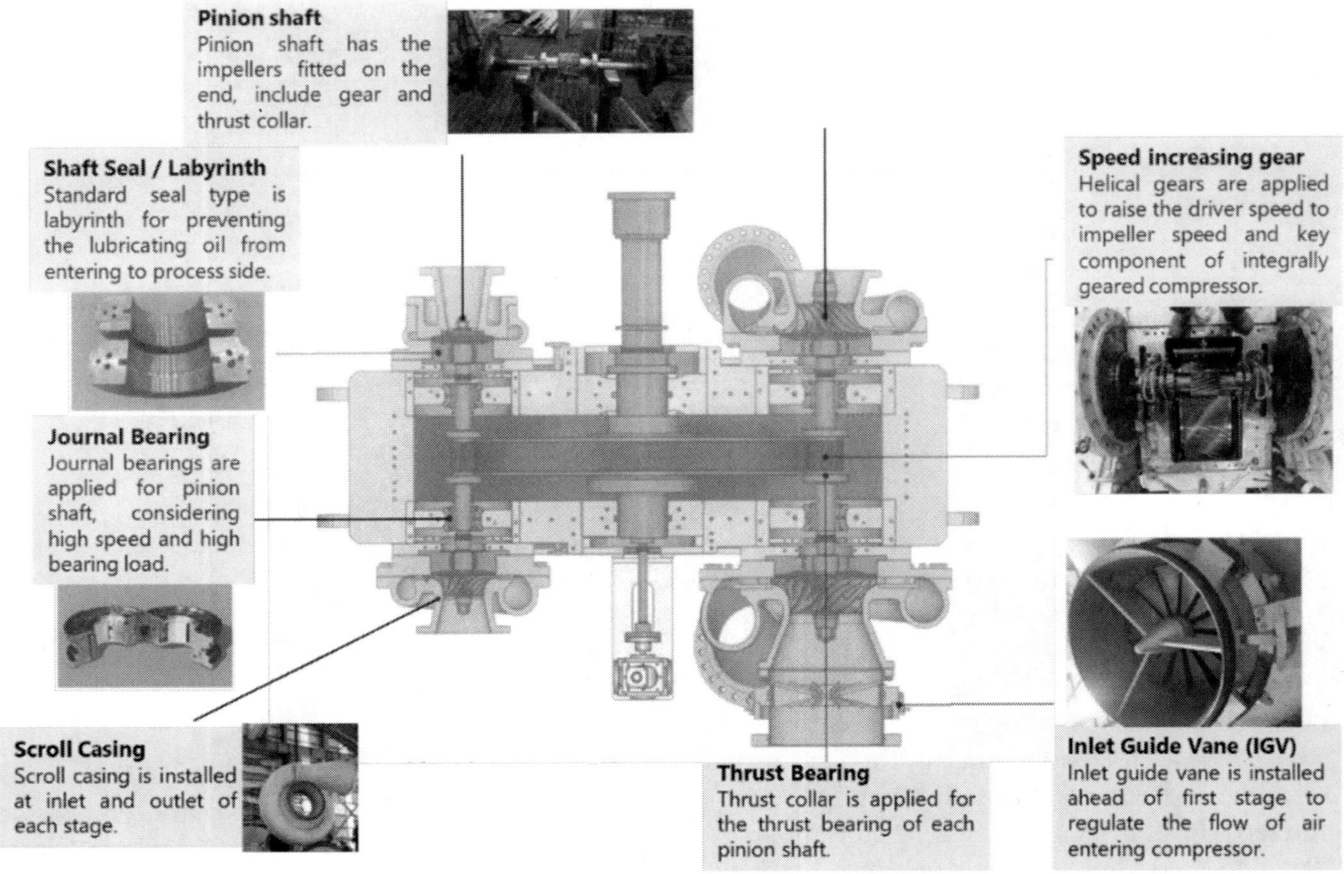

Figure 2. Component of Integrally Geared Centrifugal Process Air Compressor (101-J)

Chronology of Event

On April 30, 2021, at 08:06, the first vibration spike occurred at the 1st stage of the 101-J compressor. Vibration values reached VT-10132X: 54 μmPP (2,126 μinPP) and VT-10132Y: **92 μmPP** (3,622 μinPP). However, the 101-J did not trip since VT-10132X value was still below the trip setting (2oo2 voting logic interlock).

On the same day at 10:40, a second spike occurred, with vibration value observed of VT-10132X: 38 μmPP (1,496 μinPP)and VT-10132Y: **113 μmPP** (4,449 μinPP). The 101-J was still running when a third spike occurred at 10:50 with vibration value observed VT-10132X: 33 μmPP (1,300 μinPP)and VT-10132Y: **112 μmPP** (4,410 μinPP).

At 11:14, the highest vibration value was observed VT-10132X: 63 μmPP (2,480 μinPP) and VT-10132Y: **184 μmPP** (7,244 μinPP) and eventually at 11:20, the 101-J tripped due to high-high vibration on the first stage, with vibration value VT-10132X: **95 μmPP** (3,740 μinPP) and VT-10132Y: **104 μmPP** (4,095 μinPP). See Figure 3 for the chronology.

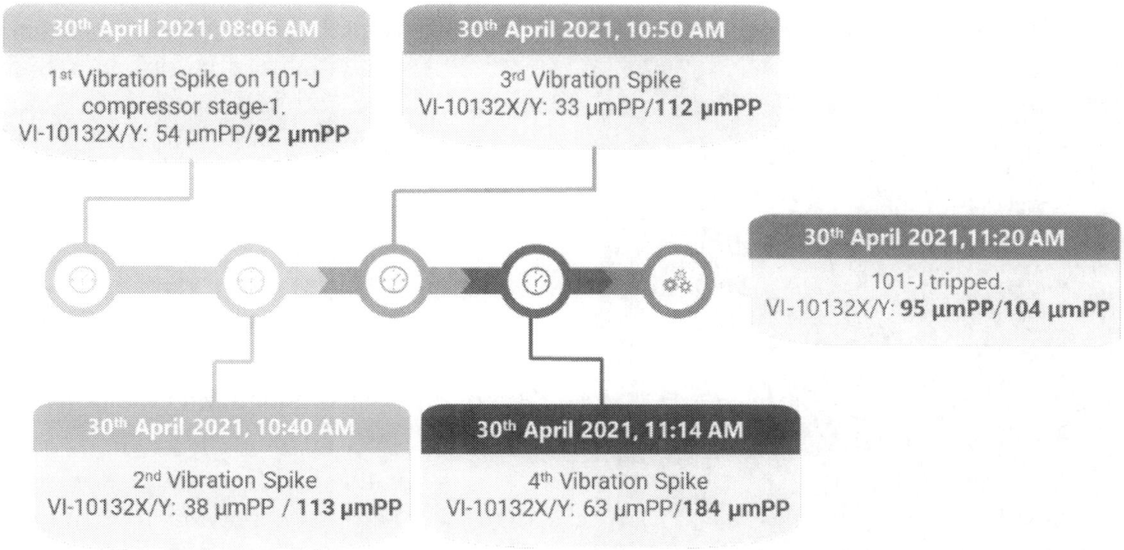

Figure 3. Infographic of chronology

After the compressor tripped, the vibration spectrum was analysed, the surging conditions were suspected as a root cause of the failure. The OEM recommended performing inspection at the following parts:

- Bearing Stage-1 (Figure 4)
- Stage-1 Intercooler (Figure 5)
- Suction Pipe of 101-J (Figure 6)
- IGV Opening and Function (Table 2)
- Vibration Probe (Figure 7)
- Anti-Surge Control Valve (Figure 8)
- Suction Filter Replacement (101-L) (Figure 9)

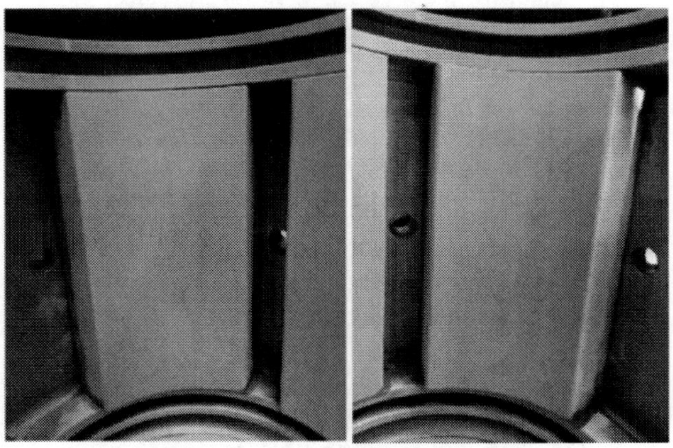

Figure 4. Bearing inspection result (Good Condition)

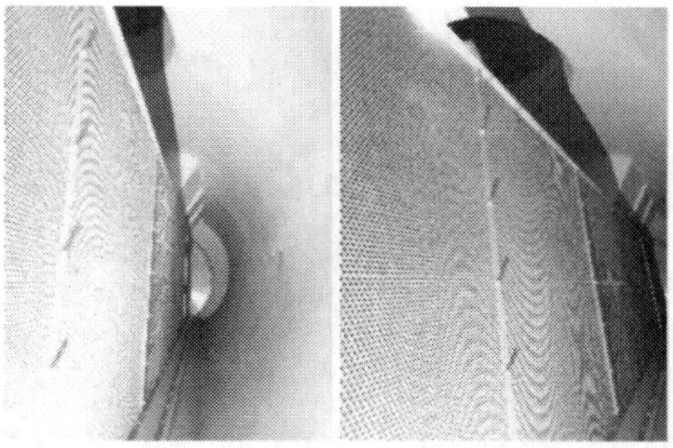

Figure 5. Internal conditions of perforated plate interstage cooler (Clean and Good Condition)

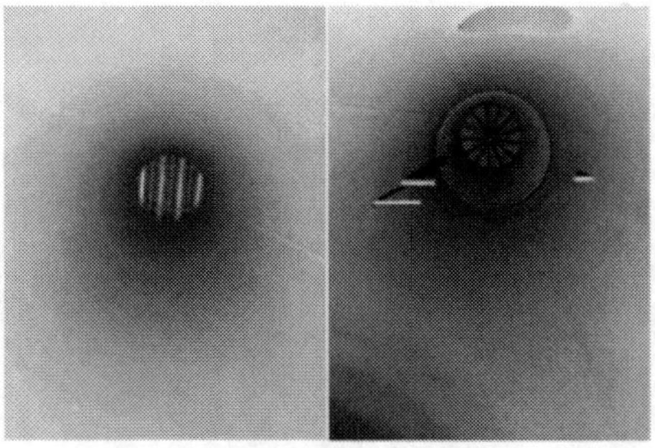

Figure 6. Internal condition of suction pipe to 101-J (Clean and Good Condition)

Reading at DCS	Actual at Field
0%	0%
25 %	24 %
50 %	49 %
75 %	76 %
100 %	100 %

Table 2. Stroke test result of IGV opening (Good Condition)

Figure 7. New probe was installed, and reading is within standard

Figure 8. Anti surge control valve inspection result (Good Condition)

Figure 9. Suction Air Filter Replacement (Good Condition)

Inspections were performed on the recommended items with results indicating that all were in good condition. No defect was found on the suspected items and the OEM recommended to re-start the Compressor.

On May 4, 2021, a restart was attempted and the 101-J tripped when the turbine speed reached 5,290 rpm with a similar problem, high vibration on 1st stage compressor (VT-10132X: **89 μmPP** (3,504 μinPP) and VT-10132Y: **123 μmPP** (4,842 μinPP)).

As per inspection result and after discussion with the OEM and BN-BH, the OEM confirmed that no mechanical abnormality was observed. The OEM and BN-BH recommended to check other possible causes, including the following items:

- Instrumentation system
- Cable and Proximitor Replacement
- Setting to prevent low frequency vibration
- Grounding system
- BN3500 cards replacement
- Electrical interference

After thorough testing, it was decided to restart the compressor again. During the compressor restart a visual check was performed at pinion shaft Stage-1 (Figure 10). The compressor tripped with the same behaviour and axial shaft movement was observed on the shaft of stage-1. Finally, it was decided to open the casing of Compresor for comprehensive inspection and overhaul.

Figure 10. Position of visual check shaft stage-1 compressor

Main Findings during Overhaul

Inspection was conducted during overhaul with main findings as follows:

- Crack on thrust collar 1st stage side (Figure 11)

- Rubbing on 2nd stage impeller (Figure 12)
- Damage on thrust collar steam turbine pinion turbine side (Figure 13)

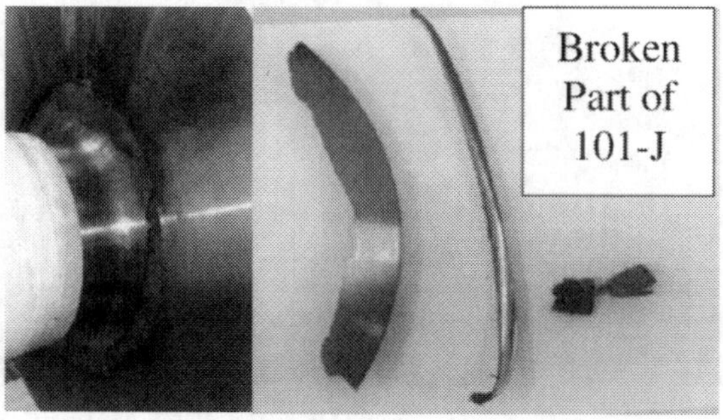

Figure 11. Crack on thrust collar 1st stage side

Figure 12. Rubbing on 2nd stage impeller

Figure 13. Damage on thrust collar steam turbine pinion turbine side

Based on the conditions above, it was decided to take some actions as follows:

- Replace all rotors with new spare set.
- Maintain 101-J operation at maximum 90% of rated capacity until finalization the RCA and corrective actions as well as improvement actions to be taken.

Root Cause Analysis

Fractography inspection and Finite Element Method (FEM) analysis were performed by the OEM to determine the failure root cause.

Fractography Inspection

Fractography inspection was performed for the specimen from the shaft with result as follows:

- Beach mark was observed, which is typically characteristic of fatigue cracking (Figure 14)
- Crack origin is from screw position, and its surface was damaged (Figure 15 and Figure 16)
- Crack origin was damaged (Figure 17)
- Fractured condition was fatigue fracture (Figure 18)

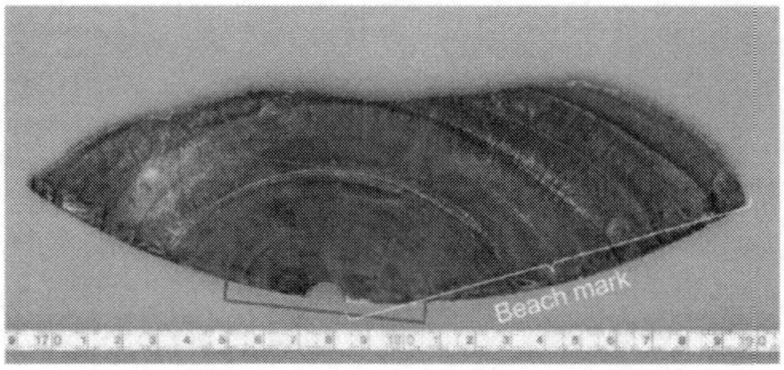

Figure 14. Overview of fractured specimen

Figure 15. Magnified of crack origin area

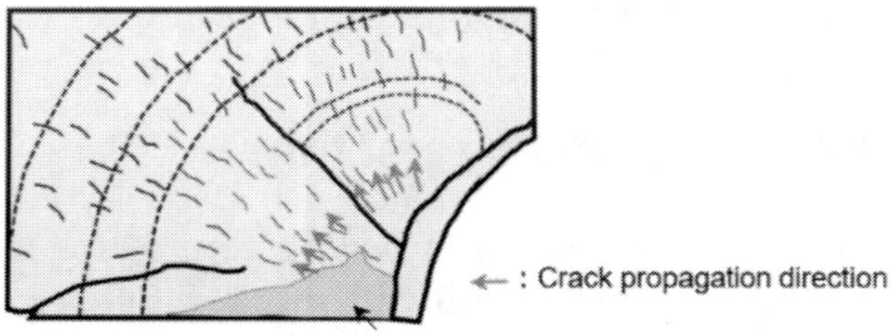

Figure 16. Crack propagation area

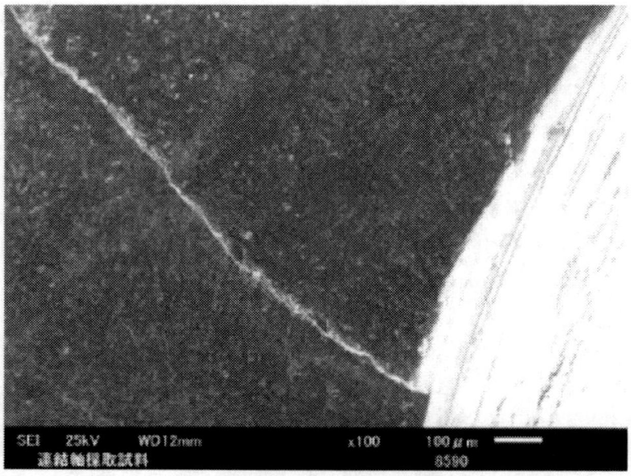

Figure 17. Crack origin was damaged

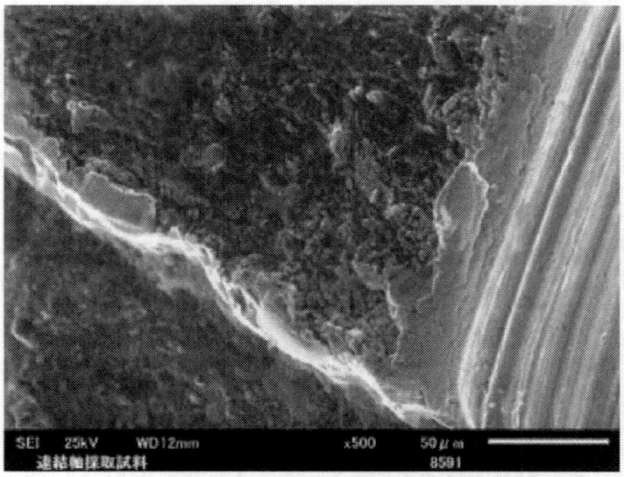

Figure 18. Fatigue cracking was suspected

Finite Element Method (FEM) Analysis

FEM analysis was performed by simulating the operating condition for both Turbine Pinion (Figure 19) and No.1 Pinion (Figure 20). Contact pressure between shaft and thrust collar was calculated with result as follows:

- Less contact pressure condition was observed in partial location
- Observed that inner side of thrust collar has lower contact pressure compared to outer side of thrust collar

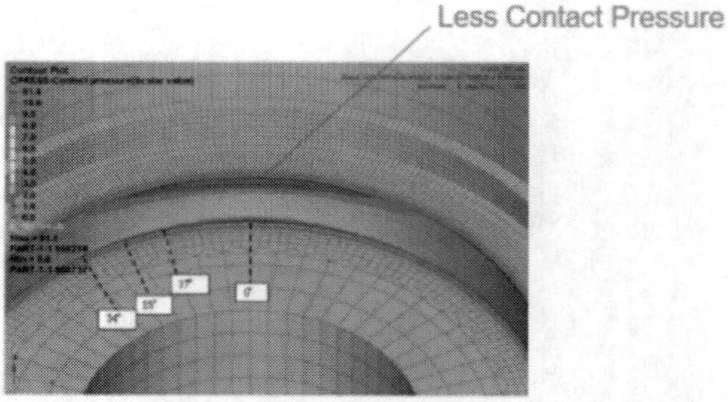

Figure 19. FEM analysis result of Turbine Pinion

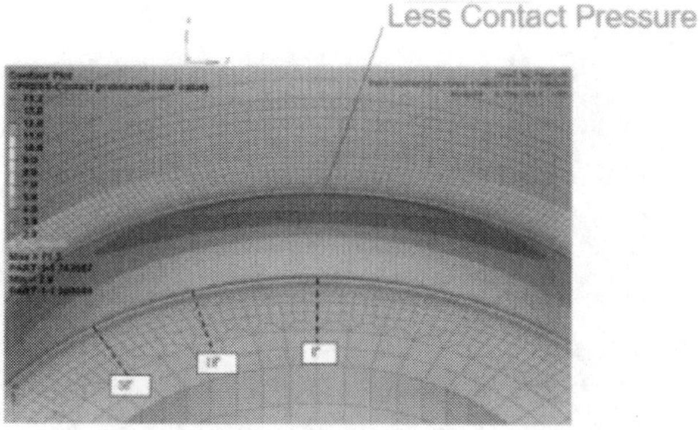

Figure 20. FEM analysis result of No.1 Pinion

Based on the investigation result, damage phenomena at the thrust collar can be concluded as follows:

- The thrust collar damage progressed gradually from the inner side to outer side at thrust collar area with metal reduction
- The thrust collar damage occurred because of lifting phenomenon of thrust collar from the shaft due to the insufficient gripping force (Turbine Pinion and No.1 Pinion) and unexpected force under the transient operating condition (No.1 Pinion)

Improvement Plan / Countermeasure

Countermeasure was developed as per the conclusion of failure investigation to avoid the failure reoccurrence with detail as follow:

- Change design collar connection from shrink fit to bolting (Figure 21). The OEM recommended a new collar connection method by bolts in axial direction and as per FEM analysis, this new design can be used under current BAP operating condition (Figure 22).
- Theoretically, thrust collar will not slip from the shaft if safety factor $\geq$ 1. After improvement, it is shown that safety factor increased to well above 1. (Table 3. Calculated safety factor at compressor rated capacityTable 3)
- Install axial displacement sensors for Turbine Pinion, No.1 Pinion, and No.2 Pinion (Figure 23, Figure 24, and Figure 25) BAP, as discussed with the OEM, decided to install additional sensors for online monitoring of axial displacement. The additional sensors are part of an online monitoring improvement for better monitoring during operation.

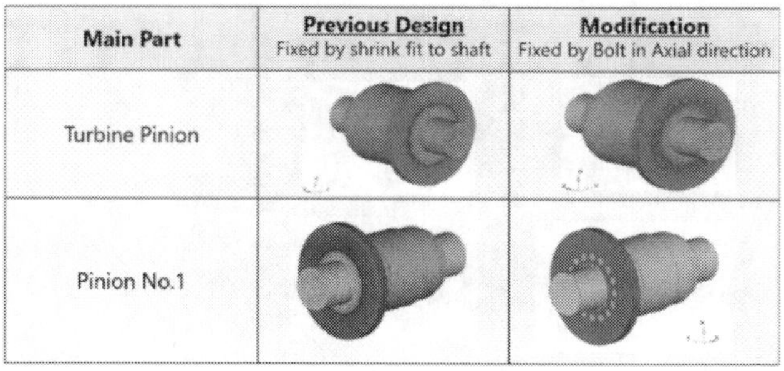

Figure 21. New design of collar connection

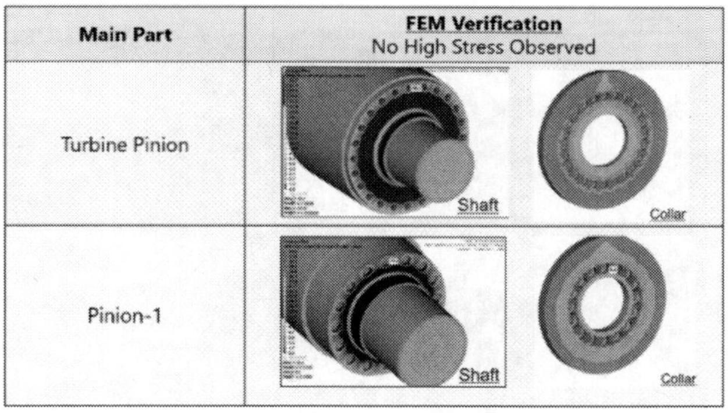

Figure 22. No high stress on BAP operating condition

Safety Factor	Turbine Pinion	No.1 Pinion
Before Improvement	1.02 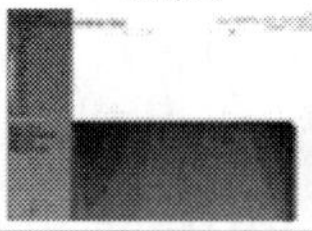	1.49 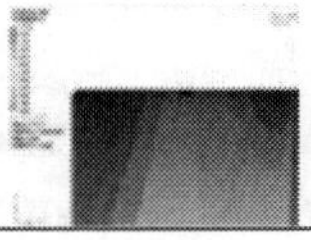
After Improvement	1.90	1.90

Table 3. Calculated safety factor at compressor rated capacity

Figure 23. Axial displacement sensor on Turbine Pinion

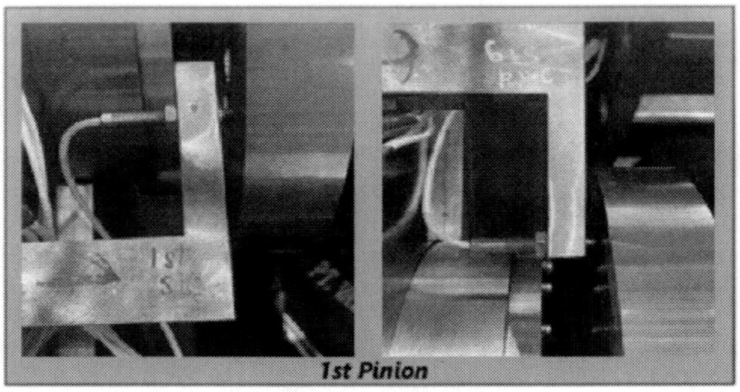

Figure 24. Axial displacement sensor on No.1 Pinion

Figure 25. Axial displacement sensor on No.2 Pinion

The above-mentioned improvement was implemented and installed in February 2023. In May 2024, an internal inspection was performed to check the gap of thrust collar No.1 Pinion after improvement with the result no gap was observed which means that there is no axial movement on thrust collar after improvement (Figure 26).

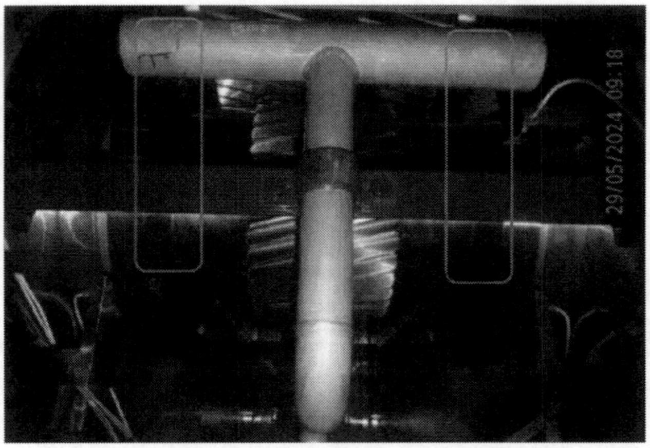

Figure 26. Inspection result (No gap observed on thrust collar)

Conclusion

In 2021, the 101-J integrally geared process air compressor tripped due to high vibration on stage-1 compressor. Inspection results showed that the thrust collar at turbine pinion and No.1 pinion was damaged. A failure investigation was performed by the OEM and concluded that the damage to the thrust collar was due to less gripping force leading to fatigue cracking. Improvements were performed to prevent reoccurrence by improving the installation of the thrust collar to the shaft with bolting connection, and installation of additional axial vibration sensors for online monitoring.

Based on this compressor failure experience, it is important that to have a detailed understanding of equipment components and the functions. Understanding possible failure modes on equipment is also important.

Innovation, as part of continuous improvement, is essential to improve equipment reliability and achieve operation excellence.

Avoiding Fatigue Load of Equipment in Green Ammonia Plants

A typical challenge for design and operation of green ammonia plants is the fluctuating availability of renewable energy. The electrolysis unit for hydrogen generation usually can cope relatively well with these fluctuations so that the hydrogen availability follows these fluctuations. If the ammonia plant follows these ups and downs, the synloop pressure would fluctuate in the same manner. This type of fluctuation would result in the regime described by ASME Sec. VIII Div. 2 paragraph 5.5.2.3 as fatigue load. One option to reduce the pressure fluctuations in the ammonia synthesis loop is to let it operate at nearly constant load by eliminating these fluctuations. This can be achieved by storage of energy (batteries) or hydrogen as an intermediate product. However, this is extremely expensive compared to the option of designing the plant to be able to operate under different loads at constant operating pressure while ramping up and down. This can be achieved through additional control elements like bypasses in the synthesis loop, operated by a master controller optimized by dynamic process simulations. An additional challenge is the resulting temperature fluctuation in the hot parts of the synthesis loop, caused by different production rates. Using FEM (Finite Element Method) analysis, it has been verified that theses stresses do not exceed the maximum permissible stresses. This article shows the results of the developments and examples for pressure and temperature profiles under different production rates.

Bernd Mielke, Klaus Nölker, Felix Florian Bensing
thyssenkrupp Uhde

Green Ammonia Process

In a green ammonia plant, hydrogen is produced by electrolysis of water, using electricity from renewable sources. This hydrogen is combined with nitrogen from an air separation unit to form ammonia (see Fig. 1). The ammonia synthesis loop, including synthesis gas (hydrogen and nitrogen) compression and refrigeration unit is fairly similar to a conventional ammonia synthesis.

The challenge is the adaption of this process to the often fluctuating availability of renewable electricity. While an electrolysis unit can relatively quickly follow the changes in electricity input, this is not the case for the ammonia synthesis. In the case of a reduced supply of hydrogen at its battery limits, a conventional ammonia synthesis loop as shown in Fig. 2, typically operating at a pressure range between 15 and 22 MPa (2,200 to 3,200 psi), would result in a reduction of the operating pressure.

In a conventional ammonia plant (based on hydrogen from steam reforming), this challenge of frequent and / or rapid load changes rarely exists. However, it has been demonstrated that operation at low load (~20% of nameplate) is possible using the control mechanisms present in a conventional plant. However, these turndowns are the result of isolated events that occur occasionally under special circumstances like temporary shortage of natural gas. This type of upset does not represent the type of fluctions that would result in fatigue.

Fluctuating Energy Input

Example for Energy Availability Profile

Availability of solar power follows the natural day and night cycle. This fluctuation can be lowered by combining it with wind power. An example for the annual energy availability at a potential green ammonia site relying on these two sources is shown in Figure 3.

The corresponding ammonia production is also shown, assuming load changes from 40 to 100% are permitted.

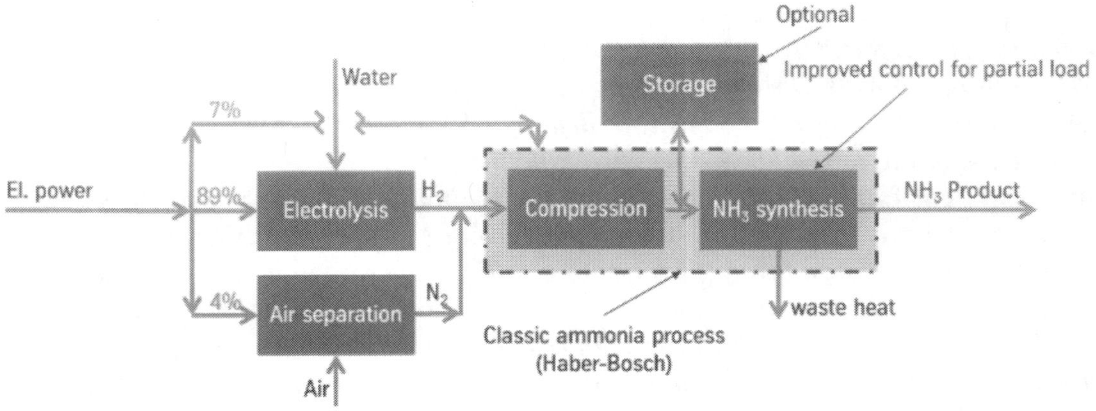

Figure 1: Green ammonia process (simplified).

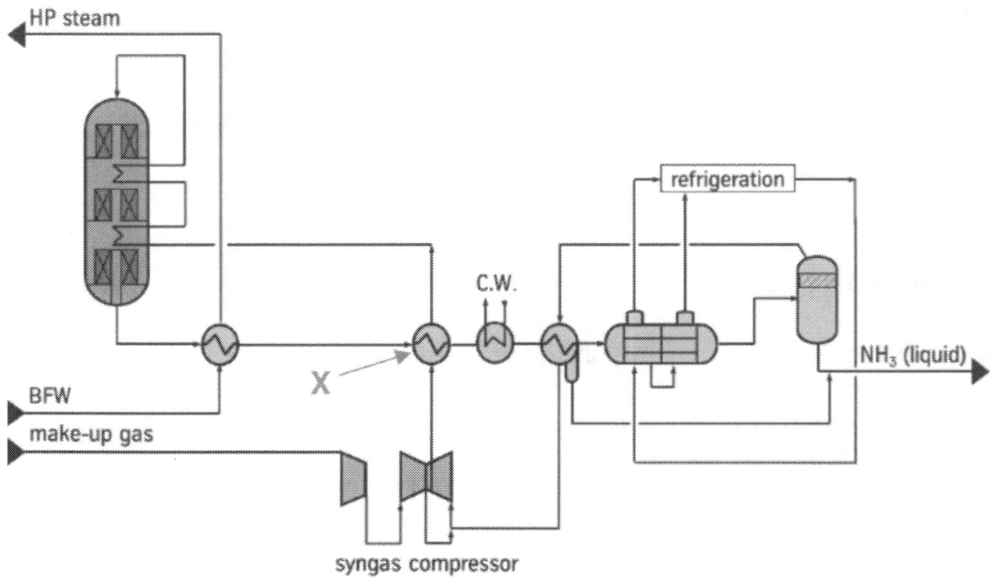

Figure 2: Ammonia synthesis loop suitable for green ammonia (clean feed, no purge stream) (simplified). "X" marks the point investigated for temperature stress in Figure 8.

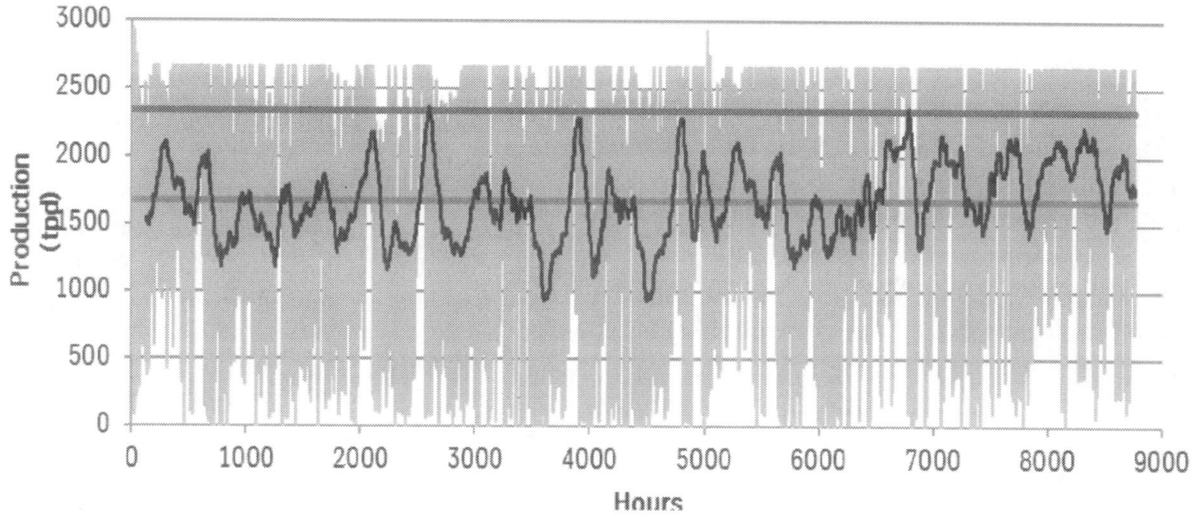

Figure 3: Light grey: Example for the availability of energy over one year, combined from solar and wind power for a potential green ammonia site.
Dark grey: Ammonia production from this energy and using energy storage for reduction of peaks.
Red: Design capacity of the plant. Green: Average output of the plant.

Effect of Fluctuating Energy Input on Ammonia Synthesis

The natural reaction of an ammonia synthesis on reduced availability of hydrogen is as follows:

Generally speaking, less hydrogen supply means less ammonia production. This means, the provided catalyst volume is oversized for the actual production, and reactions come closer to equilibrium. It might happen that there is no more unreacted gas left for reaction in the last catalyst bed. This brings disorder to the temperature profile over the catalyst beds. Due to the higher ammonia concentration and the smaller molar flow after the reactor, the pressure goes down in this case [1].

The effect also depends on the control strategy applied to the nitrogen feed. In a conventional ammonia plant, the air to the secondary reformer would be reduced approximately proportional to the primary reformer load. (This is a requirement because otherwise combustion would result in exceeding permissible temperature in the secondary reformer). In a green ammonia plant, the amount of nitrogen from the air separation is in principle independent from the amount of hydrogen, but is still linked to the hydrogen flow via the H/N ratio of about 3. However, in both cases, less feed would lead to lower synthesis pressure.

In the event of a pressure change, the auxiliary systems around the synthesis would behave as follows:

The synthesis gas compressor, if being a turbocompressor, has to process a certain minimum flow rate in order to avoid surging. Therefore, the automatic anti-surge control opens the anti-surge valve, allowing gas to spill back from the outlet to the inlet in order to maintain the required minimum volumetric flow rate. This ensures safe operation and avoids damage to the machine, but the effect is that shaft power in partial load operation does not go down proportionally to the flow but gets higher per ton of ammonia produced, meaning the efficiency of the system drops.

If one or more piston compressors are installed, there is less efficiency loss at low turndown but this comes at the cost of other known disadvantages of piston compressors, like high maintenance requirements or pulsation of the suction flow. But also here, the automatic load control itself is standard.

The refrigeration unit reprocesses the ammonia vapor formed in the chillers upstream of the ammonia separation. At this point in the process, standard control aims at maintaining a certain ammonia separation temperature (typically in the range of -10 to 0°C (14 to 32 °F) by keeping the pressure on the vapor side of the chiller constant. This works very well and has the effect, that the vapor flow to the refrigeration compressor follows the ammonia production. Also, the compressor has an automatic anti-surge protection, ensuring safe minimum flow, so that no additional control mechanisms are needed to enable low partial load.

Avoiding Risk from Pressure Changes

Risk by Pressure Changes

Pressure vessels in conventional ammonia plants are typically designed for steady conditions, with the exception of a limited number of shut-downs and start-ups. Due to the more fluctuating operating conditions in green ammonia plants, more frequent temperature and pressure cycles must be considered for a proper design of pressure vessels. Failure caused by cyclic services often begins at points of locally high stresses, like welds or sharp edges, or where different materials are combined, and propagates further. Such failures can cause considerable harm as ammonia plants operate at relatively high pressure and temperature and the media are flammable and toxic.

Fatigue failure can be caused if the number of pressure variations is above an acceptable limit. Whether or not a detailed fatigue analysis must be performed, can be determined by an assessment according to ASME BPVC VIII Division 2 [2].

Treatment of Pressure Changes in ASME Code

For the fatigue analysis according to ASME BPVC VIII Division 2 the part of a vessel can be categorized as an integral or non-integral part. Integral parts are typically components that are attached to the pressure vessel through a full penetration weld. Typical examples are shells, welded heads, welded nozzles and flanges.

Non-integral components, such as flanged vessel covers and other flanged connections, can be attached to the vessel by other methods than full penetration welds like bolting or partial welds. These components do not contribute to vessel structural integrity.

ASME BPVC VIII Division 2 offers a method A or method B for fatigue analysis screening. For both methods full range pressure cycles during start-up and shutdown and partial pressure cycles are evaluated based on a load histogram.

For method A all full range pressure cycles and pressure cycles above 20% of the design pressure for integral parts and above 15% of the design pressure for non-integral parts are counted. The limit of 15% is applied to the ammonia synthesis in order to cover all sorts of present components.

For method B a key figure is calculated representing the cumulated effect of cyclic loading, which can lead to the initiation and growth of cracks. The ASME BPVC VIII Division 2 refers to that key figure as "damage". The "damage" is calculated for all full range pressure cycles. Based on the construction of the component, the stress and the stress amplitude, derived from the design fatigue curves of annex 3-F of ASME BPVC VIII Division 2, partial pressure cycles are categorized as significant pressure cycles and counted accordingly. The "damage" associated to the significant pressure cycles is calculated.

Method A is considered to be more conservative and less complex in evaluation and is therefore preferred.

Measures to Maintain Stable Pressure

Fluctuating hydrogen production, caused by the fluctuations in the quantity of renewable energy, poses a risk to the safe operation of the ammonia plant. In addition to the challenge of fatigue load on the synthesis equipment, if the operating parameters of the operating plant are not adjusted to the different load, the reaction in the converter could be snuffed out. In case of unstable reaction temperature and finally loss of reaction, the catalyst can be damaged by thermal cycling.

All these topics have to be addressed during the plant design phase by a dedicated dynamic green ammonia process technology.

The key technology blocks that should be taken into consideration are: power generation from renewables, an optional battery for energy supply in case of curtailment of renewable energy, potential energy top-up from the grid, hydrogen production by electrolysis, hydrogen storage in case of less or no hydrogen production and the ammonia plant itself.

RHAMFS®, a computer program developed by thyssenkrupp Uhde, is the system sizing tool which carries out an optimal sizing of these key technology blocks. It should be used early in the project development phase to optimise the concepts for the lowest levelized cost of ammonia (LCoA).

The varying hydrogen feed flow requires an adjustment of the ammonia production rate. This dynamic load management has been specially developed for green ammonia plants. Dynamic simulations, encompassing the full range of power generation to ammonia production, have been carried out to develop new design features like the so called "Master Controller" [3], [4]. A simplified sketch of the Master Controller is shown in Figure 4.

Energy is provided by renewables such as wind or solar or from a power grid. The electrical energy is used by the electrolysis system to produce hydrogen. The hydrogen is compressed in a hydrogen compressor. The pre-compressed hydrogen is fed to the suction side of the synthesis compressor, but a portion can be diverted via the hydrogen storage compressor to the hydrogen storage vessel. Hydrogen has to be deoxygenated according to the requirement of the synthesis catalyst. The hydrogen is mixed with nitrogen from the air separation unit (ASU), the H_2/N_2 mixture is compressed by the synthesis gas compressor and fed as make-up gas into the synthesis loop. There ammonia is formed in the converter and is condensed by the refrigeration system to obtain the final product ammonia.

The Master Controller needs as input the amount of hydrogen produced by electrolysis, or even better a forecast for the generation of renewable electric power. Although the renewable energy, which can be a mix of solar and wind, is fluctuating, a reliable four-hour forecast is possible. Even if the plant is connected to a power grid, this does not mean that hydrogen production is constant. The price for electrical power can change and adjusting the production rate for arbitrage may be advantageous. It may also be necessary to adjust production in the event of load shedding. The Master Controller actively manages the hydrogen storage. In the event of high hydrogen production, it fills the hydrogen storage vessel; in the event of low hydrogen production, it empties it. Knowing the current amount of hydrogen produced and the amount of hydrogen stored, the Master Controller determines the amount of hydrogen to be fed into the ammonia plant, thereby also determining the ammonia production. The Master Controller also controls the ammonia plant. It adjusts several operating parameters of the synthesis loop to the varying hydrogen feed flow rate. The Master Controller controls the inlet temperature to the ammonia converter and to the catalyst beds and maintains it above set-off temperature of the catalyst so that the reaction is not snuffed out. This is achieved by reducing the steam production so that less heat is extracted from the synthesis loop but remains available for reheating the cold synthesis gas. An electric preheater must be switched on at very low partial loads. The pressure is controlled by reducing or increasing the recycle flowrate through the ammonia converter via a spill-back valve in parallel to the anti-surge valve of the recycle compressor. The loop pressure is also controlled by a bypass around the cooling train of the synthesis loop. Hereby the ammonia concentration at the inlet of the converter can be increased, inhibiting the ammonia formation, so that the amount of hydrogen and nitrogen converted matches the flowrate of hydrogen and nitrogen fed as makeup. By this means the Master Controller and its associated control elements ensure the safe operating envelope as per ASME BPVC VIII Division 2 is not left.

Process Simulation Results

Several dynamic simulations of the synthesis loop has been carried out to study the pressure fluctuations. Some results are shown in Figure 5.

Using a real power profile, the plant production rate (tons per day) has been varied every four hours in the range from 10 to 100% of the plant's design capacity of 675 mtpd (744 stpd) (blue line in Figure 5), while the Master Controller controls the loop pressure (orange line). The set point for the desired loop pressure is at 100%. The diagram shows that the loop pressure does not deviate more than 5% downwards and 6% upwards from its setpoint. This proves that the developed Master Controller concept can cope with fluctuations imposed by the renewable energy.

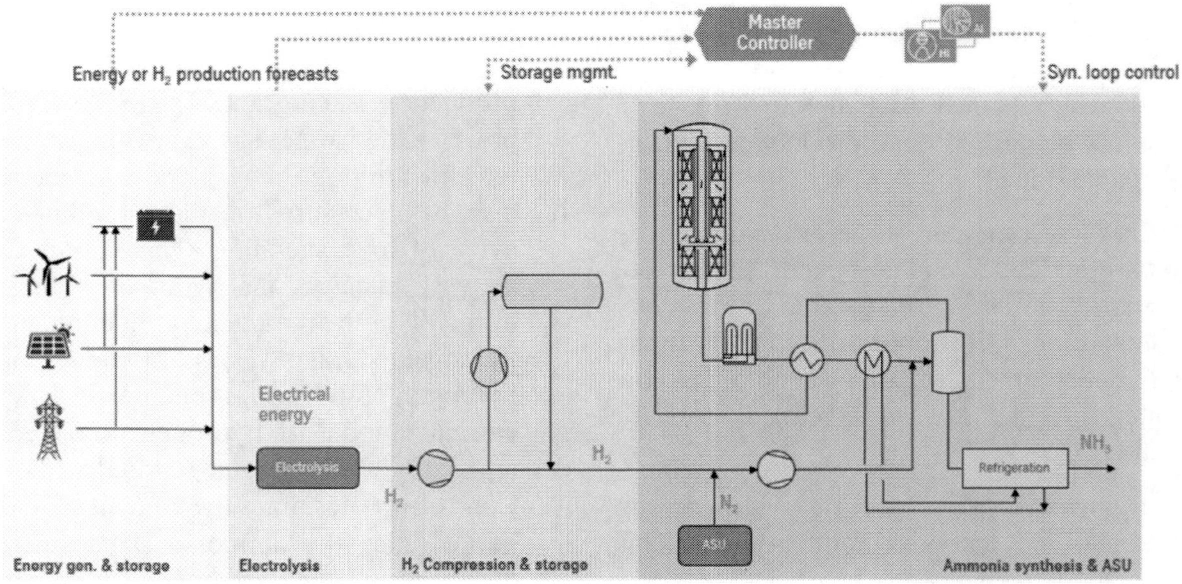

Figure 4: Master Controller for dynamic green ammonia plant load management [4].

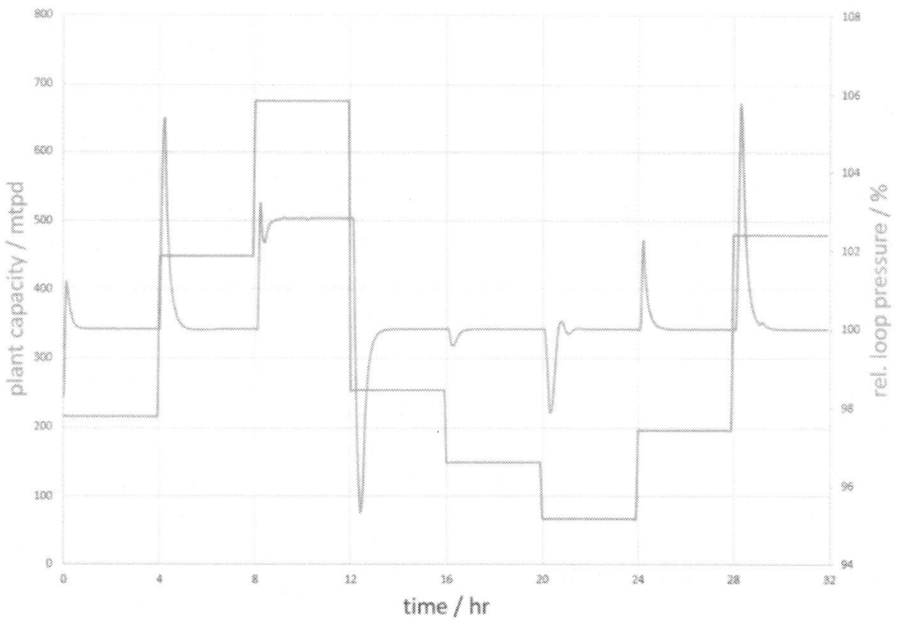

Figure 5: Dynamic simulation of green ammonia plant with variation in ammonia production rate (blue) and controlled synthesis loop pressure (orange), using the Master Controller.

Avoiding Risk from Temperature Gradients

Risk of Temperature Changes and Gradients

Beside pressure cycles, also temperature differences can cause fatigue failure due to increased local stresses. Especially susceptible parts are the components with high differential temperatures on opposite surfaces within two adjacent points on the surface. ASME Code [2] defines a limit for adjacent points within which certain temperature gradients must be considered as contributing to fatigue loading. This can be for example a radius on a surface or the distance through a component's wall.

Especially for thick-walled and non-insulated components transient effects must be considered, as the heating up and cooling down of the material by the varying operation conditions is time-delayed for the surfaces in contact with the media and the other surfaces.

Furthermore, welded joints between materials with different coefficients of thermal expansion can be susceptible to fatigue loading.

Treatment of Temperature Gradients in ASME Code

Following ASME BPVC VIII Division 2, cyclic temperature differences shall be considered based on a load histogram. For both method A and method B, temperature differences between two adjacent points and through the wall must be considered.

Method A defines a multiplier based on the range for each individual cycle. As an example, the multiplier for temperature differences of 28°C (50°F) or less is zero and the related cycles are not counted.

Following method B, the "damage" for all temperature differences during start-up and shut down must be calculated. Based on the construction of the component, the elastic modulus and the stress amplitude, derived from the design fatigue curves of annex 3-F of ASME BPVC VIII Division 2, temperature cycles are categorized as significant temperature cycles and counted accordingly. The "damage" associated with the significant temperature cycles will be calculated.

Method A can be utilized for a more conservative fatigue analysis. The more complex method B can be used to receive more accurate results.

Furthermore, for both methods, temperature cycles for welds between materials having different coefficients of thermal expansion must be counted. In method A there are specific limits beyond which cycles must be counted to consider different coefficients of thermal expansion. The same applies to method B to calculate the "damage".

Measures for Stress Reduction by Design

A detailed fatigue analysis shall be performed if:

- Either the total number of countable cycles is higher than the limit according to ASME BPVC VIII Division 2 method A
- Or the cumulated damage as per method B is greater than 1.

To reduce the effect of fatigue loading some general design details can be considered. Some examples include the use of integral constructions or avoiding fillet welds for attachments, reinforcement pads, threaded connections, partial penetration welds, stud bolt connections and nozzles in knuckle regions of heads [5].

Furthermore, the design must focus on reducing local stresses by abrupt changes in thickness, discontinuities or sharp edges. Points of locally higher peak stresses must be identified.

Further estimation of the life span of components under fatigue loading can be done by FEM analysis.

FEM Simulation of Stress by Temperature Gradient in Nozzle at Load Changes

As an example, the influence by cyclic operating conditions on the hot inlet nozzle of the Gas/Gas Heat Exchanger (marked "X" in Fig. 2) has been investigated. This nozzle is an integral part located in a hemispherical head on the tube side of the heat exchanger. The aim was to check the temperature gradient of the nozzle to identify significant temperature cycles during the typical operating conditions.*Figure* Figure 6 shows typical operating temperatures for this nozzle in the load scenario described by Figure 5 in a green ammonia plant.

Per this 24 hour interval, a total of 4 cycles of a critical temperature difference have been identified. If this load pattern repeats daily over the expected lifetime of 20 years, this will result in approximately 29,000 cycles, which is far above the limits of ASME BPVC VIII Division 2. Therefore, a detailed fatigue analysis is required.

To minimize the effort for fatigue analysis of the components, the transient effect of the changing conditions was analyzed by FEM. Figure 7 shows the temperature on the inside surface and the maximum temperature of the nozzle wall. Considering the transient temperature effects, the temperature difference through the nozzle wall is at maximum 27°C (49°F). Hence, due to the enhanced analysis of the temperature gradients, a detailed fatigue analysis can be omitted in this case.

Figure 8 shows the nozzle set up analyzed by FEM at the point just before the ramp up at 12.2 hours of operation.

The location at the gas / gas exchanger hot inlet nozzle can be seen as a critical part in the ammonia synthesis loop as it is directly affected by rapid temperature changes due to the direct contact to the flow. This leads to relatively fast increasing temperature on the inside surface. Other components may be heating up or cooling down slower due to dead zones or lower thermal conductivity.

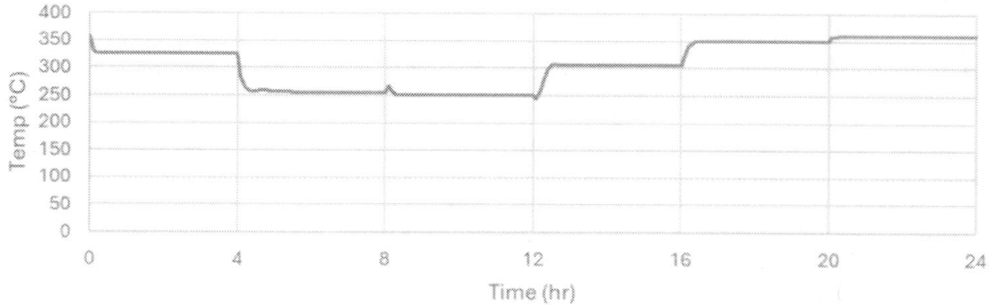

Figure 6: Process temperature of hot gas inlet nozzle of gas/gas heat exchanger marked „X“ in Figure 2.

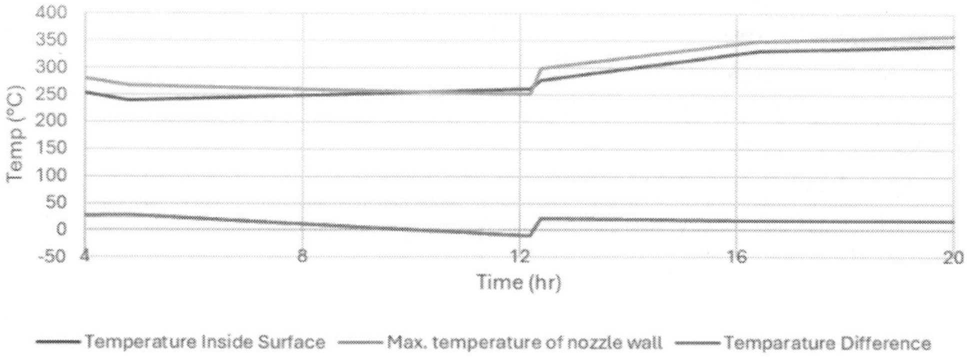

Figure 7: Temperatures at hot gas inlet nozzle of gas/gas heat exchanger at critical time interval.

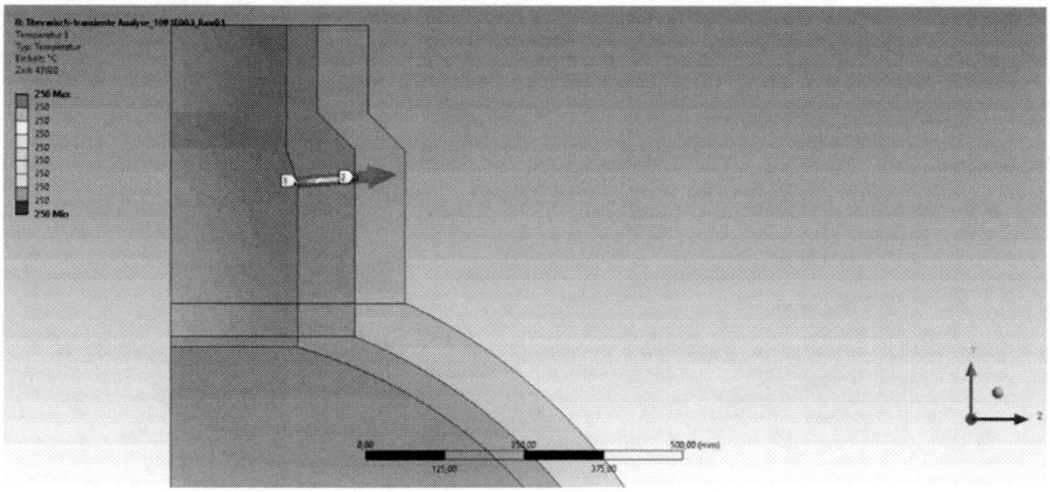

Figure 8: Hot gas inlet nozzle of gas/gas heat exchanger at 12.2 hours operation.

Summary

In a green ammonia plant, load changes are expected. Technology has been developed which enables a low turndown of 10% and a ramp rate of 2% per minute. Therefore, the plant has high flexibility to respond to changes in hydrogen supply so that the hydrogen storage can be minimal and ammonia production follows the availability of renewable energy. The Master Controller ensures the temperatures and pressures in the synthesis loop remain within an acceptable range defined by ASME code, even if the hydrogen supply fluctuates. Thus, loss of containment due to pressure cycling fatigue does not occur, and catalyst and equipment is not damaged by thermal cycling.

References

[1] European Patent EP 3 426 601 B1, Method and System for producing a product gas under changing load conditions; 2017

[2] ASME BPVC VIII Division 2, The American Society of Mechanical Engineers, 2023

[3] Patent application WO 2023/099743 A1; Method for operating an ammonia plant, and plant for producing ammonia

[4] B. Keil, C. Renk, B. Mielke, K. Bagga; Beyond the Trend: Challenges in Green Ammonia Economy; Nitrogen & Syngas Conference 2025, Barcelona, 2025

[5] M. Basic, D.R. Moss; Pressure Vessel Design Manual; Elsevier, 2013

Ammonia Converter Basket and its Catalyst Replacement Job at CFCL, Gadepan (Kota)

Chambal Fertilizers and Chemicals Limited (CFCL) Kota, India, operates three ammonia plants. The Ammonia-II plant is based on KBR's Technology and was commissioned in October 1999. In more than 20 years of plant operation above the name plate capacity, the ammonia converter basket and catalyst were not even replaced once. As basket has approached end-of-life, CFCL has decided to replace the ammonia converter basket and catalyst for reliability and process performance improvement. This article gives an insight into the installation of an ammonia converter basket designed with state-of-the-art and cutting-edge technology. It covers aspects of design innovation, equipment reliability, overall energy savings in a syn gas loop system, site safety, and planning and preparation in anticipation of unforeseen problems during site job execution.

The Ammonia Converter is the heart of ammonia plants, and ensuring its reliability is of the utmost importance. CFCL's careful planning prior to the commissioning ensured a remarkably smooth and safe installation of the new basket. Safe and smooth Implementation of such a large-scale project within the short time available during plant turnaround has always been a challenge.

KBR supplied the ammonia converter basket with all the latest upgrades as Proprietary Equipment. With the basket upgrade, the ammonia-II plant is able to achieve higher sustainable capacity.

Mr. Amit Goyal,
CFCL, Kota, India

Akhil Nahar,
KBR, Houston, USA

Vishvajeet Joshi,
KBR, Pune, India

Introduction

Chambal Fertilizers and Chemicals Limited is a Nitrogenous Fertilizer complex for production of Urea at village Gadepan, 35kms from Kota (Rajasthan). It has three plants viz. Gadepan-I, Gadepan-II and Gadepan-III, each with correspondingly numbered Ammonia and Urea plants. Each Urea plant has 2 units. The Gadepan-II Ammonia plant is based on KBR (USA) technology and Urea-II plant is based on ACES process of TEC, Japan. The Ammonia-2 or Gadepan-II plant was commissioned in October 1999. The name plate capacity of Ammonia-2 plant was 1350 MTPD and pre-upgrade capacity was 1710 MTPD.

In the Gadepan-II plant, high-pressure synthesis gas at about 163 Kg/cm2g (2,318 psig) from the synthesis gas compressor is fed to the ammonia synthesis converter (ADC-701), passing through its internal basket, for converting synthesis gas to Ammonia. The converter shell and the internal basket are designed by KBR. The converter pressure shell is multilayer (Total thickness 107mm) and horizontal in construction. The existing basket has two catalyst beds - bed 1, and bed 2, with bed 2 containing two compartments (bed 2A & bed 2B), all operating in series. One feed/effluent interchanger is installed in combination with bed-1.

CFCL planned to replace the existing converter basket with new upgraded basket in financial year 2023-24 considering the reliability of equipment as it was in operation for more than 24 years. Although there were no significant deviations in process parameters, the complete catalyst was also replaced with fresh catalyst during the basket replacement activities.

Key stakeholders:
Plant owner: CFCL,
Licensor: KBR,
Converter Manufacturer: L&T
Basket manufacturer: ISGEC
Site job contractor: Plant-Tech
Converter inspection: L&T

Key improvements in the new basket design

The Two-Bed Ammonia Converter design was introduced in the early 1970s with an installation in Japan, followed by another installation in Canada in the early 1980s. Here, KBR used the first Horizontal Converter with a two-bed, one-interchanger configuration. This configuration has been used in several Horizontal Converters installed worldwide and has proven to be extremely successful and effective.

The temperature of the pressure shell is maintained at a lower temperature than the reaction temperature, by passing a cold feed gas stream through the annulus space between the pressure shell and basket shell using the main sweep feed stream.

The cold-wall converter design includes an internal stainless-steel basket where reaction takes place over a catalyst. The main feed gas stream is heated in the interchanger before going to Bed-1. Effluent from Bed-1 is cooled in the interchanger and passed through Bed-2A and Bed-2B in series. Preheated feed gas is passed through beds utilizing a proven robust support system holding highly active catalyst for maximum conversion to produce ammonia most efficiently. All the feed gas flow is passed through the entire catalyst volume without using any quench. The thermodynamic bed 1 intercooler system thus enables maximum possible conversion at a minimum synthesis loop operating pressure.

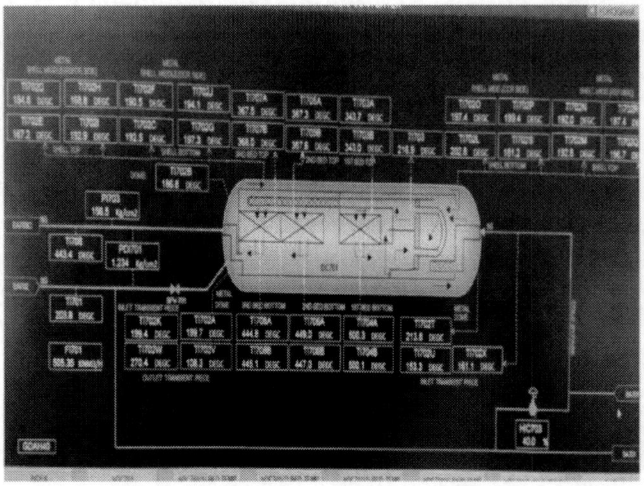

Figure 1 DCS display of ammonia converter with parallel operation of Bed 2A and Bed 2B

Over time, KBR has made significant developments in the design of the converter basket to enhance reliability and operational efficiency, as mentioned below.

A. Parallel configuration for Bed 2A & 2B

In all prior Converter designs, Beds 2A and 2B were arranged in series, i.e., all gas flows first through Bed 2A and then through Bed 2B. In the modified configuration, Beds 2A and 2B are arranged in parallel, i.e., half of the gas passes through each Beds simultaneously. In this configuration, the overall pressure drop across the Converter is reduced (Refer Figure 2).

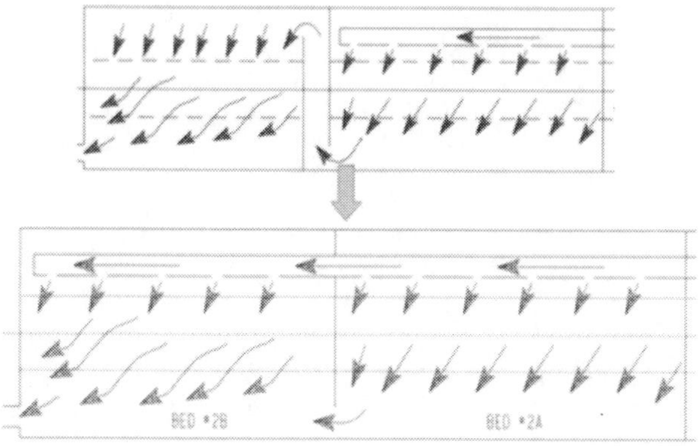

Figure 2 Bed 2A & 2B in series (top) vs parallel (bottom) arrangement

The estimated pressure drop savings due to the change in bed configuration from series to parallel is 0.2 - 0.3 kg/cm^2 (3-4 psi). Reduced differential pressure across the converter will help reduce power consumption or increase the synthesis loop capacity.

The approximate pressure drop in the Normal Case (SOR-1750MTPD), including the pressure drop in the interchanger, estimated in the new design basket, is 1.3 kg/cm^2 (18 psi).

B. Upgrade of Catalyst Support

KBR upgraded the Catalyst Bed support arrangement to the latest design, which is a floating screen design. The design provides more robustness especially in transient conditions like start-up/shutdown. The new support design results in marginal movement of the screen, unlike the old design where the screen panels were needed to move up to ~ 20mm for hot conditions.

Construction of new basket in brief:
Dimensions: Ø2548 x L28135 mm
Material of construction: SS304
Minimum metal design temperature: -12°C
Fabricated weight: 113 MT
Weight with catalyst: 396 MT
Total Catalyst volume: 85.8 m^3
Basket pneumatic leak test pressure: 1 kg/cm^2 g

Planning and scheduling of the job

The complete job execution was challenging for CFCL as the converter basket was opened for the first time in 25 years since plant commissioning in 1999. Moreover, there were no past records of such work experience by the site contractor and CFCL for horizontal converter baskets. Meticulous planning, including identifying each involved activity and its associated risks, was crucial for successfully commissioning the new basket. With multiple roundtable discussions and site visits of equipment licenser and site contractors, a schedule was prepared to complete the mechanical job in 28 days and production to production within 36.5 days. The bar chart is attached in Table 1 for reference. Along with the progress of work, CFCL, KBR and Plant-Tech made various optimizations to complete the mechanical work much before the scheduled time. By doing some of the activities in parallel and collaborative planning, the job was completed in **23 days**.

Project execution sequence in brief:

1. Prefabrication work: The work included inspection and readiness of the trolley and puller arrangement, ground area levelling, assembly of two crawler cranes (LR1400 and CKE-2500), blinds fabrication with nitrogen tapping, inspection of new basket along with supplied loose items, nitrogen header laying and winch machine installation.

2. Converter Isolation for manpower entry into the converter: After converter cooling and depressurization, nitrogen purging was done until the hydrogen concentration reached an acceptable limit. All three external piping: gas inlet to converter annular space, cold shot piping and gas outlet piping, connected to the converter, were positively isolated by temporary blinding. For outlet piping, the 60mm thick (SS304) pipe was cut by a cold cutting machine and a blind was welded to the pipe with a blind & sleeve arrangement.

Figure 3 Cold cutting of 24" outlet pipe cutting

3. Converter basket isolation: The manpower entered the converter in a nitrogen atmosphere wearing breathing apparatus and taking all safety precautions after opening the manholes located at both ends of the converter. The Basket was positively isolated from both cold shot piping and outlet piping by providing welded blinding. A 1" nitrogen hose was also connected to the 1" Greyloc connection in the basket to maintain the inert atmosphere over the pyrophoric catalyst beds.

4. Removal of AT/BT thermowell: After the instrument cables were removed from the thermocouples junction box, the thermowell was pulled out with the help of a moving trolley arrangement and cranes as shown in Figure 4.

Figure 4: Removal of Thermowells

5. Dome cover removal: The moving trolley specifically designed to remove the dome cover and basket was positioned below the dome cover flange and aligned with the help of four 70 MT hydraulic jacks to ensure complete contact of saddle plate to flange outer diameter. The Hydraulic tensioners (supplier Tentec) were utilized to remove the 32 studs of size M160 X6P. The dome cover was pulled back with the help of fixed winch machine as shown in Figure 5.

Figure 5 Removal of Dome cover using trolley and winch machine

6. Basket removal: Before basket removal, a gasket and stud protection ring and flange extension structure were placed at the shell

flange end. This activity includes attaching the trolley to the basket at end plate 1 side and providing installation wheels at end plate 2 side with the help of hydraulic jacks (see Figure 6). The provision of installation wheels in a nitrogen atmosphere was strenuous and required extreme safety precautions.

Figure 6: Removal of basket on installation wheel

7. Catalyst removal: The catalyst removal activity was a very specialized job as it had to be done under a Nitrogen atmosphere. The deputed contractor Plant-Tech mobilized complete breathing apparatus which was connected to bank of oxygen cylinders placed outside the converter shell.
   Meanwhile, the converter was handed over for internal inspection under normal atmospheric conditions in parallel to catalyst unloading activities within 5.5 days.

8. Replacement of old basket with new basket: The empty old basket was lifted by two crawler cranes, 400MT and 250MT, using nylon belts of the requisite capacity, as per the approved rigging plans.

Figure 7: Lifting of new basket

9. Basket insertion trial and internal piping work: For this activity, the empty new basket was inserted into the converter shell (see figure 8), and the dome cover was tightened with 50% studs at 100% tensioning load. After that, both side cold shot and gas outlet piping alignment and welding work were completed, followed by NDT. The Basket was again removed from the converter shell for catalyst loading.

Figure 8: Insertion of empty basket in shell

10. Catalyst loading: As per KBR procedure, the catalyst in beds 2B and 2A was loaded under normal atmosphere and the catalyst in bed 1 was loaded under nitrogen atmosphere. The details of the catalyst. loaded in bed 2B, 2A and 1 are as below table-2.

Bed No.	1	2A & 2B
Brand	KM-1R	KM-1
Particle Size, mm	1.5~3/3~6	1.5~3/3~6
Volume, m3	25.58/3.18	51.16/6.36
Quantity, kg	68961	165438

Table: 2 Catalyst details bed wise

The total quantity of catalyst is 86.28 m3.

Figure 9 Catalyst loading in basket

11. Basket insertion and stop pin insertion: After catalyst loading of basket and welding of basket shell manhole, the basket final inspection was carried out for external surface cleaning. The basket was then pushed into converter shell by trolley and winch machine arrangement. The alignment of basket was done at the Greyloc flange faces and by measuring circumferential gap between basket outer diameter and converter shell inside diameter. The lifting jack in rear end was utilized for removal of installation wheel. In parallel, the trolley jacks were lowered and barrel removal done. The final gap measurements were taken after insertion of basket stop pin in brackets.

12. Dome cover and manhole fixing: The Dome cover with assembled new double conical gasket with graphite was mounted on the trolley and pulled towards the basket with the winch machine. With proper alignment, the dome cover was tightened in a multiple pressure cycles with hydraulic tensioners. The final pressure cycle was carried out at 7500PSI to maintain the flange gap 17.5mm. Meanwhile the dome cover was boxed up and the thermowells were inserted into the basket with the trolley and crane arrangement. A fabricated spreader beam was utilized to prevent undesirable sagging in thermowell pipes.

    Similarly, the manhole at both ends of the converter were tightened up after final cleaning inspection up to pressure cycle 14000PSI with desired gap of 54mm.

13. External piping normalization: While the dome cover is tightened , the converter outlet piping was de-blinded and rewelded. Besides this, the cold shot flange and annular gas inlet pipe flange blind was removed and boxed up. All piping supports were also normalized as per drawing.

    Mechanical handover and commissioning: Before the mechanical handover, the site was cleaned, and unnecessary items were removed to ensure smooth movement of manpower during plant start-up activities. The insulation on the dome and piping was restored.

Commissioning

The commissioning of the converter started with N2 purging pressure up to 12 KG/cm2 (170 psig), followed by Syn Gas charging up to 120kg/cm2 (1700 psig) for leak checking. As no leak was observed in the Syn loop system, the pressure was reduced to 85kg/cm2g (1200 psig) and the start-up heater was taken in line for activation of the catalyst in bed-1. With controlled heating and regulating the gas feed into the converter, the reduction of the catalyst in the remaining two beds was completed. It took almost three days for the complete catalyst reduction of all three beds and on-stream production was declared after achieving 96.72% purity of Ammonia.

Process Performance

With the new design basket and catalyst, the syn loop pressure was reduced by 7-10 kg/cm2 (100-140 psi) , and the pressure difference across the basket was reduced from 1.7 kg/cm2 to 1.3 kg/cm2 (24 to 18 psi). This reduction in loop pressure and converter differential pressure resulted in an overall energy savings in the syn loop system for the same plant load, 1740 MTPD. Moreover, CFCL can now increase throughput to 1820 MTPD ammonia production.

Converter inspection & repair work

Immediately after pulling the catalyst basket from the converter shell, the converter returned to normal atmospheric working condition. Catalyst removal activities were going to happen outside the converter shell. Therefore, CFCL took advantage of this duration and planned to inspect the converter shell in parallel to catalyst unloading activity within 5.5 days. In that way, the overall schedule of the site job was reduced.

Material of Construction (MOC) of the shell for the outer two layers is SA 533 GR B CL2, inner shell MOC is SA 302 GR A and dished end is SA 302 GR B. To check the integrity of the equipment, Ammonia Converter A-DC-701-V was offered for fitness for service assessment as per API 579-1 / ASME FFS-1 to TCR Advanced.

The process /design parameters are shown in Table 3 and general arrangement drawing is shown in Figure 10.

Parameter	Unit	Value
Code	-	AD MERKBLATTER, 1995 EDITION
Design pressure (Int.)	kgf/cm^2g	188
Design temperature	°C	290 Except nozzle B & C
Hydro test pressure	Kgf/cm^2g	244
Working pressure (Int.)	kgf/cm^2g	177.38
Working temperature	°C	231
Corrosion allowance	mm	1.5

Table 3: Design parameters of the Ammonia converter shell

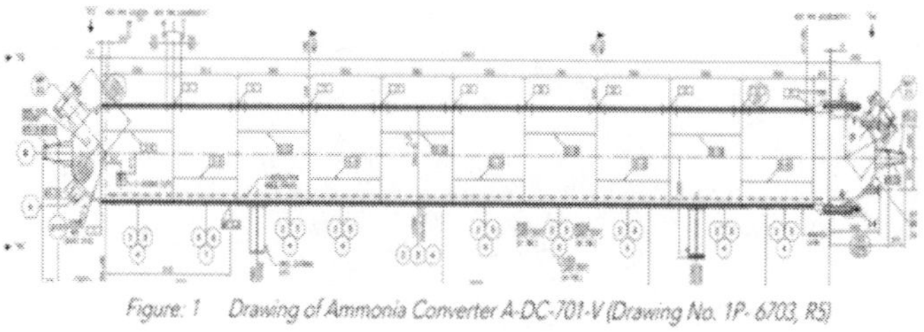

Figure 10: General arrangement of Ammonia converter shell

L&T was awarded the job for fitness for service assessment as per API 579-1 / ASME FFS.
For the FFS assessment, the following inspection was carried out by L&T.

1. Visual examination, thickness measurement
2. Dimensional measurements of different components,
3. Positive material identification (PMI),
4. Dye Penetrant Testing (DPT),
5. Wet fluorescent Magnetic Particle Testing (WFMPT),
6. Ultrasonic Flaw Detection (UFD),
7. Phased Array Ultrasonic Testing (PAUT) of all accessible weld joints to detect subsurface flaws,
8. Inspection of dished ends for HTHA detection,
9. In-situ microstructural examinations (replication metallography)
10. Hardness measurements.

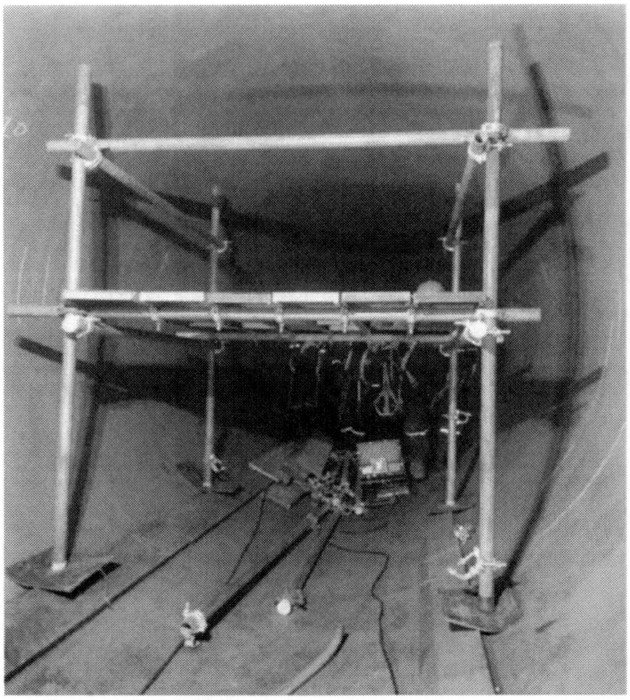

Figure 11 Scaffolding provision for internal inspection of the converter

Using wet Fluorescent Magnetic Particle Inspection, minor linear indications were identified on the inner surface of the shell. These indications were removed by smooth grinding with proper contouring and dehydrogenation treatment.

SR. NO	NDT METHOD	LOCATION	TYPE OF DEFECT	DESCRIPTION/REMARKS
1	WFMPT	L SEAM 38 930mm from C seam 37 towards 270°	1 Linear Indication in HAZ parallel to weld joint.	Length: 18 mm approx. Depth: 3 to 4 mm approx.
2	WFMPT	C SEAM 37 1000mm from L seam 38 toward 0°	1 Linear Indication in HAZ perpendicular to weld joint.	Length: 45 mm approx. Depth: 3 to 4 mm approx.
3	WFMPT	C SEAM 21 700 mm towards 270° from sliding rail	2 Linear Indication in HAZ perpendicular to weld joint.	Length: 25 mm approx. Depth: 3 to 4 mm approx.
4	WFMPT	L SEAM 18 260mm from C seam 21 towards 0°	3 Linear Indication in HAZ parallel to weld joint.	Length: 25 and 15 mm approx. Depth: 3 to 4 mm approx.

Table 4: Locations of defect indication in pressure shell

Figure 12 Converter inspection by MPI

Safety at site

Safety was given the utmost importance during the job execution. Some of the safety highlights are mentioned below:

a) Safety briefing marked the beginning of the job by the work permittee in all shifts.
b) State-of-the-art fire protection equipment was installed to avert any untoward fire incident.
c) A key safety feature was the use of an online camera to record video inside the vessel during catalyst unloading activity in a nitrogen atmosphere.
d) In case of a sudden increase in catalyst bed temperatures, an arrangement for monitoring bed temperatures was provided for quick evacuation of technicians working inside the inert atmosphere.
e) An ambulance was always present at the site during the execution of project.

Conclusion

The successful replacement of the ammonia converter basket with the catalyst reflects CFCL's commitment to operational reliability and efficiency. Executing a project of this magnitude within the limited time available during the plant turnaround poses significant challenges for planning and execution, especially in the absence of previous references and records. The project underscores the importance of meticulous planning, teamwork, and quick problem resolution, enabling completion ahead of schedule. The basket replacement has helped the Gadepan-II plant in increasing the plant capacity and improving energy efficiency. Such endeavours are infrequent and critical for the longevity of fertilizer plants, marking significant milestones in their operational life.

Split Flow Reactor - Optimized Catalyst Converters for High-Capacity Ammonia Plants

The increasing global demand for ammonia, driven by its critical role in agriculture and emerging low-carbon applications, has necessitated the development of scalable and efficient production technologies. The continuous upscaling of ammonia production facilities requires the development of innovative engineering solutions to address the limitations of traditional reactor designs. Traditional ammonia production facilities, with capacities ranging up to 3,000 metric tons per day (MTPD), comprise of conventional reactor designs, which rely on single catalyst beds, face numerous challenges using the same design approach for mega scale plants exceeding 3,000 MTPD. These challenges hinder the scalability, cost efficiency, and environmental compliance of high-capacity ammonia plants, necessitating innovative reactor designs to address these limitations. First, the reactor diameters increase significantly, complicating manufacturing, transportation, and site installation leading to logistical issues. Then, the larger catalyst beds result in increased pressure drops. Third, the expanded reactor sizes require larger plot areas, which may not be feasible in existing facilities and results in space constraints and finally, there are material challenges resulting from increased shell thickness for larger reactors creating difficulties in material procurement and fabrication. This paper introduces the Split Flow Reactor, an innovative solution developed by KBR to address these challenges in catalyst converters, including High Temperature Shift (HTS) converters, Low Temperature Shift (LTS) converters, and Methanators. The Split Flow Reactor divides the catalyst bed into two parallel sections within a single reactor shell, achieving several benefits.

Umesh Kudav, Partha Pratim Chowdhury, Sudesh Kumar, Shailesh Mudaliar
KBR

Introduction

The global ammonia market is set to grow threefold in the coming decades, with almost all growth driven by low-carbon ammonia. Clean ammonia produced from natural gas (blue) is expected to dominate the long-term supply, as more projects are nearing investment decisions compared to green ammonia. In addition, existing operating units are working towards enhanced efficiency and greater yield. The continuously growing demand is driving the need for scalable and efficient production technologies suitable for construction of mega scale ammonia plants.

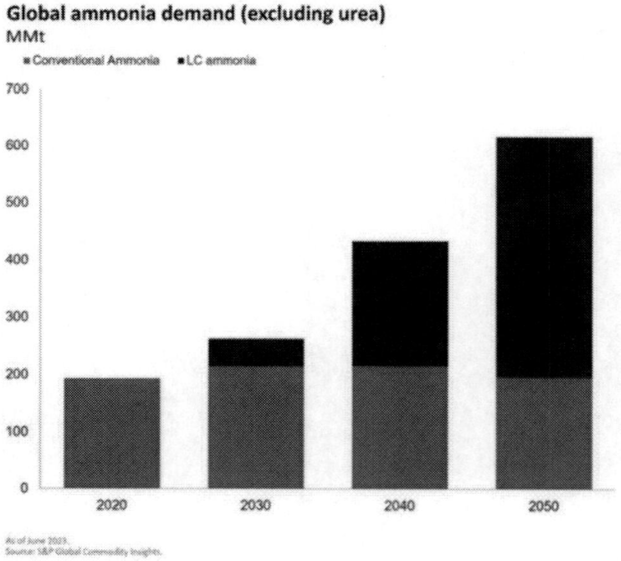

Figure 1. Global ammonia demand forecast [5]

Globally, ammonia production relies on Steam Methane Reforming (SMR), a process where natural gas is used to produce hydrogen, which is then combined with nitrogen to form ammonia. Natural gas (methane) reacts with steam to produce hydrogen and carbon monoxide. Hydrogen is then combined with nitrogen from the air in the presence of a catalyst to form ammonia. Carbon monoxide produced in steam methane reforming is a catalyst poison for further process steps as well as a significant environmental pollutant. Hence, its removal is critical to the ammonia production process. This is accomplished through a series of reactions including high-temperature and low-temperature shift conversion followed by methanation. In conventional ammonia plants these reactions occur in pressure vessels housing a fixed catalyst bed downflow reactor. Scaling of the units for larger capacities to meet the rising global demand involves addressing significant engineering challenges from the reactor design perspective. If conventional design approaches are applied to reactors in mega-scale ammonia plants, this will result in significantly larger diameters, necessitating thicker reactor walls and complicating fabrication. Deeper catalyst beds will further increase pressure drop across the reactor. Additionally, the larger equipment size poses logistical difficulties, increases on-site construction requirements, and may adversely affect fabrication quality. The increased footprint due to larger diameter also makes a case for more sustainable and cost competitive design.

The challenges mentioned above can be addressed through an innovative Split Flow Reactor design, while enhancing operational efficiency and maintaining high conversion rates.

Conventional Carbon Monoxide Removal Process

Shift Conversion: Water-gas shift reaction plays a key role in SMR effluent treatment converting carbon monoxide into carbon dioxide taking oxygen from water and in the process generating additional hydrogen. High temperature shift reactors operate at 310°C-450°C (590°F-842°F) to accelerate reaction rates, while low temperature shift reactors, operating at 200°C-250°C (392°F-482°F), maximize equilibrium conversion. Hence, the high temperature shift converter is followed by the low temperature shift converter. These reactions take place in conventional single bed down flow reactors, which increase in diameter and length as capacity and flow rates increase.

$$CO + H_2O \rightarrow CO_2 + H_2 + \text{Heat}$$
(Equation 1)

Methanation Reaction: Residual carbon monoxide and carbon dioxide post shift conversion are further converted to methane through methanation reactions after the CO_2 removal section.

$$CO + 3H_2 \rightarrow CH_4 + H_2O + \text{Heat}$$
(Equation 2)

$$CO_2 + 4H_2 \rightarrow CH_4 + 2H_2O + \text{Heat}$$
(Equation 3)

These exothermic reactions favor low temperatures for equilibrium and high temperatures for kinetics, demanding precise operational control.

Figure 2 is a snapshot of a typical KBR syngas process highlighting shift conversion and methanation sections.

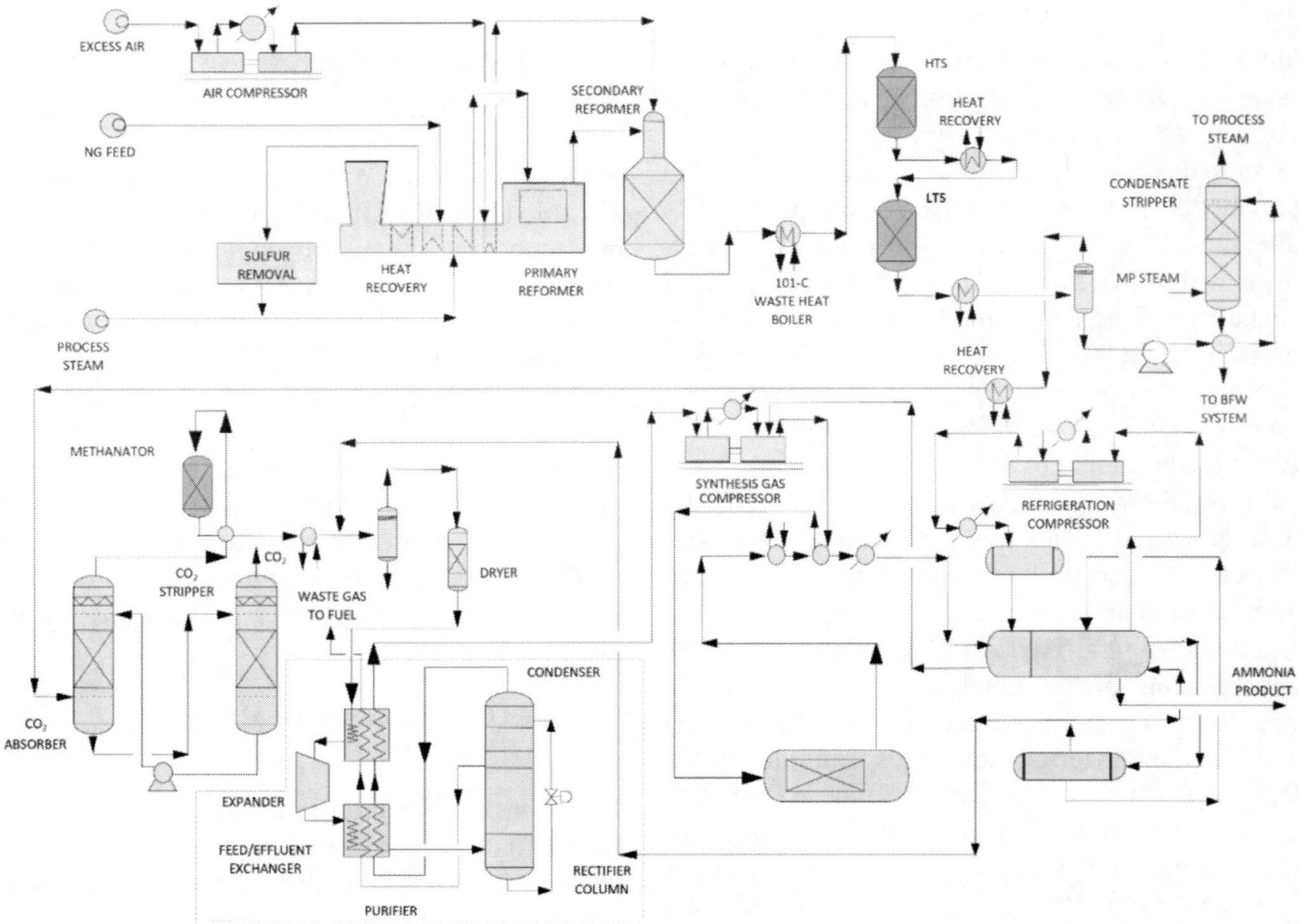

Figure 2. Typical KBR ammonia process scheme

Conventional Reactor Design

Traditionally, high temperature and low temperature shift converters and methanators are designed as single catalyst bed, down flow vertical reactors. (Figure 3)

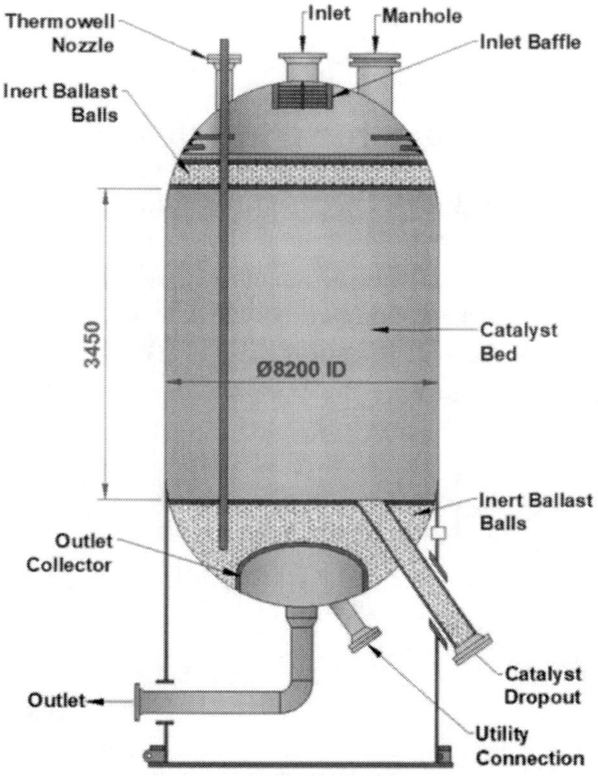

Figure 3. Conventional reactor schematic diagram

A comparison of the conventional down flow high-temperature shift reactor design for typical plant capacities from 1,200 MTPD through 6,000 MTPD and challenges involved are presented below in Figures 4, 5 & 6.

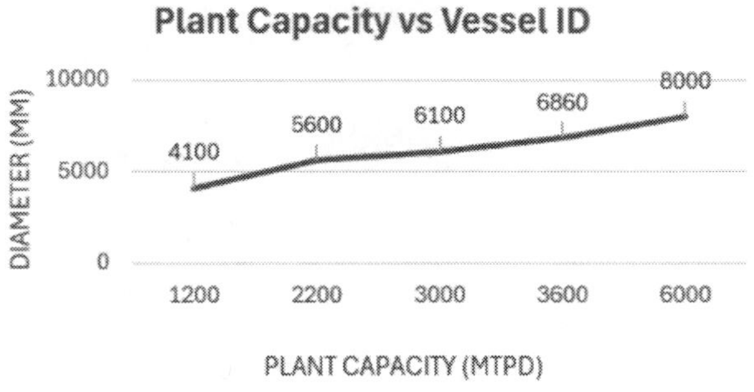

Figure 4. Ammonia plant capacity vs equipment diameter comparison

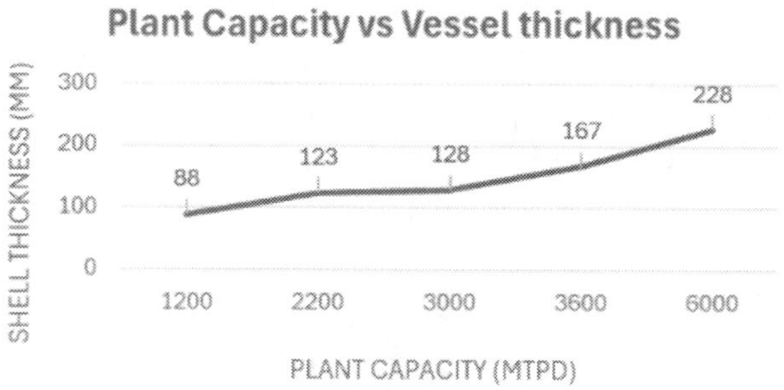

Figure 5. Ammonia plant capacity vs equipment shell thickness comparison

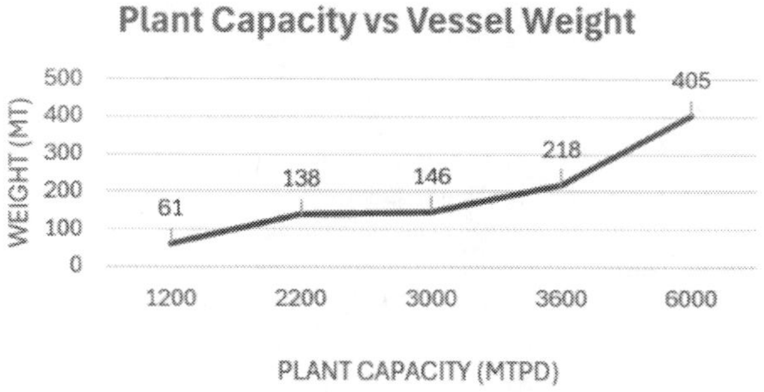

Figure 6. Ammonia plant capacity vs equipment weight comparison

For single-train large ammonia plants (~6,000 MTPD), HTS converter shell diameters are close to 8,000 mm, resulting in proportionally thicker shells. This poses challenges in material procurement, fabrication, and significantly raises costs. Oversized dimensions complicate road transport, often necessitating site fabrication, which increases complexity, schedule risk, and may compromise quality. Larger vessels also demand more plot space.

The Split Flow Reactor Design [3, 4]

The Split Flow Reactor design innovatively divides the catalyst bed into two parallel sections within a single reactor shell. The schematic diagram (Figure 7) below illustrates catalyst bed arrangement and bypass piping.

The split flow reactor introduces a **parallel flow mechanism** within a single reactor shell adding two catalyst beds vertically, one above the other.

The catalyst beds are supported by a robust mechanical support system. The bottom head here is empty unlike the single bed down flow reactor, where it is filled with inert ballast that helps achieve lower pressure drop.

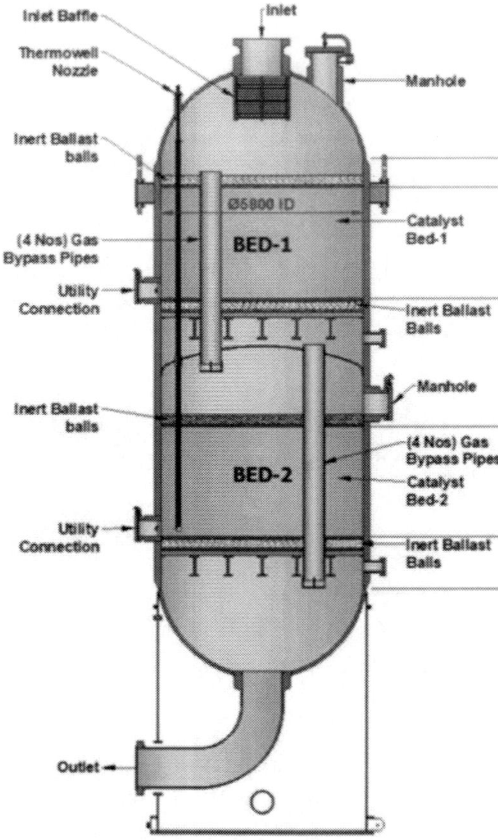

Figure 7. Split flow schematic diagram

Multiple bypass pipes are uniformly positioned in each bed to ensure uniform distribution and minimize flow imbalances. 50% of the flow bypasses the first bed and directly enters the second bed. The remaining 50% flows through the first bed and bypasses the second bed to mix at the outlet. Complete conversion for the full flow is accomplished within the reactor beds operating in parallel.

With parallel bed configuration, the reactor diameter is significantly lower than traditional down flow design, with increased height (tan-to-tan length).

Design Features and Advantages

Uniform Flow Distribution and Optimized Pressure Drop

Computational Fluid Dynamics (CFD) [1,2] simulations were conducted to validate velocity profiles & flow distribution (Figure 8), temperature profile (Figure 9) and pressure drop (Figure 10) across the reactor in bypass configurations. Results indicate uniform flow distribution, desired temperature profile and pressure drop across both beds, maintaining consistent reaction conditions.

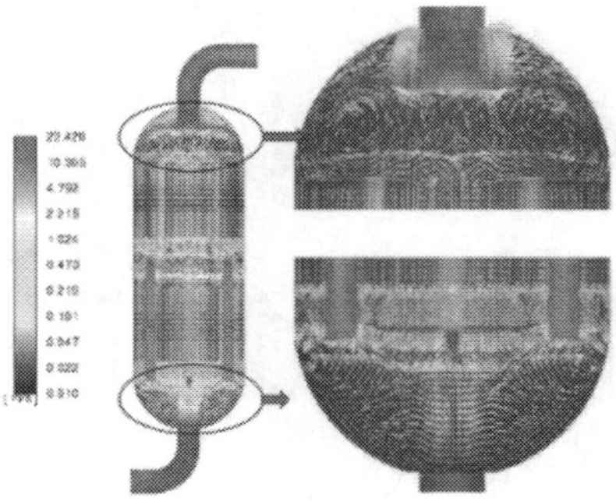

Figure 8. Velocity profile and flow distribution

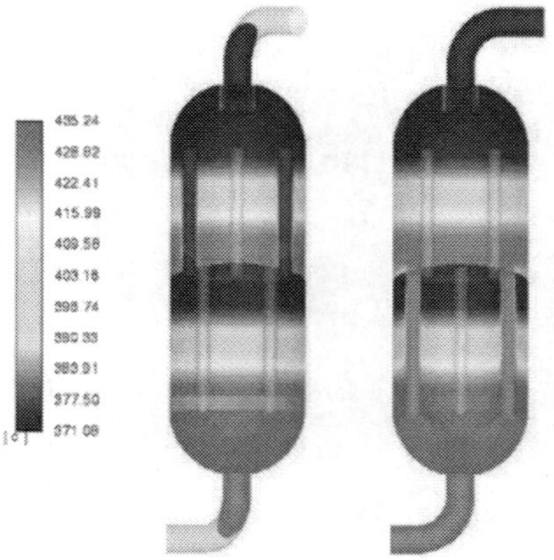

Figure 9. Temperature profile

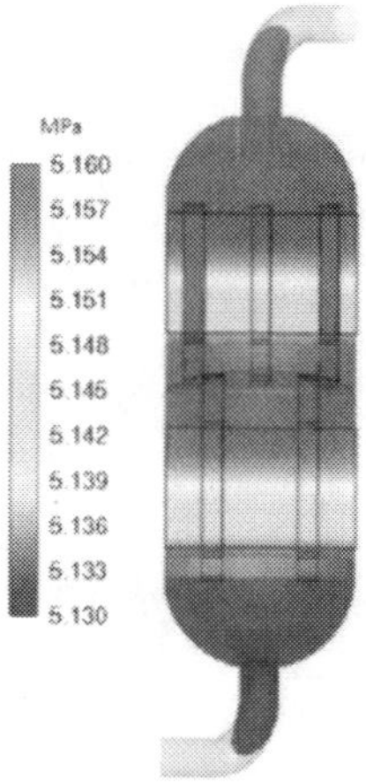

Figure 10. Pressure drop profile

Space Efficiency

Eliminates the need for parallel reactors. Overall plot space is reduced due to single lower diameter equipment.

Manufacturing and Transportability

Smaller diameters reduce fabrication and logistical challenges.

Cost-Effective Solution

Low capex due to the reduced diameter and thickness with less expensive internals.

Results and Discussion

A comparison of the conventional down flow vs split flow HTS reactor design is presented below (Figure 11 & 12) for a 6,000 MTPD plant capacity. With split flow reactor design it is possible to achieve significant reduction in equipment diameter thus reducing the thickness requirements. This translates to easier material procurement, shop fabrication with improved quality. Logistics have become simplified and there is no need for site fabrication due to a lack of logistical challenges.

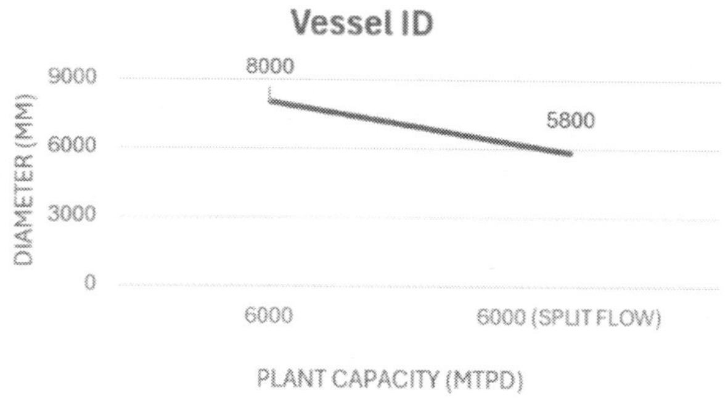

Figure 11. Equipment diameter comparison

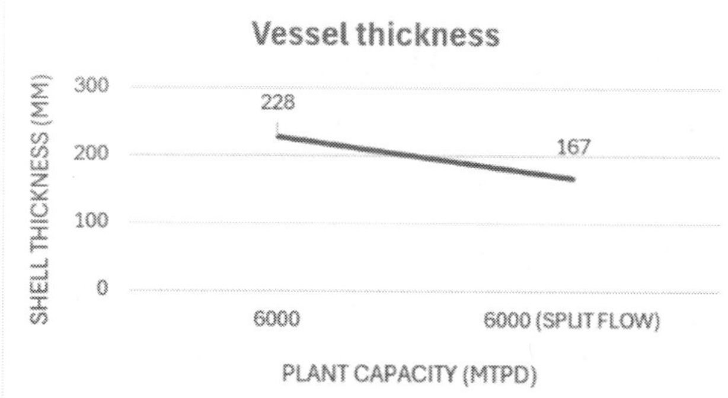

Figure 12. Equipment thickness comparison

A comparison of 6000 MTPD conventional versus split flow designs HTS, LTS and Methanator is presented in Table 1 below focusing on the following parameters:

- Internal diameter
- Height of the reactor
- Catalyst and Inert ballast volumes
- Pressure drops across internals

The split-flow reactor design offers a substantially reduced diameter while maintaining the same catalyst volume as conventional reactors. This compact configuration also eliminates the need for parallel reactors, thereby reducing overall plot dimensions and achieving material and fabrication cost savings estimated at 20–25%.

Parameter	6000 MTPD (Conventional)	6000 MTPD (Split Flow)	Remarks
HTS Converter			
Inside diameter mm (ft)	8000 (26.25)	5800 (19.0)	~28% lower diameter
Catalyst volume m3 (ft3)	181 (6392)	181 (6392)	Same catalyst volume
Inert ballast volume m3 (ft3)	93 (3284)	7.6 (268.4)	~92% reduction in inert ballast requirement
T.L. to T.L mm (ft)	2100 (6.9)	11250 (36.9)	
No of beds	1	2	
Total bed height mm (ft)	3600 (11.8)	7160 (23.5)	
Pressure drop kPa (psi)	42 (6.09)	35 (5.08)	~17% reduction in pressure drop
LTS Converter			
Inside diameter mm (ft)	8700 (28.5)	6200 (20.3)	~29% lower diameter
Catalyst volume m3 (ft3)	282 (9958.6)	282 (9958.6)	Same catalyst volume
Inert ballast volume m3 (ft3)	127 (4484.9)	8.7 (307.2)	~93% reduction in inert ballast requirement
T.L. to T.L mm (ft)	3450 (11.3)	13800 (45.3)	
No of beds	1	2	
Total bed height mm (ft)	4950 (16.2)	9660 (31.7)	
Pressure drop kPa (psi)	44.5 (6.45)	40 (5.8)	~10% reduction in pressure drop
Methanator			
Inside diameter mm (ft)	6300 (20.67)	4600 (15.0)	~27% lower diameter
Catalyst volume m3 (ft3)	96.6 (3411.4)	96.6 (3411.4)	Same catalyst volume
Inert ballast volume m3 (ft3)	43 (1518.5)	2.45 (86.5)	~94% reduction in inert ballast requirement
T.L. to T.L mm (ft)	1900 (6.23)	9900 (32.5)	
No of beds	1	2	
Total bed height mm (ft)	3100 (10.1)	6000 (19.7)	
Pressure drop kPa (psi)	31.5 (4.57)	30 (4.35)	~5% reduction in pressure drop

Table 1. Equipment size comparison

Optimized reactor dimensions further contribute to reduced land footprint, supporting sustainable site utilization and facilitating seamless integration into existing industrial infrastructures with minimal environmental disruption.

Efficient carbon monoxide (CO) conversion ensures compliance with environmental emission standards. This contributes to reducing the global environmental footprint of ammonia production facilities.

Future Innovation Development

The principles of split flow design could be easily applied to other catalytic processes, such as hydrogen production, petrochemical refining, and carbon capture systems, fostering innovation across the chemical industry (diversified applications).

Conclusion

The Split Flow Reactor design is a significant advancement in CO removal for ammonia production, offering a scalable, cost-effective, and sustainable solution. By splitting flow through parallel catalyst beds within a single reactor shell, this innovation minimizes spatial, manufacturing, and operational challenges associated with large-capacity plants and revamps.

Bibliography

1. Tridiagonal Solutions, CFD analysis of Split Flow Reactors, 2022
2. KBR Internal Analysis, 2024
3. United State Patent, US 8,197,785 B2, Split Flow Contactor, 2012
4. United State Patent, US 7,081,230 B2, Split Flow Vertical Ammonia Converter, 2006
5. Figure 1 : Ammonia Market to Triple by 2050 with Nearly All Growth Coming from Low-Carbon Supply - Jul 11, 2023

Safety Analysis of PT Green Ammonia Indonesia: Determination of Threat Zone Using ALOHA

Mandate for reducing CO2 emissions, as well as increasing demand for ammonia in the future has encouraged PT Kaltim Parna Industri (PT KPI) to establish a subsidiary called PT Green Ammonia Indonesia (PT GAI). PT GAI will be established in the Tanah Kuning industrial area located in North Kalimantan province. The plant, which is planned to be the first green ammonia plant in Indonesia, will utilize renewable energy from a power plant of the Kayan river. This Power Plant will be a Hydroelectric Power Plant with a total capacity of 9,000 MW and the construction process will be carried out in stages, with phase I expected to be completed in 2028 with a capacity of 900 MW.

PT GAI will use Alkaline Water Electrolysis technology to produce hydrogen. This paper will map out safety aspects of handling hydrogen using Areal Locations of Hazardous Atmospheres (ALOHA) software. ALOHA will help to make simulations that estimate threat zones associated with hazardous chemical releases, including toxic gas clouds, fires, and explosions.

Baskara A. Nugraha
PT. Kaltim Parna Industri, Bontang, East Kalimantan – Indonesia

Wiratni, & Cahyono, R. B.
Department of Chemical Engineering, Universitas Gadjah Mada – Indonesia

Introduction

Green Ammonia not only serves as a substitute for traditional "grey" Ammonia applications, such as those in the fertilizer industry, but also paves the way towards innovative utilization strategies in other sectors, contributing to a broader paradigm shift towards greener industrial processes. Green ammonia can serve as a chemical energy storage of renewable energy, a green hydrogen carrier for bulk transportation, and as a zero-carbon fuel, especially in the maritime and energy production sectors [1]. PT Green Ammonia Indonesia (GAI) was established to spearhead the development of Green Ammonia in Indonesia. GAI will develop a 300 Metric Ton Per Day (MTPD) Green Ammonia plant in North Kalimantan, Indonesia. The Green Ammonia plant will be built in Kawasan Industri Tanah Kuning (Yellow Land Industrial Park). The plant will utilize renewable energy from a hydropower plant of the Kayan River. The hydropower plant will consist of a five-unit cascade. The first hydropower plant will be about 900 Megawatt (MW), commissioned in the year 2028. The electricity generated will be delivered to the industrial estate area located about 124 miles (200 kilometers) from the dam.

Process Hazards Related to Green Ammonia Processes

PT GAI will apply Alkaline Water Electrolysis technology to produce hydrogen up to 2,260 kg/h of green hydrogen with 104^0F (40^0C) temperature and 580 psi (40 bar) pressure. In an electrolyzer system, oxygen and hydrogen coexist within the installation, separated solely by a membrane. This presents a risk of an explosive mixture forming inside the equipment. Major process hazards of electrolysis are associated with oxygen-hydrogen flammable mixtures that may be formed because of cell membrane degradation or failure in process operations. An uncontrolled contact between the two generates a strong potential explosive mixture that can be easily ignited.

The Haber-Bosch process has been considered converting nitrogen and hydrogen into ammonia via a catalyzed process conducted at high temperatures and pressure (GAI uses a single converter reactor at high pressure around 5,076 psi/350 bar)-significantly higher than operational pressure of conventional ammonia converter and temperature 725^0F/385^0C). Safety concerns about the synthesis loop are mainly associated with fires and the loss of containment of the hydrogen/ammonia mixture. The reactor unit, therefore, can be addressed as one of the most hazardous units apart from the electrolyzers.

Although water electrolysis was a mature technology by the mid-20th century, it gradually lost prominence in the industrial sector after the 1960s. This shift was primarily driven by economic factors:

- Steam Methane Reforming (SMR), developed and scaled during that period, became the dominant method for industrial hydrogen production due to its much lower cost, especially as natural gas became widely available and inexpensive.
- Electrolysis, in contrast, required large amounts of electricity, and in most regions, electricity was (and still often is) generated from fossil fuels, making electrolytic hydrogen economically and environmentally less attractive at the time.

Moreover, the lack of climate or environmental policy incentives meant there was no push toward low-emission hydrogen. As a result, the industry prioritized cost over sustainability, leading to the deprioritization of electrolyzers outside niche applications. Technologically, electrolyzers remained relatively stagnant for decades, with little investment in innovation or scaling, unlike SMR technologies, which continued to improve and expand. Due to the lack of historical electrolyzer data, the unique nature of hydrogen, the complexity of large-scale systems, and the need for standardized safety practices, conducting a Quantitative Risk Assessment (QRA) study is essential to model potential explosion scenarios, assess the likelihood of a fire occurring, and implement effective safety measures. Similarly, ammonia converters operate under high-pressure and high-temperature conditions, posing significant risks of explosions. A QRA for ammonia converters is crucial, especially when integrated with hydrogen production systems.

QRA was carried out using "What if" analysis and HAZOP result as inputs which has been carried out on a green ammonia plant concept [2]. Then, the magnitude of fire and explosion will be estimated based on the simulation results provided by the Areal Locations of Hazardous Atmospheres (ALOHA) software.

What if	Consequences	Recommendation
Ammonia Converter		
reactor leaks	Release of the material contained at high pressure and temperature: made of flammable species, it may be ignited readily; the contained materials have toxicological concern	- Frequent maintenance operations for control of external reactor integrity - Implement sensors for detection of hydrogen releases and loss of pressure - Alert workers and plant emergency shut-down operations should begin when the scenario occurs
Electrolyzer		
unit or pipelines leak	O_2/H_2 releases, with flammability concern (oxidant or easily ignited substance)	- Frequent maintenance and control of gaskets is suggested

Table 1. "What if" Analysis

Methodology

Data concerning location was collected from Meteoblue website. Meteorological data were generated from the Indonesian Agency for Meteorological, Climatological and Geophysics. Data concerning chemical materials, activities and processes, employee's records, and company assets were collected from design specifications. "What if" analysis and HAZOP were conducted to retrieve information about safety risks and assess possible process malfunction.

Table 1 represents only one possible accident from the ammonia converter and electrolyzer. For more detailed study of other possible causes, a "What if" analysis was also carried out. For example "What if" analysis for the converter includes: converter collapse, cooling capacity decrease, bed support breaks, and coolant leaks in reactor beds. For the electrolyzer: a blackout, overpressure, and membrane degrade. The maximum credible event (MCE) from the study is leakage of hydrogen. Therefore, both "What if" analysis and HAZOP play a significant role in constructing Event Tree Analysis, especially for the event of gas leak.

Event Tree Analysis (ETA) focuses on the determination of the frequency of a consequence based on the frequency of the initial event [3]. The frequency of the gas leak and probability are determined by using the data from Oil and Gas (OGP) Risk Assessment Data Directory [4]. By combining ETA, frequency data, and part count analysis for each equipment segment, an event frequency table will be obtained. ETA illustration can be seen in Figure 1.

All those data were evaluated then analyzed using QRA method to determine the individual risk levels. The data analysis was done by identifying the risks of fire and explosion and then verifying the field to set a fire and explosion scenarios. Individual Risk per Annum (IRPA) of individual risk was calculated from frequency analysis using ETA.

No	Guide-Word	Eleme nt	Deviati on	Possibl e Cause	Conse quenc es	Safegu ards	Comm ents	Action Requir ed
Electrolyzer – Node 1 Design Intent: Production of H_2 and O_2 from water electrolysis Operative Condition: P = 40 bar, T = 40^{0}C								
1	More	Flow	More Flow	-Faulty instrumentation and controls -Faulty valves (excessive opening) -Wrong operation of the electrolysis stack	-Wrong ratio of reactant -Compressor capacity varies -higher hydrogen content enhances safety risk	-Flowmeter	-control system redundant -control and mange compressor	-control the redundancy -safety interlock -maintenance
Converter – Node 2 Design Intent: Ammonia conversion Operative Condition: P = 350 bar, T = 385^{0}C								
1	More	Temperature	More Temperature	-Lack of cooling capacity -Loss of control on cooling unit	-Kinetic affected -catalyst damaged	-Temperature meter	-May affect compression and the integrity of reactor	-temperature control redundancy -Alarm for process deviation -safety interlock

Table 2. HAZOP Analysis – example of 2 nodes from electrolyzer & converter

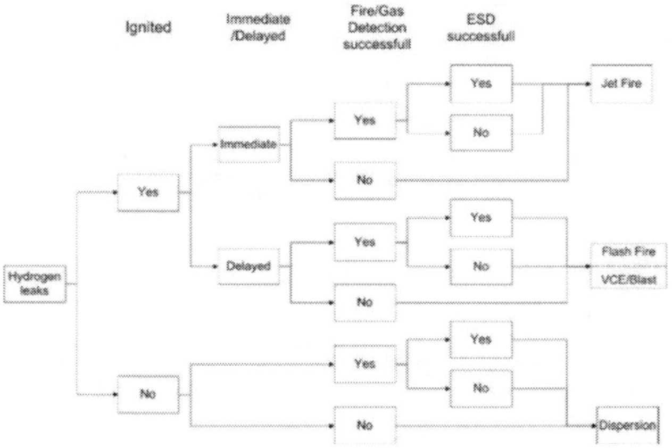

Figure 1. Event Tree Analysis of Gas Leak

Furthermore, the consequences were analyzed using Areal Location of Hazardous Atmosphere (ALOHA) software of fire and explosion based on a worst-case scenario modelling, which has been built upon the Gaussian dispersion model of continuous, buoyant air pollution flumes. The simulation model was jointly developed by organizations including the United States Environmental Protection Agency (USEPA), Chemical Emergency Preparedness and Prevention Office (CEPPO), and National Oceanic and Atmospheric Administration Office of Response and Restoration (NOAA). The model can simulate the dispersion model for over 900 chemicals and is primarily used in the simulation of accidental release of hazardous substances and the dispersion of chemical vapor [5].

Location Specific Individual Risk (LSIR)

To quantify the risk, it is necessary to calculate Location Specific Individual Risk (LSIR). LSIR calculated by a hypothetical individual who is positioned there for 24 hours per day, 365 days per year [6]. The LSIR in each respective area with regular intervals from 1E-01 to 1E-08 per year shall be produced for all scenarios.

$$LSIR = f.P.C.O.p$$

Where:

- f = Event frequency (per year).
- P = Probability of events
- C = Fatality probability
- O = Occupancy factor (fraction of time people are present).
- p = population

Individual Risk per Annum (IRPA)

IRPA in each worker group is calculated based on manning level of each worker group. Individual Risk Per Annum (IRPA) represents the risk of an individual in a specific worker group becoming a fatality while working on the facility in a single year, considering the number of hours exposed each year.

$$IRPA = \sum (LSIR.F)$$

Where:

- LSIR = Location-Specific Individual Risk at a given location
- F = Fraction of time an individual spends at that location

Aggregate risk levels are compared with risk acceptance criteria – refer to HSE UK.

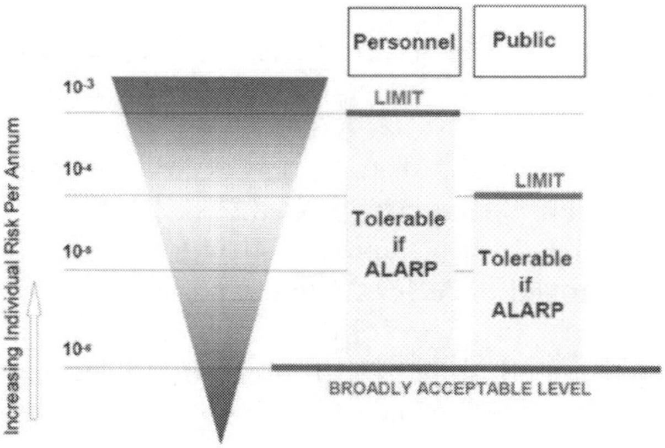

Figure 2. IRPA for Working Group Risk Acceptance Criteria, based on HSE-UK

Result

EVENT FREQUENCY & PROBABILITY (1/year)	
Jet Fire	6.46E-07
Flash Fire	9.30E-06
VCE/Blast	6.20E-06
Gas Dispersion	6.30E-04

Table 3. Event for Electrolyzer Leak

EVENT FREQUENCY & PROBABILITY (1/year)	
Jet Fire	1.10E-06
Flash Fire	1.58E-05
VCE/Blast	1.05E-05
Gas Dispersion	1.07E-03

Table 4. Event for Converter Leak

The event frequency from Electrolyzer and Converter are displayed in Table 3 and Table 4. The results are obtained from OGP Risk Assessment Data Directory with Full Bore Release Assumption.

	Personnel (Working) Group Description	Shift / Daily	No. of Personnel
Shift	Supervisor Ammonia	Shift	1
	Supervisor Utility	Shift	1
	DCS Panel Operator	Shift	2
	Field Operator - Utility Plant	Shift	1
	Field Operator - Ammonia Plant	Shift	2
	Security Officer	Shift	2
Non-shift	Plant Manager	Daily	1
	Officer (Housekeeper and front desk)	Daily	2

Table 5. Personnel Working Group

Table 5. Personnel Working Group, obtained from manpower plant data based on working hours, shifts and non-shifts. This data will later be used when calculating LSIR and IRPA, namely when determining Occupancy factor, population, and Fraction of time. The placement location of each individual personnel can be determined from the general plot plant division in Figure 3.

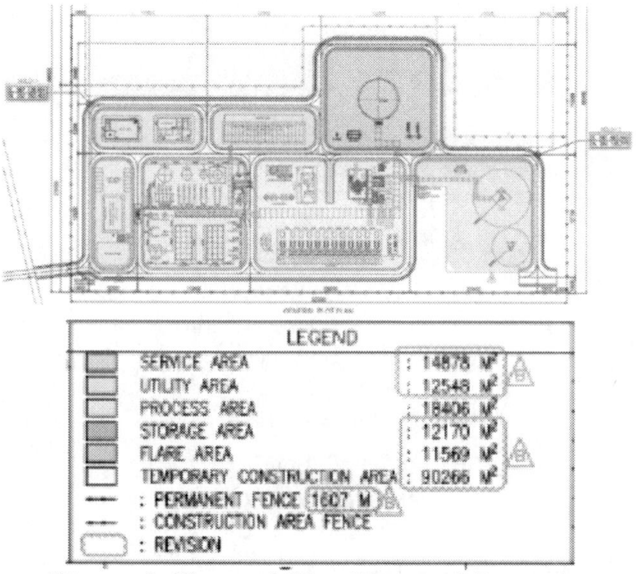

Figure 3. General Plot Plan and Legend

From the results of the ALOHA simulation and with the help of a general plot plan, threat zones and affected areas will be obtained as in tables 6 and 7. For some illustrations of the threat zone from the ALOHA simulation results, see Figures 4 – 7.

No	Release Source	Hazard to Analyze	Threat zone	Reach (m)	Effects	Affected Area
1	Direct	Flammable Vapor Cloud	Red	371	24000 ppm = 60% LEL = Flame Pockets	All Area
			Orange	1000	4000 ppm = 10% LEL	All Area
2		Overpressure (blast force) from vapor cloud explosion	Red	273	8.0 psi = destruction of buildings	Process Area, Utility Area, Storage Area, Flare Area, and part of service area
			Orange	279	3.5 psi = serious injury likely	Process Area, Utility Area, Storage Area, Flare Area, and part of service area
			Yellow	333	1.0 psi = shatters glass	All Area
3	Pipeline	Flammable Vapor Cloud	Red	296	24000 ppm = 60% LEL = Flame Pockets	Process Area, Utility Area, Storage Area, Flare Area, and part of service area
			Orange	602	4000 ppm = 10% LEL	All Area
4		Thermal Radiation (Jet Fire)	Red	14	10.0 kW/(sq m) = potentially lethal within 60 sec	Process Area
			Orange	20	5.0 kW/(sq m) = 2nd degree burns within 60 sec	Process Area
			Yellow	31	2.0 kW/(sq m) = pain within 60 sec)	Process Area

Table 6. Electrolyzer leak modelling result

No	Release Source	Hazard to Analyze	Threat zone	Reach (m)	Effects	Affected Area
1	Direct	Flammable Vapor Cloud	Red	512	24000 ppm = 60% LEL = Flame Pockets	All Area
			Orange	1400	4000 ppm = 10% LEL	All Area
2		Overpressure (blast force) from vapor cloud explosion	Red	376	8.0 psi = destruction of buildings	All Area
			Orange	385	3.5 psi = serious injury likely	All Area
			Yellow	452	1.0 psi = shatters glass	All Area
3	Pipeline	Flammable Vapor Cloud	Red	718	24000 ppm = 60% LEL = Flame Pockets	All Area
			Orange	1400	4000 ppm = 10% LEL	All Area
4		Thermal Radiation (Jet Fire)	Red	30	10.0 kW/(sq m) = potentially lethal within 60 sec	Process Area
			Orange	42	5.0 kW/(sq m) = 2nd degree burns within 60 sec	Process Area
			Yellow	65	2.0 kW/(sq m) = pain within 60 sec)	Process Area

Table 7. Converter leak modelling result

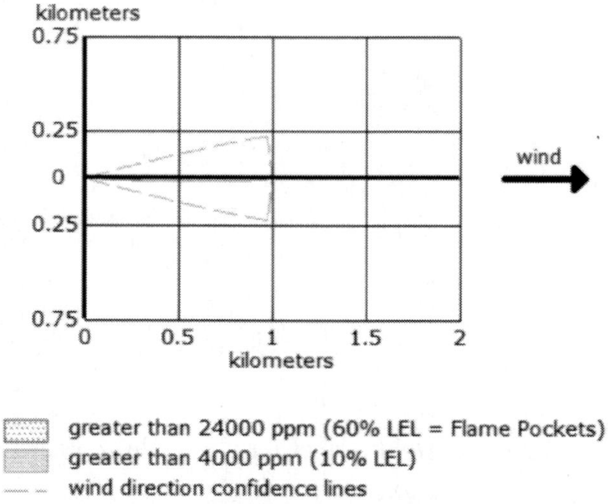

Figure 4. Flammable Vapor Cloud Threat Zone (Electrolyzer leak)

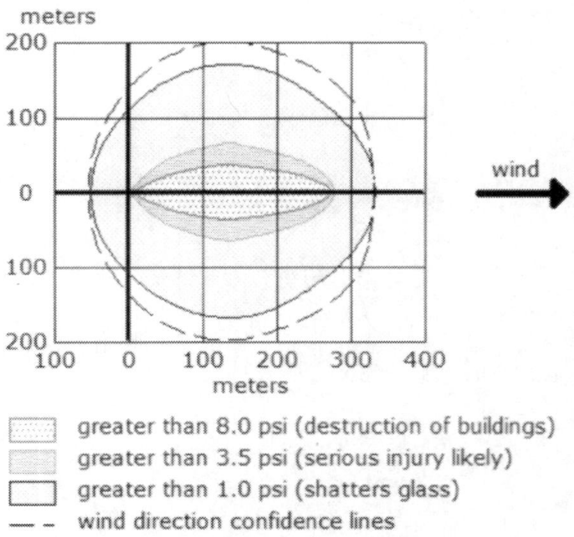

Figure 5. Overpressure Threat Zone (Electrolyzer leak)

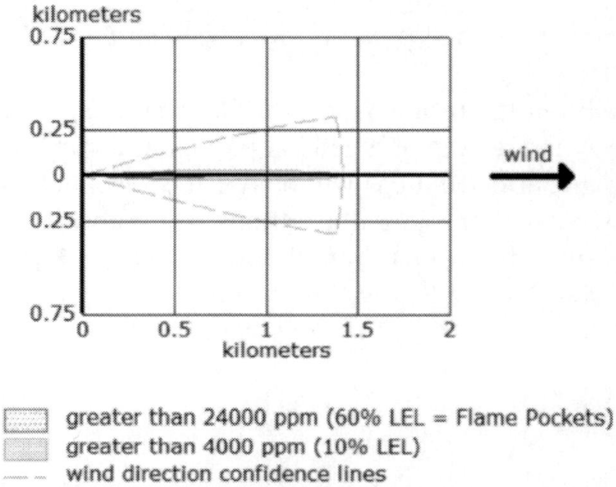

Figure 6. Flammable Vapor Cloud Threat Zone (Converter leak)

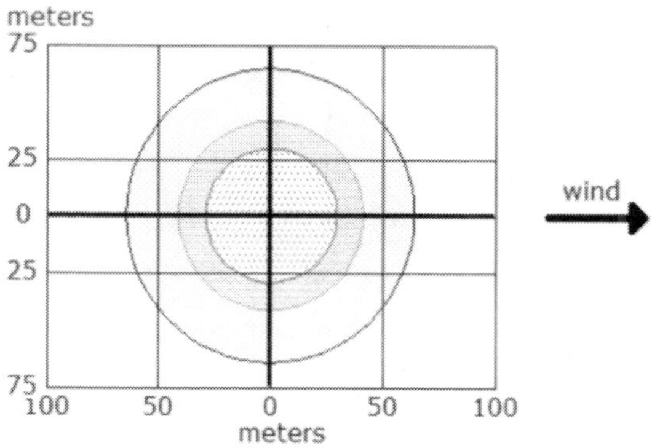

Figure 7. Jet Fire Threat Zone (Converter leak)

LSIR and IRPA

LSIR for each area (Hydrogen leakage from Electrolyzer):

Event	Freq. & Prob.	LSIR				
		Service Area	Utility Area	Process Area	Storage Area	Flare Area
Jet Fire	6.46E-07			1.29E-06		
Flash Fire	9.30E-06	1.12E-04	9.30E-05	1.86E-04	4.65E-05	1.16E-05
Explosion/Blast	6.20E-06	7.44E-05	6.20E-05	1.24E-04	3.10E-05	7.75E-06
	LSIR total	1.86E-04	1.55E-04	3.11E-04	7.75E-05	1.94E-05

Table 8. LSIR for Electrolyzer Leak

LSIR for each area (Hydrogen leakage from Converter):

Event	Freq. & Prob.	LSIR				
		Service Area	Utility Area	Process Area	Storage Area	Flare Area
Jet Fire	1.10E-06			2.20E-06		
Flash Fire	1.58E-05	1.90E-04	1.58E-04	3.16E-04	7.90E-05	1.98E-05
Explosion/Blast	1.05E-05	1.26E-04	1.05E-04	2.10E-04	5.25E-05	1.31E-05
	LSIR total	3.16E-04	2.63E-04	5.28E-04	1.32E-04	3.29E-05

Table 9. LSIR for Converter Leak

Based on CCPS [7] and UK HSE [8], it is explained that individual risk should be computed for each initiating event. So, the IRPA assessment is divided into two parts.

Personnel (Working) Group Description	IRPA
Supervisor Ammonia	1.03E-04
Supervisor Utility	8.60E-05
DCS Panel Operator	1.86E-04
Field Operator - Utility Plant	7.64E-05
Field Operator - Ammonia Plant	2.49E-04
Security Officer	1.86E-04
Plant Manager	6.20E-05
Officer (Housekeeper and front desk)	1.24E-04

Table 10. IRPA for Electrolyzer Leak

Personnel (Working) Group Description	IRPA
Supervisor Ammonia	1.76E-04
Supervisor Utility	1.46E-04
DCS Panel Operator	3.16E-04
Field Operator - Utility Plant	1.30E-04
Field Operator - Ammonia Plant	4.22E-04
Security Officer	3.16E-04
Plant Manager	1.05E-04
Officer (Housekeeper and front desk)	2.10E-04

Table 11. IRPA for Converter Leak

Conclusion

From the results of the two IRPA calculations, it is known that the highest risk value was obtained by the Field Operator Ammonia Plant personnel group. This can happen because the location of the electrolyzer and converter are both in the Process area. When compared to the HSE-UK personnel ALARP standard, both IRPA values are still acceptable. However, strategies and steps are still needed to handle risks if they arise, including:

- Monitor hydrogen leaks with online detection systems
- Install Automated shutdown systems if leaks exceed threshold integrated with the Safety Instrumented System (SIS) such as PAHH (Pressure Alarm High-High), and LAHH (Level Alarm High-High) as appropriate to ensure there is a reduction in the level of risk, so that the risk that will be caused becomes Tolerable Risk or in ALARP conditions [9]
- Design a fire and explosion-proof control room building

- Provide a hydrant system (tank and piping)
- Update the QRA study on a regular basis (every five years). This QRA update activity is part of the Safety Management System (SMS) audit. This SMS audit is very important in monitoring the performance indicators of a plant's Process Safety Management. [10]

References

[1] The Royal Society. (2020). Ammonia: zero-carbon fertilizer, fuel, and energy store.

[2] Salgari, F. (2024). Conceptual Design of a Green Ammonia Production Process: Cost and Safety Assessment. Universita Degli Studi di Padova, Italy.

[3] Institute for Sustainable Process Technology (2023); "Safety Aspects of Green Hydrogen Production on Industrial Scale". Hydrohub Innovation Program – Public Report

[4] International Association of Oil and Gas Producer (IOGP). 2010. OGP Risk Assessment Data Directory

[5] J.M. TSENG, T.S. SU & C.Y. KUO. 2012. Consequence Evaluation of Toxic chemical releases by ALOHA. Procedia Engineering 45 (2012) 384 – 389. Doi: 10.1016/j.proeng.2012.08.175

[6] Quantitative Risk Analysis Report PT Kaltim Parna Industri. 2023. KPI-QSHE-QRA-001

[7] Guidelines for Chemical Process Quantitative Risk Analysis (CCPS, AIChE, 2nd Edition)

[8] Reducing Risks, Protecting People (R2P2) – UK HSE

[9] Sihotang, K. J. & Widanarko, B. (2023). Kajian dampak Pajanan Radiasi Panas Saat Terjadi Tumpahan Minyak dan Kebakaran Tangki di Pusat Pengumpul Produksi Minyak PT. X. Jurnal Rekayasa Proses Volume 17, Nomor 1, 2023, 110–116. https://doi.org/10.22146/jrekpros.84097

[10] Hassan, C. R. C., Pitt, M. J., & Wilday, A. J. (2005). Integrating Performance Indicator into the Audit Process Safety Management System. ASEAN Journal of Chemical Engineering, 5(2), 161-168. Retrieved from https://jurnal.ugm.ac.id/v3/AJChE/article/view/7646.

[11] API RP 581 (Risk-Based Inspection Technology) 4th edition.

[12] Hastiti, L. R., Lestari, F., & Susilowati, I. H. (2018). Quantitative Fire and Explosion Risk Assessment of Fuel Tanker Truck. Indian Journal of Public Health Research & Development, December 2018, Vol. 9, No. 12. DOI Number: 10.5958/0976-5506.2018.01875.2

Lifecycle Integrity Management of Outlet Pigtails

Outlet pigtails transfer reformed gas from catalyst tubes to the outlet manifold in steam methane reformers. They are typically designed per API 530 for resistance to creep from internal pressure stress. However, additional factors—such as thermal expansion, gravity loads, and other in-service damage like stress relaxation, and nitridation—are not fully addressed by code compliance. Failures from these mechanisms can cause unplanned outages. Effective lifecycle management includes optimizing design, fabrication, and weld procedures, understanding operational risks, and applying targeted inspection and monitoring. This paper outlines damage mechanisms, mitigation strategies, and lifecycle practices to ensure safe, reliable pigtail performance.

Olivia Chung, Nicholas Goodman, Charles Thomas
Quest Integrity, New Zealand

Introduction

Outlet pigtails have a finite service life. A pigtail failure will, in the best case, force a plant outage with the associated economic loss and, in the worst case, lead to significant secondary and environmental damage. This places emphasis on active management of these components throughout their lifecycle. Such management requires a lifecycle management strategy that is informed by knowledge of the factors that impact reliability. This approach both ensures reliable operation and facilitates decision making during each stage of the lifecycle. The general lifecycle management (LCM) process for fixed, pressurised equipment is shown in Figure 1 [1] [2]. Plant operators and mechanical designers will make decisions that involve compromises between competing objectives, such as maximum strength and minimum weight. The operator should have input to such decisions as part of LCM planning; too often, the design of pigtails and its associated installation and operation are taken at face value. The objectives of this paper are to:

- Present the problems that could lead to unanticipated in-service failures
- Provide key points for enabling mechanical de-signers, manufacturers and operators to develop an LCM plan that maintains the rate of failures as low as reasonably practicable

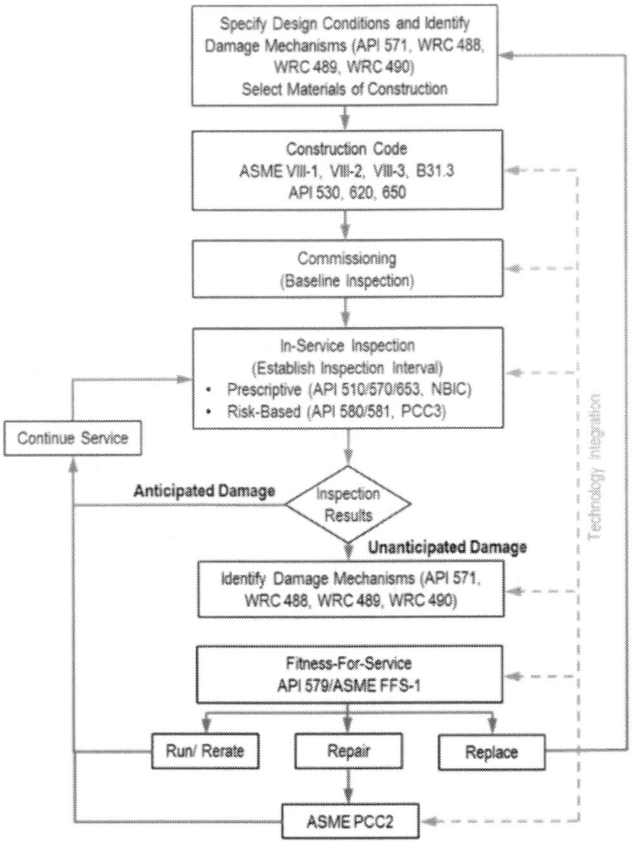

Figure 1. Overall LCM process for fixed, pressurised equipment [1] [2].

Materials of Construction

Alloys 800H (UNS N08801) and/or Alloy 800HT (UNS N08811) are almost universally selected as the materials of construction for reformer outlet pigtails. Selected nominal composition of Alloy 800H and Alloy 800HT and material specifications are provided in Table 1.

Material Property		UNS N08800 Alloy 800	UNS N08810 Alloy 800H	UNS N08811 Alloy 800HT	
Selected chemical composition (weight %)	C	0.10 max.	0.05 - 0.10	0.06 - 0.10	
	Al	0.15 - 0.60	0.15 - 0.60	0.15 - 0.60	Al + Ti 0.85 - 1.20
	Ti	0.15 - 0.60	0.15 - 0.60	0.15 - 0.60	
	Cr	19.0 -23.0	19.0 - 23.0	19.0 - 23.0	
	Fe	Balance	Balance	Balance	
	Ni	30.0 - 35.0	30.0 - 35.0	30.0 - 35.0	
Average grain size		-	ASTM 5 or coarser		
Tensile strength		520 MPa (75 ksi) (cold-worked annealed)	450 MPa (65 ksi) (all conditions)		
Yield strength		205 MPa (30 ksi) (cold-worked annealed)	170 MPa (35 ksi) (all conditions)		
Elongation		30% (all conditions)			

Table 1. Selected specified properties of ASTM B407 Alloys 800, 800H and 800HT [3].

The expected microstructure of wrought Alloy 800H and Alloy 800HT pigtails is wrought, equiaxed, austenite grains with some twinning (see Figure 2). Alloy 800H and Alloy 800HT have additions of titanium and aluminum, which can lead to the formation of carbonitrides and nitrides, as highlighted in Figure 2. As produced, the alloy typically has a fairly continuous network of chromium rich carbides along the grain boundaries, with very fine titanium carbides/ carbonitrides distributed within the grains. This high density of very fine secondary precipitates is a significant factor in the high creep strength of the alloy. Larger, blocky titanium carbonitrides may also be present. After service at high temperature, the grain boundary carbides agglomerate and coarsen, leading to a discontinuous network of larger carbides on the grain boundaries. The network of titanium carbonitrides in the grains also similarly coarsen and agglomerate (see

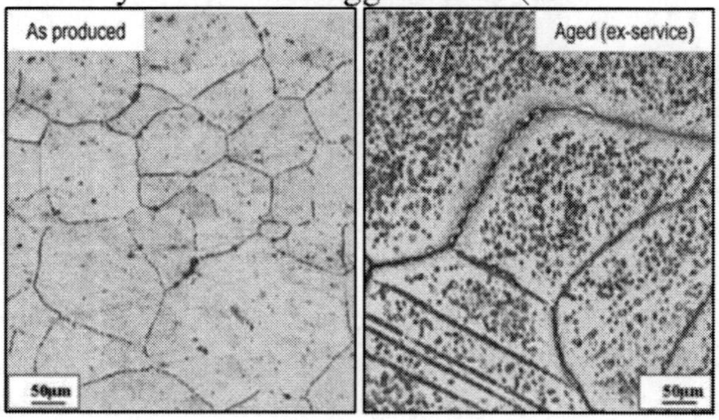

Figure 3).

The degree of coarsening is a function of both time and temperature and is often referred to as aging. Aging of this material is normal and expected. Aged material is not inherently defective or unfit for service. Material aging typically leads to a small drop in hardness. More aged material generally has lower creep strength but potentially higher creep ductility.

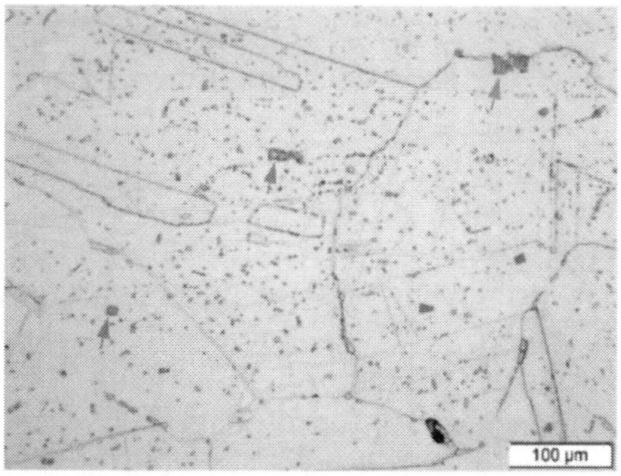

Figure 2. Representative microstructure. Arrows show examples of Ti(C,N) precipitates [4].

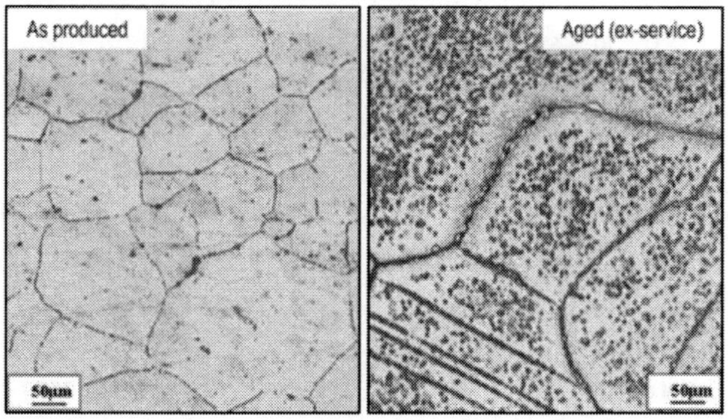

Figure 3. Micrographs of Alloy 800 showing change of microstructure with aging [5].

Grain size is typically controlled to ASTM size 5 or coarser for creep strength. Table 1 depicts the trivial differences in composition between the various grades. However, the required differences in grain size result in massive differences in creep strength. Generation of the correct grain size involves heat treatment designed to generate grain growth. Pigtail fabrication typically involves cold forming and drawing of the tube and cold bending to form the shape of the pigtail. Annealing and stabilizing heat treatments are required after these mechanical processes. The degree of cold working will influence recrystallisation and grain growth during heat treatment. This cold working and heat treatment can lead to wide variation in grain size, even within individual pigtails.

For the outlet pigtail connections, ERNiCr-3 (Inconel 82) weld filler material was used extensively until creep failures were experienced at these welded connections. Current industry practice is to weld Alloy 800H/HT with an overmatched nickel-based filler material such as ERNiCrFe-3 (Inconel 182) or ERNiCrCoMo-1 (Inconel 617) [6]. It should not be assumed that there is a "standard" filler metal for pigtail connections. In general, the industry has appeared to migrate from Inconel 82 to higher strength alloys.

Damage Mechanisms

Damage that occurs in outlet pigtails can be generally categorized as creep, creep fatigue, environmental attack, and stress relaxation cracking (SRC) [4] [7]. A general overview of these damage mechanisms is provided below. The onset of damage can be minimized during the design and procurement phases by giving additional consideration to the specified fabrication and construction processes, particularly for in-service damage not readily accounted for in design codes and standards.

Creep

Outlet pigtails operate at elevated temperatures, typically between 750 °C and 875 °C (1382 °F and 1607 °F), and at internal pressures between 1 MPa and 3 MPa (145 psi and 435 psi). As such, creep is the dominant damage mechanism considered during design. The design life of an outlet pigtail is based on creep rupture from internal pressure (i.e., hoop stress) to a prescribed time (e.g., 100,000 hours) using API 530. This aspect of design requires selection of material of construction and dimensions to provide adequate strength to resist creep.

Creep Fatigue

In addition to creep rupture from hoop stress, creep rupture from axial loading can also occur during service. Axial loads can be from transient thermal stresses from start-up/shutdown, bending, and other system loads. Additional fabrication factors, such as ovality in the pigtail after forming and/or bending, may also contribute to locally higher axial stresses. In addition, any cyclic operation due to thermal stresses will contribute to creep-fatigue damage.

Neutral Axis Cracking

There have been a number of cases of cracking at the neutral axis of bends [6] [7] [8]. The reason for this is a combination of localized high stress due to ovality caused by the forming operation and the response to post-bend solution anneal heat treatment. The issue of ovality caused by flattening of the tube during operation should be controlled by the fabricator and specified by the operator.

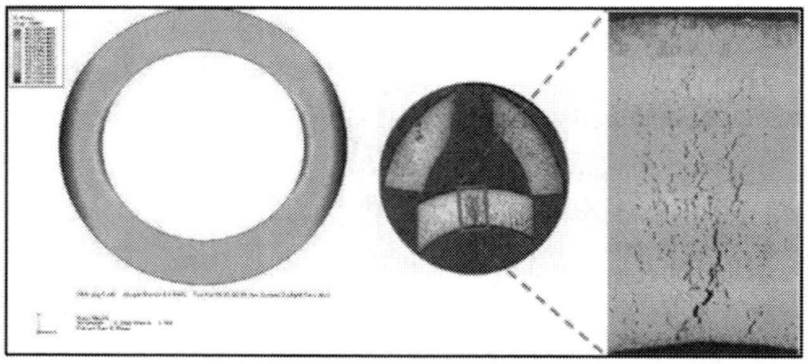

Figure 4. Stress redistribution due to ovality and internal pressure and observed creep damage associated with ovality at neutral axis. [7].

Environmental Attack

Pigtails are exposed to two environments during service:

- High-temperature syngas (i.e., hydrogen, carbon monoxide, and small amounts of methane and steam) at the internal surface. Intergranular oxidation, decarburization, and carburization can occur.
- High-temperature air at the external surface. Intergranular oxidation and nitridation can occur.

Presence of creep voids and cracking can provide a path for environmental attack to occur [7].

Perhaps the least understood environmental attack mechanism is nitridation, particularly that formed during in-service operation. The formation of nitride-rich acicular precipitates in service has been observed around creep cracks and has contributed to pigtail failures due to a reduction in creep ductility that results in embrittlement. Although, published literature validating the effect of in-service nitridation on mechanical and creep properties of Alloy 800H and 800HT is very limited [4].

Damage from in-service nitridation is confirmed by the presence of acicular and blocky aluminum and chromium nitrides near surfaces exposed to air (i.e. a nitrogen-containing gas) (see Figure 5). The depth of formation is controlled by the diffusion of nitrogen, so these nitrides are typically only found very near an exposed surface. The passive chromium oxide film that naturally forms on the pigtails provides some degree of protection from nitridation. However, in areas of high strain, this film can fracture and allow for higher nitridation rates (e.g., at locations of creep damage). An initial model of the kinetics that enable nitridation to occur in-service for Alloy 800H has been provided by Young et al. [9].

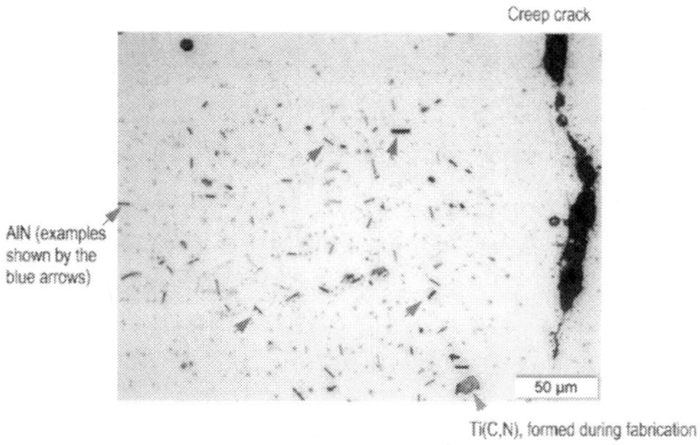

Figure 5. Micrograph showing nitride-rich precipitates. A creep crack is also shown [4].

Stress Relaxation Cracking

Stress relaxation cracking (SRC) can occur either during fabrication or in service; this damage mechanism is also known by other names, as summarized in Table 2 [4] [6]. Material cracking occurs due to post-weld heat treatment or service at elevated temperatures between 500 °C to 750 °C (932 °F to 1382 °F), resulting in grain boundary strain [6] [7].

Fabrication	In-Service
Reheat cracking Stress relief cracking Stress relaxation cracking	Stress relaxation cracking Creep embrittlement cracking Stress induced cracking Stress assisted grain boundary oxidation (SAGBO cracking)

Table 2. Names for SRC mechanisms [6].

Critical factors that increase susceptibility to SRC include:

- Composition and impurity elements
- Large grain size
- High residual stresses from fabrication (i.e., cold working, welding)
- Heavy section thickness (which controls restraint and stress state)
- Notches and stress concentrators
- Weld metal and base metal strength
- Welding and heat treatment conditions

Industry experience has indicated that Alloy 800H/HT is susceptible to SRC. Sudden in-service cracking and failure can occur as SRC is a brittle fracture mechanism [6].

Although this damage mechanism is generally associated with thick-walled components, it can also be observed in thin-walled pipework such as outlet pigtails. The observed damage is either:

- Intergranular cracking with an iron-nickel-rich layer sandwiched between chromium/oxide layers (see Figure 6, left); or
- A band of creep voids in the parent near the weld fusion line due to mismatch in material performance between the parent and over-matched weld filler (see Figure 6, right).

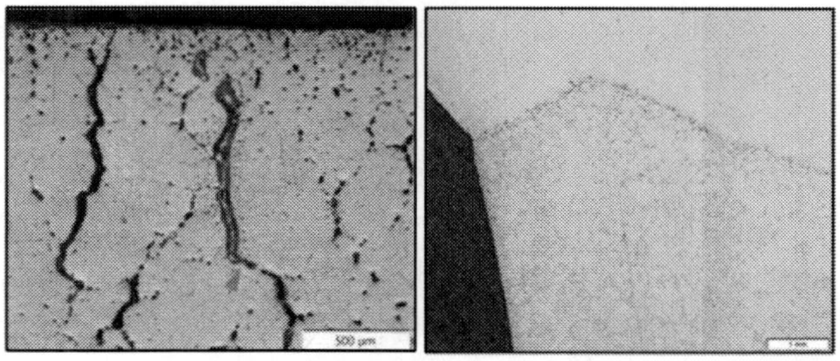
Figure 6. Left – SRC in an Alloy 800HT pigtail, highlighted by the arrow. Creep voiding and cracking is also shown. Right – Creep voiding along the fusion line of a pigtail weld joint [4].

Challenges and Potential Solutions – Design

There is no standardized code available for designing outlet pigtails, and many decisions are determined by the designer. Guidelines in ASME B31.3 [10] provide an approach for defining the design criteria for process piping based on the fluid service, design of appropriate supports and materials of construction as well as supporting information relating to fabrication, assembly, inspection, and acceptance criteria. The selection of materials to minimize in-service damage is not within the scope of ASME B31.3. In addition, the use of other design standards such as API 530

[11] can supplement ASME B31.3 in defining the minimum wall thickness based on the desired design creep life, which will affect the resultant weight of the piping. Determination of the minimum wall thickness will also require a diameter input, which will be based upon a trade-off between pressure drop of the process along the length of the piping and available syngas compressor throughput to optimize the process. The dimensions of outlet pigtails for existing installations are not likely to change. However, for new installations, the decisions relating to pigtail geometry would need to be justified with respect to the desired design life and any constraints around self-weight and the process.

Despite the design of outlet pigtails being primarily based on ensuring adequate creep performance from hoop stress, damage caused by system bending loads or system stresses acting on the pigtails should also be considered in the design. It is accepted that pigtails are required to accommodate movement of the system, hence the convoluted shape used to intentionally impart flexibility. Although inherent flexibility is a key aspect in the performance of outlet pigtails, bending loads from operation are unlikely to be fully eliminated. Use of tools such as finite element stress analysis would help the designer understand the nature of these complex axial loads and identify locations potentially susceptible to circumferential cracking (e.g., terminal weld at the manifold/hot collector end shown in Figure 7). However, it is noted that detailed analysis of the outlet pigtail above a generic pipe model based on design and anticipated operating conditions is rarely a significant input to the overall design.

The system design includes the choice of butt welds or socket welds for the pigtail connections. This is the designer's prerogative, but the operator must live with the consequence of that decision. A butt weld is often the default but does have downsides. Among other things, the material above the catalyst tube weld is part of the catalyst tube and failure here would require a tube replacement. Having a life-limiting component of the pigtail system dictating catalyst tube replacement would be undesirable. On the other hand, socket and fillet welds require care and attention to avoid stress concentrations at the toe of the weld and excessive restraint, as these can contribute to the onset of SRC. This avoidance could be achieved, for example, by specifying some form of automatic welding. Socket weld fit up is easy and no purge is required, so they are somewhat easier to maintain.

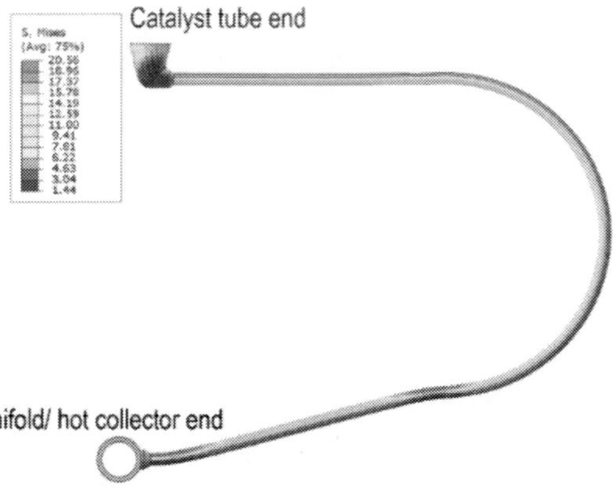

Figure 7. Creep analysis of an outlet pigtail, highlighting high-stress regions on the pigtail and at the weld joint toward the outlet manifold.

Challenges and Potential Solutions – Materials of Construction

Chemical Composition

Operators should, perhaps, be involved in selecting the composition of the alloy. The two variants considered for the construction of outlet pigtails have only minor variations in composition, mainly in the total content of titanium and aluminum. Indeed, it is not uncommon for material to qualify as both 800H or 800HT based on the similarities in composition. The presence of aluminum and titanium additions can promote the onset of nitridation and SRC damage. In particular,

- Nitriding due to absorption of nitrogen from the external atmosphere. The consequence of which is embrittlement due to the formation of internal nitride precipitates, chief among which is aluminum nitride [4] [7].

- The risk of SRC increases if the base metal and weld filler material have a combined aluminum and titanium content greater than 0.7 % [6] [7].

Hence, specifically controlling the content of aluminum and titanium (to a lesser extent) in the alloy may be considered a reasonable course of action. Indeed, for sufficiently large orders, it may well be preferable to procure a single heat of material of known and controlled composition even at a premium price.

Grain Size

Of prime importance for Alloy 800H and Alloy 800HT is the material grain size. The impact of grain size on the creep performance of Alloy 800 has long been established and has resulted in the very large differences in creep strength between the basic Alloy 800 and the grain-size controlled Alloy 800H and Alloy 800HT grades. The general requirement for 800H and 800HT is a grain size coarser than ASTM grain size number #5, as noted in Table 1. Examples of outlet pigtails that generally meet this grain specification are shown in Figure 8. (All grain size measurements shown were analyzed using Electron Back Scatter Diffraction [EBSD] in accordance with ASTM E2627 [12]).

As shown in Figure 8, variations in the grain size and grain size distribution can exist in the final product that will affect the resultant creep behavior. In particular [4],

- Coarse grain size can lead to poor ductility despite high nominal creep strength. Ductility controls the flexibility and the ability to withstand (thermal) cyclic loading.
- Fine-grained regions have better creep ductility and can deform with less voiding.

The material grain size and the distribution of grain size is controlled by the manufacturing process (i.e., drawing, forming or pilger), any bending, and heat treatment(s) following the mechanical processes. Based on the observed variations in the final product, a need to improve the control of grain size and resultant microstructure to manage the risk of creep damage has been recognized, as demonstrated by Imizcoz, Tait, and Kral [13]. Recrystallisation and grain growth is controlled by the degree of deformation and cold work induced during forming and bending operations and the subsequent heat treatment. However, this aspect has not been widely studied in the literature [14]. Based on an assumed cold work of 40% to 60% of Alloy 800 H after pilger forming, recrystallisation was noted at 850 °C (1562 °F), with temperatures above 1100 °C (2012 °F) producing larger grains. A heat treatment of 1075 °C (1967 °F) for at least one hour was estimated to obtain the ASTM grain size requirement of 5 or coarser [15]. It is noted that an ASTM grain size of 3.5 or finer shows greater resistance to SRC [6]. The requirement to maximize creep ductility by limiting damage due to thermal loading can, therefore, be at odds with the requirement to maximize creep strength. Specifying a tighter range/tolerance above those specified in ASTM B407 will ensure the best compromise between creep strength and creep ductility (e.g., minimizing the presence of excessively coarse grains).

The importance of grain size control requires a consistent and repeatable measurement and analysis approach. Intercept methods, such as those described in ASTM E112-24 [16], are prone to error due to challenges in distinguishing between grain boundaries and twin boundaries and the limited area covered in the analysis. A methodology based on grain size analysis using EBSD, as described in ASTM E2627 [12] and/or ASTM E1181 [17], is considered to provide a standardized analysis approach less prone to human bias. However, care must be taken in selecting appropriate EBSD parameters and associated algorithms to optimize the accuracy of the grain size measurement [13]. Defining and incorporating a grain size measurement approach as part of the qualification process, analogous to qualification of a weld procedure specification, would provide additional certainty to the operator with respect to the final product condition prior to installation.

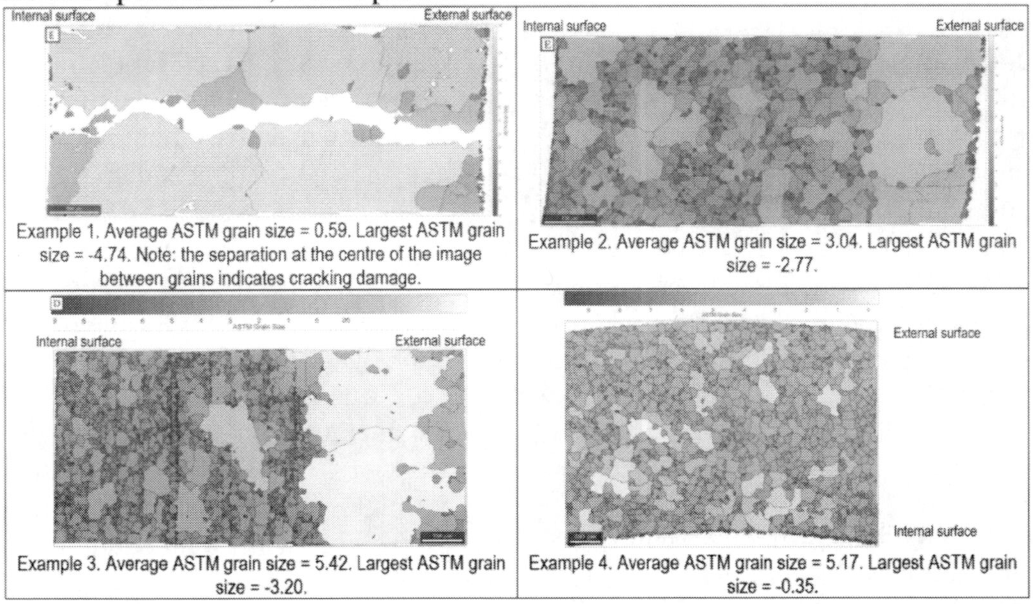

Example 1. Average ASTM grain size = 0.59. Largest ASTM grain size = -4.74. Note: the separation at the centre of the image between grains indicates cracking damage.

Example 2. Average ASTM grain size = 3.04. Largest ASTM grain size = -2.77.

Example 3. Average ASTM grain size = 5.42. Largest ASTM grain size = -3.20.

Example 4. Average ASTM grain size = 5.17. Largest ASTM grain size = -0.35.

Figure 8. Examples of outlet pigtail grain size distributions measured with EBSD.

Bending

The bending process has contributed to a number of examples of neutral axis cracking occurring at bends remote from the terminal welds. The reasons for which have been attributed, at least in part, to the grain size issues discussed previously. After cold work, solution anneal heat treatment has multiple effects. Cold work is recovered, deformed grains recrystallize, and grain growth occurs. At the neutral axis, the degree of cold work is low; hence, the response to heat treatment is dominated by grain growth that leads to the formation of low ductility extremely coarse grains at this location. In addition, any ovality and wall thickness reduction caused by the bending process can lead to a localized stress variation [4] [8]. Dimensional control and limits on ovality and wall thickness reduction should also be included in the procurement specification.

Selection of Weld Consumable

The issue of material selection extends to weld consumable. The current industry practice of using over-matched, nickel-based weld filler consumables increases the susceptibility to SRC. The risk of SRC can be reduced by using an intermediate or matching alloy, where available and appropriate [4] [6]. As in the case of socket or butt welds, the weld consumable would typically be specified by the designer. The operator should give input to this decision and be assured that a suitable weld consumable is specified.

In Service Management

Managing the life of large numbers of outlet pigtails installed in the reformer outlet system is challenging. It is possible, however, to identify the high-risk locations and emphasize inspection during planned outages. Available inspection techniques and technologies that can be applied for detecting damage for outlet pigtails include [6] [7] [18]:

- Visual examination of the general condition and distortion in its cold condition.
- Use of Go/No-Go gauges set to predetermined amounts/percentage of allowable diametral expansion. A typical limit of 3% diametral growth is often used.
- External crawlers to measure outer diameter along the length of the outlet pigtail.
- Dye penetrant testing at the pigtail connection welds for detecting surface-breaking features (e.g., cracking) (see Figure 9).
- Metallurgical replication at the external surface to identify the surface condition.
- Radiography to detect creep cracking at bends.

The extent of available in-service inspection techniques will have limited scope and ability to identify all known damage mechanisms or contributing factors. For example, there is no known screening tool for grain size distribution; this information would require destructive laboratory analysis. Furthermore, the consideration for appropriate access for inspection, including scaffold structures, removal of insulation, and surface preparation can add additional challenges and constraints for the in-service inspection program during planned outages.

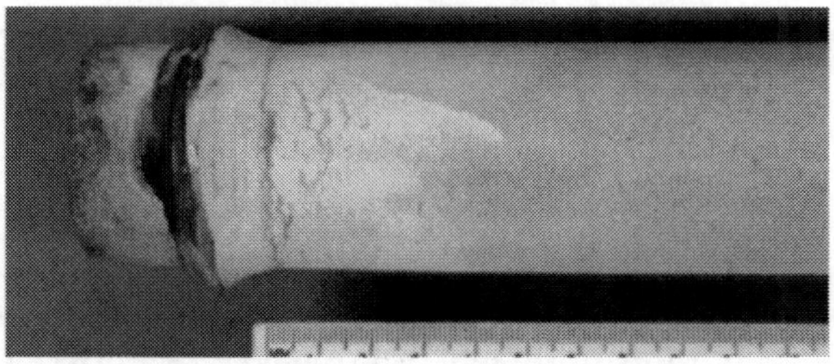

Figure 9. Linear indications detected by dye penetrant testing at a weld and along the pigtail.

Other engineering activities can be considered to supplement inspection during planned outages, including:

- Monitoring of pigtail temperatures based on the outlet temperature of the catalyst tube they are attached to. Detailed trending and reviews of infrared temperature data from the catalyst tubes have value for understanding at-risk pigtails in addition to integrity management of the catalyst tubes themselves. Thermal tube growth monitors are available to identify hot and cold regions of the reformer.
- In-service position monitors can identify movement of the system from start-up/shutdown cycles as well as daily temperature variations. Experience has shown such movement on a consistent basis to be significant. Understanding the dynamics of the system allows the operators to estimate damage rates and identify inspection locations.

This information can be used as inputs for a global finite element analysis of the outlet system, highlighting the locations of maximum thermal movement, determining the conditions at which they occur, and identifying the locations most susceptible to creep/creep-fatigue damage, particularly if axial loads are significant. Such an analysis may be available from original design. In addition, laboratory metallurgical analysis and testing of outlet pigtails can be carried out to further understand the material condition and extent of any in-service damage observed at high-risk locations. The information can be used to validate any finite element analysis and provide insight on the likely condition of outlet pigtails in service, particularly to inform the extent of inspection scope for future planned outages.

Where unanticipated damage is observed for equipment operating in-service, there is a need to consider any remedial actions by undertaking a fitness-for-service (FFS) assessment. The LCM process would typically involve an FFS assessment undertaken using a standard such as API 579-1/ASME FFS-1 [18] to make run, repair, or replace decisions for damage discovered during inspections. However, FFS technology may be used to optimize the decision-making process at any stage in the LCM process [2]. If the option to run with the observed damage is not acceptable, the following would need to be considered for repair/replacement:

- The weld procedure specification for any repairs, noting that the components the pigtails are connected to will have aged and will not be of the same material condition as at the time of commission (e.g., manifold/hot collector). Additional guidance is available in ASME PCC-2 [19].
- Any mechanical modifications to the pigtail tube that may affect its response to bending/axial loads (e.g., addition of pup pieces or collars).
- Any "nipping" of outlet pigtails.
- The number of spare components available to undertake isolated replacements, if needed.

The outcome of these decisions will affect the remaining service life of the outlet pigtails and, hence, the time for consideration of full replacement. At that time, the LCM framework can be re-visited based on learnings and experience from operation.

Conclusions

In many cases, operators accept the pigtail design provided by the Original Equipment Manufacturers, Licensors, and/or designers and rely on in-service inspection to identify damage. As discussed, comprehensive inspection of pigtails would be an extremely expensive exercise with no guarantee of success. Relying primarily on turnaround inspection, while undoubtedly of value, is not likely to optimize pigtail reliability, particularly when unanticipated failures or damage occurs. Implementing a comprehensive LCM strategy will significantly enhance the operator's ability to maximize the service life of their outlet pigtails whilst ensuring reliability for operation. Figure 10 depicts the LCM process, as applied to outlet pigtails, including specific aspects related to the damage mechanisms, materials of construction and in-service inspection described above.

In the case of pigtails, many of the factors and questions that ultimately control life and in-service reliability arise at the time of design and procurement, before they enter service. Such issues to be challenged and addressed as part of the life management strategy have been described above. Examples include (but are not limited to):

- **Design:** Has self-weight been considered when the wall thickness has been selected based on pressure stresses? Self-weight may impact the propensity to sag and generate bending loads.
- **Material selection:** Has grain size requirement be specified adequately? Variable grain size may cause poor flexibility. It is recommended to laboratory test batch samples as part of the procurement specification.
- **Weld consumable:** Has the most appropriate weld consumable been specified? High-strength, over-matched consumables may lead to enhanced risk of SRC.
- **Weld design:** Has the most appropriate weld design been specified? The use of sock-o-let or butt welds is a balance between ease of fabrication and ease of maintenance.
- **System design:** Pigtails are designed to be flexible. Has the thermal movement of the system been analyzed and the risk of creep fatigue damage been assessed?

Once in service, the life management strategy will overlap with the risk-based inspection program, which describes the in-service inspections that can reasonably be undertaken and the limitations of such inspections. A detailed thermal analysis that highlights the locations of maximum thermal movement and hence creep-fatigue loading will significantly aid in determining high-risk locations.

It is acknowledged that design codes may not fully account for all damage mechanisms for outlet pigtails. Despite this, the onset of in-service damage can be minimized at the time of design and procurement, particularly with respect to understanding how the material composition and fabrication processes contribute to creep damage, environmental attack, and possible stress relaxation cracking. Active collaboration between the operator and designer at this early stage of the lifecycle process will improve the lifecycle management of outlet pigtails once constructed and commissioned, as well as minimize the occurrence of unanticipated failures and damage.

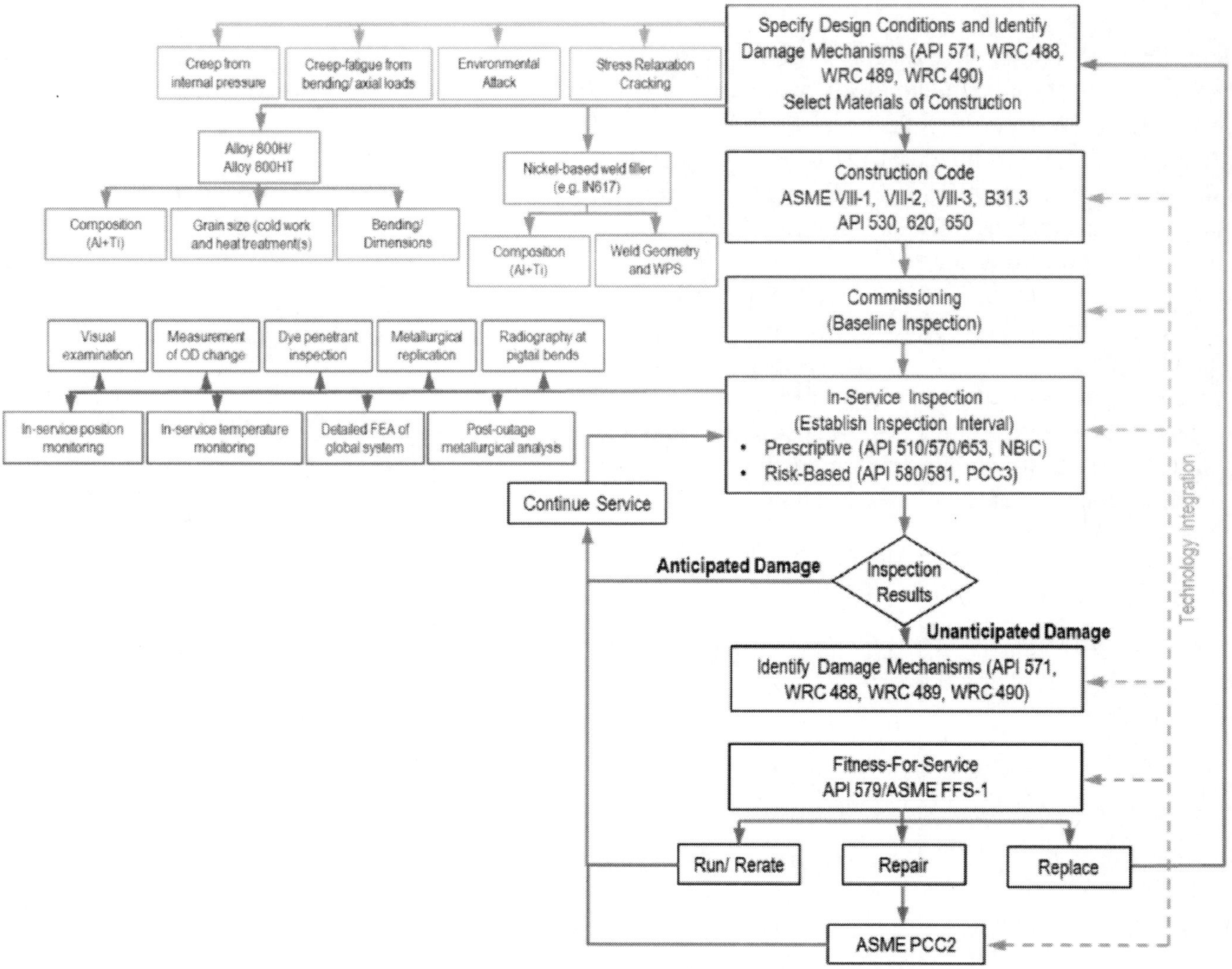

Figure 10. Overall LCM process, as applied for outlet pigtails.

References

[1] D. Osage, D. Thornton and P. Henry, "Life-Cycle Management of Pressurized Fixed Equipment (PVP2011-57903)," in *Proceedings of the ASME 2011 Pressure Vessels & Piping Division Conference (PVP2011)*, Baltimore, 2011.

[2] WRC, "WRC BULLETIN 584 - Overview of API579-1/ASME FFS-1 With An Emphasis on Data and Inspection Requirements," Cleveland, 2020.

[3] ASTM, "ASTM B407 - Standard Specification for Nickel-Iron-Chromium Alloy Seamless Pipe and Tube," Philadelphia, 2022.

[4] O. Chung, N. Goodman and C. Thomas, "Reliability of reformer outlet systems," *Nitrogen + Syngas 2024 Technical Preprint,* p. 269–280, 2024.

[5] R. Dehmolaei, M. Shamanian and A. Kermanpur, "Microstructural changes and mechanical properties of Incoloy 800 after 15 years service," *Material Characterisation,* vol. 60, no. 3, p. 246–250, 2009.

[6] API, API Technical Report 942-A - Materials, Fabrication and Repair Considerations for Hydrogen Reformer Furnace Outlet Pigtails and Manifolds, Washington DC, 2014.

[7] C. Thomas and M. Smillie, "Failures of pigtails: Why they happen and how to avoid them," *Ammonia Technical Manual,* vol. 52, p. 101–111, 2011.

[8] M. Smillie, O. Olatte, C. Thomas and T. Levi, "Investigation of in-service damage accmulation of Alloy 800HT methanol reformer outlet tubes," *Materials at High Temperatures,* vol. 24, p. 241–248, 2007.

[9] A. Young, M. Kral and C. Bishop, "Time-Temperature-Precipitation Relations for Nitrides and Evaluation of Internal Oxidation Theory for Nitridation of Austenitic Stainless Steel," *Metallurgical and Materials Transactions A,* vol. 51, no. 9, p. 4456–4470, 2020.

[10] ASME, ASME B31.3 - Process Piping, New York, 2022.

[11] API, API 530 - Calculation of Heater-tube Thickness in Petroleum Refineries, Washington DC, 2021.

[12] ASTM, ASTM E2627 - Standard Practice for Determining Average Grain Size Using Electron Backscatter Diffraction (EBSD) in Fully Recrystallized Polycrystalline Materials, Pennsylvania, 2019.

[13] C. Bishop, S. Mucalo, M. Kral and D. Lewis, "Experiences During Manufacturing of Reformer Outlet Pigtails," *Ammonia Technical Manual,* vol. 52, p. 213–224, 2013.

[14] C. Bishop, S. Mucalo, M. Kral and D. Lewis, "Continuous Grain Size Gradients in Austenitic Incoloy 800H: Design, Processing, and Characterization," *Metallurgical and Materials Transactions A,* vol. 51, p. 1719–1731, 2020.

[15] A. Beardsley, C. Bishop and M. Kral, "EBSD Characterization of Pilgered Alloy 800H After Heat Treatment," *Materials Performance and Characterization,* vol. 5, no. 5, p. 717–739, 2016.

[16] ASTM, ASTM E112 - Standard Test Methods for Determining Average Grain Size, Philadelphia, 2024.

[17] ASTM, ASTM E1181 - Standard Test Methods for Characterizing Duplex Grain Sizes, Philadelphia, 2023.

[18] API, API RP 579-1 / ASME FFS-1 - Fitness for Service, Washington DC, 2021.

[19] ASME, ASME PCC-2 - Repair of Pressure Equipment and Piping, New York, 2022.

[20] ASM, ASM Speciality Handbook: Nickel, Cobalt and Their Alloys, Cleveland, 2000.

AUTHOR INDEX

Agrawal, Mukesh....256
Ahmad, Sheikh A....56
Al Mehairbi, Saeed....180
Al-Sakran, Abdulaziz N....230
Al-Shehhi, Abdul A....180
Alyami, Khalid....49
Anderson, Brian....44
Archer, Taylor....1, 208
Ariharnomo, Novi....256
Awan, Hafiz A. R....144
Aymans, Benjamin....131
Azevedo, Cristiano....110
Balasubramanian, Ganesan....180
Banu, Faritha....110
Baratto, Francesco....88, 171
Bashir, Maham....56
Benny, Rajindra....218
Bensing, Felix F....265
Blennerhassett, Daragh....198
Braden, Karen....69
Cahyono, R. B....292
Cardin, Pablo....131
Carlucci, Marco M....88, 171
Chaudhry, Kamran A....56
Chowdhury, Partha P....284
Christensen, Thomas S....76
Chung, Olivia....301
Cousins, Matt J....230
Dahl, Per J....76
Dharmavaram, Seshu....1
Ejaz, Tayyab....189
Ezzat, Abdulrahman....49
Goodman, Nicholas....301
Goyal, Amit....275
Grills, Gabrielle....208
Guang, Sun....88
Hosein, Dane....218
Imizcoz, Pedro....131
Jakobi, Dietlinde....131
Jamal, Faisal....56
Joshi, Vishvajeet....275
Kalatrakkal, Sachin....198
Kimpel, Alexis....208
Knudsen, Lari B....76
Kudav, Umesh....284
Kumar, Sudesh....284
Latorre, Felipe R. F....171
Lee, Jackson....69, 95
Luckman, Brad....44
Manikandan....49
Masciarelli, Enrico....208
McMurphy, James....246
Mehta, Bhupen....165
Michael, Davies....44
Mielke, Bernd....265
Modi, Kamleshkumar....105
Mohammed, Taiebi....49
Moncalvo, Davide....44
Moolchan, Mikyle....34
Mudaliar, Shailesh....284
Nahar, Akhil....275
Nölker, Klaus....265
Nugraha, Baskara A....292
Oliynyk, Volodymyr....198
Ormann, Heinrich....171
Pattabathula, Venkat....1, 121
Potter, Dean....69
Powell, Stuart....25
Rampat, Kiran....34
Raza, Syed H....144
Redaelli, Luca....208
Richardson, Jim....121
Sarrack, Anthony....25
Scheerder, Alex....154
Schroeder, Mark....246
Shaffer, Dorothy M....25
Somers, Marcel A. J....131
Thomas, Charles....301
Vaughen, Bruce K....121
Villa, Matteo....131
Wiratni....292
Wohlgeschaffen, Ken....240

AIChE
120 Wall Street, FL 23
New York, NY 10005-4020

CURRAN ASSOCIATES INC.
proceedings.com

ISBN 979-8-3313-2886-3
9 798331 328863